Routledge Revivals

Change and Development in the Middle East

The Middle East is a region of great traditional diversity, which has been characterized by immense political, social and economic changes, still developing over thirty years after the title's original publication. A group of oil-rich countries have achieved great political significance and some of the highest *per capita* incomes in the world. Much modern development has been spatially polarized, accentuating the concentrations of rapidly growing populations and posing severe problems for planners. Cultivation and pastoralism, the main traditional activities, have often suffered from neglect and insufficient investment, and both require re-evaluation.

These are the issues addressed by this volume, first published in 1981, which contains a series of overviews and case studies written by present or former members of staff and research students of the Department of Geography in the University of Durham in honour of W.B. Fisher. *Change and Development in the Middle East* provides an interesting and relevant geographical and demographic analysis of this diverse and volatile region.

Change and Development in the Middle East

Essays in Honour of W.B. Fisher

Edited by John I. Clarke and Howard Bowen-Jones

Routledge
Taylor & Francis Group

First published in 1981
by Methuen & Co.

This edition first published in 2013 by Routledge
2 Park Square, Milton Park, Abingdon, Oxon, OX14 4RN

Simultaneously published in the USA and Canada
by Routledge
711 Third Avenue, New York, NY 10017

Routledge is an imprint of the Taylor & Francis Group, an informa business

Publisher's Note
The publisher has gone to great lengths to ensure the quality of this reprint but points out that some imperfections in the original copies may be apparent.

Disclaimer
The publisher has made every effort to trace copyright holders and welcomes correspondence from those they have been unable to contact.

A Library of Congress record exists under LC control number: 80041649

ISBN 13: 978-0-415-82081-3 (hbk)
ISBN 13: 978-0-203-38395-7 (ebk)

Change and development in the Middle East

WILLIAM BAYNE FISHER
Professor of Geography, University of Durham, 1956–81

Change and development in the Middle East

ESSAYS IN HONOUR OF

W. B. Fisher

edited by
John I. Clarke and
Howard Bowen-Jones

METHUEN
London and New York

First published in 1981 by
Methuen & Co. Ltd
11 New Fetter Lane, London EC4P 4EE
Published in the USA by
Methuen & Co.
in association with Methuen, Inc.
733 Third Avenue, New York, NY 10017

Printed in Great Britain at the
University Press, Cambridge

British Library Cataloguing in Publication Data

Change and development in the Middle East.
1. Near East – History – 20th century
I. Clarke, John Innes II. Bowen-Jones, Howard
III. Fisher, William Bayne
956′.04 DS62.4 80–41649

ISBN 0–416–71080–8

Contents

List of figures

Notes on contributors

TAIBA A. AL-ASFOUR is a Lecturer in Geography in the University of Kuwait, and was formerly a research student in Geography in the University of Durham.

J. A. ALLAN is a Senior Lecturer in Geography in the School of Oriental and African Studies in the University of London, and was formerly a student in the University of Durham.

PETER BEAUMONT is a Professor of Geography at St David's University College, Lampeter, and was formerly a Senior Lecturer in Geography in the University of Durham.

BRIAN W. BEELEY is a Staff Tutor (Social Sciences) in the Open University and was formerly a Research Fellow in Geography in the University of Durham.

J. STACE BIRKS is working with the World Bank, and was formerly a Research Fellow in Geography in the University of Durham.

GERALD H. BLAKE is a Senior Lecturer in Geography, University of Durham.

HOWARD BOWEN-JONES is a Professor of Geography, University of Durham, and Director of the Centre for Middle Eastern and Islamic Studies (1968–80).

BRIAN D. CLARK is a Senior Lecturer in Geography, University of Aberdeen, and was formerly a Lecturer in Geography in the University of Durham.

JOHN I. CLARKE is a Professor of Geography, University of Durham.

JOHN C. DEWDNEY is a Reader in Geography, University of Durham.

WILLIAM J. DONALDSON was formerly a Research Fellow in Geography in the University of Durham.

MICHAEL P. DRURY is a Lecturer in Geography, University of Durham.

RODERIC W. DUTTON is a Research Fellow in Geography, University of Durham.

MOHAMMED EL MEHDAWI is a Lecturer in Geography, Garyounis University, Libya, and was formerly a research student in Geography in the University of Durham.

ALLAN M. FINDLAY is a Lecturer in Geography, University of Glasgow, and was formerly a research student in Geography in the University of Durham.

THOMAS H. GREENSHIELDS is an Assistant Map Research Officer at the Directorate of Overseas Surveys, Tolworth, and was formerly a research student in Geography in the University of Durham.

ALLAN G. HILL is a Senior Research Fellow in the Department of Medical Statistics and Epidemiology at the London School of Hygiene and Tropical Medicine, and was formerly a research student in Geography in the University of Durham.

RICHARD I. LAWLESS is Assistant Director of the Centre for Middle Eastern and Islamic Studies, University of Durham.

KEITH McLACHLAN is a Senior Lecturer in Geography in the School of Oriental and African Studies in the University of London, and was formerly a research student in Geography in the University of Durham.

NASSIR A. SALEH is Dean of Admissions and Registration at King Abdul Aziz University, Mecca, Saudi Arabia, and was formerly a research student in Geography in the University of Durham.

JOHN H. STEVENS is Senior Agricultural Advisor, Knight, Frank and Rutley, International Estate Agents and Property Consultants, and was formerly a Senior Lecturer in Geography in the University of Durham.

Preface

The study of change is a great intellectual challenge, whether it concerns the past, the present or the future. Few would deny that the Middle East, however defined, has experienced momentous change during the twentieth century, especially since 1950 when Professor W. B. Fisher's book *The Middle East, a Physical, Social and Regional Geography* was first published by Methuen.

Although not one of the more populous parts of the world, the Middle East comprises immense diversity arising from many factors: its crossroads location between three continents; its physical diversity; its prolonged human occupation; the spread of empires and their conflicts; the emergence of great monotheistic religions and lesser sects; the differential impact of the west; the striking political fragmentation; and also the numerous contrasts between the desert and the sown, between town and country, between ruler and ruled, between landlord and peasant, and between rich and poor.

In addition to this more or less traditional diversity, the second half of this century has been characterized by immense political, economic and social changes. The Middle East has been in a constant condition of political flux with the strengthening of nationalism and the achievement of independence by many countries; decolonization and the nationalization of European assets; the diversification of external political influences; the growth of Arab unity; the persistence of the Palestine problem; and changes in the forms of government. A group of oil-rich countries have achieved great political significance and some of the highest per capita incomes in the world. Oil has become the main catalytic agent, introducing ultra-modernity, enabling industrialization, encouraging urbanization and provoking vivid contrasts between the 'have' and 'have-not' countries. Much modern development has been spatially polarized, accentuating the increasing concentration of the rapidly growing populations and posing severe problems for planners. Cultivation and pastoralism, the traditional activities, have often suffered from neglect and insufficient investment, and both require re-evaluation. Traditional as well as modern activities all rely critically on the fundamental basis of life – water resources – without which much of the Middle East remains a desert.

Change and development are the themes of this volume, which contains a series of overviews and case studies written by present or former members of staff and research students of the Department of Geography in the University of Durham, as a *festschrift* in honour of Professor W. B. Fisher who retires in 1981 after twenty-seven years as head of department. He has done so much to develop the department, and since 1965 he has also been Principal of the Graduate Society. In particular, he has stimulated research interests in the Middle East, work which has been partly facilitated by the growth of the University of Durham Centre for Middle Eastern and Islamic Studies, of which he was the first Director in 1961–4. Consequently, it seemed most appropriate to focus this *festschrift* on the Middle East, rather than incorporate chapters representative of the research interests of the department as a whole. The editors and publishers wanted to produce a coherent, useful volume which combined macro and micro views and which would in some way complement Professor Fisher's own major volume, now in its seventh English edition.

All but one of the chapters were written specially for this book, but grateful acknowledgement is made to Croom Helm for permission to republish Chapter 9. The editors were faced with a dilemma in selecting authors, because so many Durham geographers have researched in the Middle East – there are more than a hundred higher degree theses alone. Inevitably, many former researchers are no longer academically involved in the region while others are strewn all over the world. The chapters therefore represent to some extent a sample, but we hope a valuable amalgam of generalization and uniqueness.

In the preparation this book, the editors have been greatly supported by the work of the secretarial and technical staff of the Department of Geography and the Centre for Middle Eastern and Islamic Studies of the University of Durham, notably Barbara Tranter, Joan Dresser, Hilary Smith, George Brown, Arthur Corner, Derek Hudspeth, Ian Middlemass, John Normile, Dennis Ewbank and Marlene Crighton, to all of whom thanks are given. We are also grateful to Tim Unwin for reading through the text for inconsistencies, a near impossible task for a multi-authored text on a region where names are so variously transliterated. Indeed, in at least one chapter where numerous Arabic names occur it was felt better to retain the author's transliterations rather than make them conform; it is hoped that this does not detract from the value of the book. In addition, our sincere thanks are due to Peter Atkins for kindly preparing the index at short notice.

Finally, it is appropriate to mention that royalties accruing from the sale of this book will go to a W. B. Fisher Fund within the University of Durham, which will be used to assist students undertaking geographical work in the Middle East.

J. I. Clarke
H. Bowen-Jones
Durham, November 1979

PART I

Overviews

CHAPTER ONE

Development in the Middle East

HOWARD BOWEN-JONES

During the last thirty years or so the countries of the Middle East not only have experienced profound changes but, increasingly, have consciously engaged in changing themselves, through deliberate and planned development policies. These policies themselves have changed in ethos and emphasis, so that, whilst the first generation of plans were strongly orientated towards production projects and largely based on physical feasibility and economic viability, recently they have become more and more concerned with social desiderata – with values and with concepts of justice, welfare and equality. This becomes immediately apparent from a perusal of the longest sequences of development plans – in Iran, Pakistan and Turkey extending from the 1950s to the late 1970s – as well as in relatively short sequence changes, for example the First and Second Development Plans of the Kingdom of Saudi Arabia.

This trend, of course, reflects changes not only derived from experience but also those which have occurred within intellectual and theoretical approaches to planned development. On one plane it may be said to represent a change in emphasis from mechanical engineering to social engineering and, on another, a belief that what determines the course of economic development 'is not primarily economic forces' (Sayigh 1978a). Such a change in intellectual attitude can easily become transformed into development-agency philosophy, e.g. 'planning that is confined to the control of strictly economic variables might not be adequate to generate a development process having genuine spontaneity and continuity in the long term' (UNIDO 1976). At this stage two potential dangers appear. The first of these is the loss of known control disciplines and a reluctance to establish new ones, so that, for example, 'investment' with strict connotations becomes replaced by 'expenditure'. This tendency is well illustrated in the Gulf Emirates (Al Kuwari 1978) but is not confined to them. The second danger is that apparently inadequate technical determinisms may be replaced by newer and even more stultifying universalist doctrines. These factors become of critical importance as development policies grow from the youthful basis of simple state interventionism to the full apparatus of centralist planning.

In this short survey attention is concentrated on the need, within the Middle East in this case, for attaining a reasonable balance between regional realities and social desires. Development, also, is 'the art of the possible'.

Homogeneity and heterogeneity

However we areally define the Middle East, this, as with all other regions, can only be given the status of 'region' because within it can be found a significant number of interrelated phenomena which together – and only together – confer upon it some distinctiveness. This is not the place to argue in detail for any particular areal delimitation of the Middle East but it must be emphasized that, whichever is used, the implied regional distinctiveness neither connotes uniformity or total homogeneity *within* the region nor demands total dissimilarity to areas *without*. These are or should be truisms, yet their logical consequences are worthy of further examination in the present context.

Socio-political attitudes to development

The preamble to the Saudi Arabian Development Plan (Saudi Arabia 1975) contained an illuminating statement of relevance to us:

> Planning implies the efficient use of a country's resources . . . for the attainment of nationally-cherished goals. Since goals are culturally, historically and politically orientated, a country's development plan essentially reflects its fundamental values and principles.

Planned development as a deliberate process necessarily must always have a *terminus ab quem*, a starting point. It is also and necessarily a process designed and implemented by some authoritative body; today, in the Middle East as elsewhere, this implies the sovereign state. The Saudi Arabian Plan statement, cited above, frankly and clearly recognizes this latter fact – the national basis of most development planning. Moreover, in its identification of the cultural and other roots from which spring the goals of national development, it predicates that each plan will start from a base peculiar to the country concerned. This is not a simple assertion but an affirmation of human heterogeneity within the region. Relevant to the field of development what can we further add? At a macro-political level all of the twenty-four sovereign states which we now identify in the region extending from the Hindu Kush in south-west Asia to the Atlantic coast of north-west Africa, and from the Black Sea to the Sahel and the shores of the Indian Ocean, can be described as 'emergent'. The manner of their emergence during the last half-century, however, has been extremely varied. Of the six states which in 1929 were politically independent, two, Turkey and Iran, were then being driven through secular revolutions by modernizing autocrats, and one, Saudi Arabia, was being forged into a unified kingdom out of tribal fragmentation. Oman had to wait for a new national will until 1970, and Yemen (now the Arab Republic of Yemen) and Afghanistan might be said still to be finding themselves.

All other states either had their sovereignty limited to some extent, as with countries as diverse as Egypt, Qatar and Iraq, or were effectively colonies of European powers. In many cases complete independence was achieved by the relatively painless withdrawal of the controlling power, as in Sudan and the Gulf Emirates; in some, such as Libya, the final succession was orderly but preceded by major upheavals. In others, however, the achievement of independence was associated with bitter conflict, notably in the Maghreb and Southern Yemen (People's Democratic Republic of Yemen), whilst the genesis of Israel added to the multiple birth-pangs of the old Levant.

For the most part these transformations were brought to term as a consequence of the trauma of the Second World War and the 'winds of change' which blew thereafter. In the nascent states, nationalism, étatism, radicalism, – for some even a sense of revolutionary liberation – in various combinations formed the climates in which new leaders and new élites formulated development goals.

Already, then, the sub-regional variables begin to appear even at this general level. The new decision-makers themselves were and are highly varied in origin, type and attitude, ranging from Islamic monarchs to revolutionary councils, from uneasy *modus vivendi* between interest groups to elected representative governments, from aristocratic paternalism to total radicalism – 'The Popular Revolution will not be realized until the masses seize power and until they are given complete control over all the sectors' (Gaddafi 1973).

It should not be surprising that development planning has been approached with different degrees of alacrity and is extremely varied in ethos. For one group of countries, in particular those which effectively had a colonial experience, such as Algeria, the political motivation was extremely strong – 'a five year plan became almost as much a requirement for asserting independence as a new flag and a national anthem' (Stevens 1977). In others, such as Oman, one finds a more sober appraisal of the economic basis for attaining development objectives within an existing continuum (Oman 1976). This first distinction, between those countries and peoples which reacted vigorously against a colonial past dominated by 'the West' and those who experienced no such need, is of vital importance to any understanding of recent and contemporary developmental attitudes and processes. Tibawi (1969), in his examination of the French Mandate, 'The Sacred Trust of Civilization', in the context of Syrian modern history, illustrates how in some lands the offensiveness of alien imperialism could lead to a Hegelian reaction against the West including much that could have been valuable for development. It was for much the same reasons that Somalia in 1970 turned to a strange blend of Leninist 'scientific socialism' and Islam. Thus also has been reinforced a drive to diminish 'dependence' on alien economies and political balances.

Analogous in its effects in contributing to an undiscriminating hostility expressed by many Arab leaders and thinkers towards Western tutelage was the creation of the state of Israel. Here we must limit comment on what is essentially a separate theme but two points at least are relevant here. The first is that not only did this creation and its continuing consequences even more strongly drive many Arab countries to turn from Western paths but it also exacerbated regional disunity and suspicion. King Abdallah of Jordan in *Al-Takmilah* (1978) provides evidence, remarkable for its absence of rancour, for what some felt was lost in opportunities for Arab regional collaboration at the turn of the half-century. It may be noted, in parenthesis, that Fisher in 1948 contemporaneously postulated alternative paths for the Middle East based on either 'narrow regionalism' or 'the organization of large parts of the Middle East as a single unit' (Fisher 1950). It is certain, *inter alia*, that the geopolitical consequences for the Middle East of the formation and manner of survival of Israel have, on balance, been discordant and destructive rather than unifying.

Secondly, of course, there has been a vast diversion of time, talent and treasure to war and preparations for war. At peak, between one-quarter and one-third of Egyptian national income during the last two decades went on military expenditure. In 1976, the proportion of the male population between the ages of 18 and 45 in the armed forces of Syria and Israel was 24 per cent and 19 per cent respectively. Quite apart from normal defence needs, the Arab–Israeli hot and cold wars have consumed excessive quantities not only of capital and foreign currency but of even more precious resources of management and technical skills so critically necessary for development. This also has had a non-uniform impact on the countries of the Middle East.

This last point concerning manpower and skills is then further complicated by the fact that part of the development demand for trained manpower, particularly in the non-frontline states, can only be met by employing displaced Palestinians and Jordanians. It is a sad fact that while many of these are wanted, for their expertise as workers, by the richer developing Arab countries, they are often less welcome as residents because of the political embarrassments which may be associated with their presence.

Whilst these and other factors may be identified in fundamentally creating a heterogeneity of national socio-political attitudes to development, there are other forces which, on the same general plane, strengthen the crystallization of regional and sub-regional community interests. None of the much-heralded attempts towards political union or federation between states has succeeded, with the exception of the formation of the United Arab Emirates. Most recently, the small-scale but open hostilities between Libya and Tunisia illustrate the fragility of plans to unify very dissimilar countries. However, when issues less sensitive than political sovereignty or nationality are involved then we find expressions of supra-national consciousness. This is of growing significance to development on a variety of scales and types of activity. The Arab League has almost a score of major agencies relevant to our theme, ranging from the Arab Fund for Economic and Social Development to the Arab Centre for the Study of Arid Regions. The Organization of Arab Petroleum Exporting Countries (OAPEC) has been associated with or has funded collaborative action by its members in, for example, the Arab Shipbuilding and Repair Co., and the Arab Petroleum Investments Co.

These are but illustrations; what becomes apparent on more detailed examination is that, within the Middle East as a whole, exclusive as well as inclusive sub-regional groupings are especially strong in a developmental context. Arab organizations, by definition, exclude Iran, Israel and Turkey; CENTO's development activities, important particularly in the 1960s and early 1970s, involved Iran, Pakistan, Turkey and Britain. The composition of a group as apparently coherent as the Gulf states is in practice inconstant, changing with variations in sectoral development interest; the vicissitudes of attempts at Maghrebian sub-regional co-operation have been considerable.

One of the most significant questions to emerge from a consideration of the dynamic and complex scene sketched above is this: to what extent does the diversity of national and sub-regional socio-political approaches to development reflect a maturing recognition of the diversity of geographical reality within the Middle East? Superficially, this query may appear trite but its implications are of considerable import.

If diversity is regarded primarily as the result of human perversity; if dissatisfaction with existing situations leads to preoccupation with searches for universal institutional panaceas; then we may fear for the future. This not only because critically important determinants particular to each state or sub-region may be then under-weighted, but also because certain equally critical combinations of factors common to all or many Middle-Eastern countries may be virtually ignored. Thus, both in Sayigh's examination of *The Determinants of Arab Economic Development* (1978b), and in Amin's castigation of *The Modernization of Poverty* (1974), valuable as they are, we find implied a dangerously exaggerated sense of freedom of human manoeuvre. Here, as elsewhere in this volume, it is recognized that development is for people and by people. Nevertheless, development is also place and time specific: man is only a free agent within a particular space–time continuum.

What then matters is the nature of human response to the limited but still considerable opportunities which remain within the broad constraints of life on earth and, more specifically, life within the Middle East. It is to an examination of the implications for development of some of the non-institutional elements in the balance between challenge and response that we now turn.

Resource availability and resource use: the rural scene

Idealists in the affluent world may deplore the emphasis placed on materialism by development planners in the Middle East and elsewhere, and we may remind ourselves that 'man does not live by bread alone' – a basic doctrine of the great monotheistic religions born in the Middle East. Nevertheless, it remains inescapably true that any desirable quality of life, however we define it, requires an appropriate supportive production system. As the Turkish proverb succinctly states, 'A sack must be full to stand upright'. What emerges when we examine even a few of the implications of this statement?

One starting point is that of food supply. Here, and in much of what follows, the strength of the argument relies, in part at least, on the use of statistics which, in terms of the orders of magnitude involved and as indicators of trends, are incontrovertible despite various detailed qualifications which may have to be made for other purposes.

In Table 1.1, column A would seem to indicate that, for most of the countries selected to illustrate a range of national conditions, the mean calorific nutritional level is adequate, but only if we ignore not only inequalities of distribution but also qualitative dietary deficiencies. Both of these latter, as any close observer of the Middle East knows, are almost as important today as forty years ago: 'Food supplies were insufficient in quality rather than in quantity and any improvement tended to increase the numbers to be fed rather than raise the nutritional level' (Food Research Institute 1956). Even so, significant increases in life expectancy and decreases in child death rates (Table 1.1, columns B and C) must provide some corroborative evidence for, at least, a maintenance of the nutritional *status quo* in spite of considerable increases in populations.

How has this been achieved? Very largely by an increase in food imports rather than by growth in domestic production. Even data concerning the proportion of merchandise imports taken by food show some disturbing facts (Table 1.1, column D). Two out of the ten countries named experienced an increase in proportional import dependence between 1960 and 1976, i.e. Egypt and Libya, whilst only three, Turkey, Iran and Saudi Arabia, were significantly better off in this respect than even those industrialized nations notoriously reliant on food imports, Japan and the United Kingdom. In

Table 1.1 Selected indicators of food supply and health

	A Daily per capita calorie supply % requirements	B Life expectancy at birth, years		C Child death rates (ages 1–4)		D Food as percentage of merchandise imports	
	1974	*1960*	*1977*	*1960*	*1977*	*1960*	*1976*
Egypt	113	46	54	31	18	23	28
Morocco	108	47	55	30	17	27	20
Syria	104	48	57	29	14	24	17
Algeria	88	47	56	30	16	26	18
Turkey	113	51	61	24	10	7	3
Iraq	101	46	55	31	17	*	15
Iran	98	46	52	24	14	14	11
Saudi Arabia	102	38	48	48	28	*	12
Libya	117	47	55	30	17	13	15
Kuwait	100	60	69	12	2	*	*

Source: World Bank, *World Development Report* 1979
* Not available

absolute terms, the position is even worse. During the 1970s Middle-Eastern countries as a whole have been forced to import over 20 per cent of staple wheat requirements; Syrian efforts to increase cotton production by 1975 contributed to an export of textile fibres which did little more than pay for imports of sugar and honey and left her in large deficit in cereals, fruit and vegetables. Algerian imports of foodstuffs are four times larger than equivalent exports and, in 1975 alone, Libyan subsidies on imported foodstuffs were equal in value to one-quarter of the total expenditure allocated to agriculture and agrarian reform in the whole 1976–80 development plan.

Data presented in Table 1.2 allow this analysis to be taken still further. Kuwait exemplifies the special condition of the Gulf city-states, one in which domestic agricultural production is virtually insignificant. When we examine trends within the territorial states further disturbing features appear. In every case the proportional contribution of agriculture to the Gross Domestic Product has fallen, but that of manufacturing has only risen significantly in Egypt and Turkey (column C). This could be explained in some cases by the overwhelming dominance of the oil sector, but this in itself can hardly be regarded as a healthy indicator of true development. In column D we find that those few countries which have achieved statistically significant growth rates

Table 1.2 Selected indicators of economic change

	A		*B*		*C*				*D*				*E*			
	GNP per capita		*Population: annual average growth %*		*GDP sectoral distribution %*				*Sectoral average annual growth rates %*				*Percentage of labour force in two sectors*			
	Average annual growth % 1960–70	*1977 US $*	*1960–70*	*1970–7*	*1960*		*1976**/77*		*1960–70*		*1970–6**/7*		*Agriculture*		*Industry*	
					Agri.	*Manuf.*	*Agri.*	*Manuf.*	*Agri.*	*Manuf.*	*Agri.*	*Manuf.*	*1960*	*1977*	*1960*	*1977*
Egypt	2.1	320	2.3	2.2	30	20	28	24	2.9	4.7	3.1	5.7**	58	51	12	26
Morocco	2.2	550	2.7	2.8	29	12	21	12**	4.2	4.0	0.6	5.7**	62	53	14	19
Syria	2.3	910	3.2	3.3	25	16	17	11?	4.4	4.4	6.4	7.5**	54	49	19	23
Algeria	2.1	1,110	2.1	3.5	21	10	8	11	0.4	7.7	0.2	6.9	67	35	12	18
Turkey	4.1	1,110	2.5	2.5	41	13	28**	20**	2.4	10.7	3.4	*	78	62	10	14
Iraq	3.8	1,550	3.2	3.4	17	10	8**	7**	5.7	5.9	−1.5	11.5**	53	43	18	25
Iran	7.9	2,160	2.8	3.0	29	11	10	13	4.4	12.0	5.8	16.7	54	41	23	32
S. Arabia	6.7	6,040	2.6	3.0	*	*	1	5**	*	*	3.7	4.1	71	63	10	14
Libya	6.6	6,680	4.0	4.1	14	9	3	3	*	*	14.0	15.3	53	22	16	27
Kuwait	−3.1	12,270	10.3	6.1	†	*	†	*	†	*	†	*	1	2	34	34

Source: World Bank, *World Development Report* 1979

* Not available

** 1976 † Insignificant

in agriculture are also those already identified as still being highly food-import dependent. Most critical of all are the facts and trends displayed in column E.

Even though the proportion of the labour force deployed in agriculture has fallen everywhere since 1960, it remains remarkably high in the vast majority of countries. We therefore must conclude that the situation in Iraq, for example, where, with 43 per cent of the labour force still employed in agriculture, an average decline in production has been achieved in the 1970s, imports of food and live animals in 1975 being ten times the value of equivalent exports, is but an extreme example of a general malaise.

Of course considerable differences exist between and within the non-agricultural sectors of Middle-Eastern states and the variety of the impact on agriculture and rural life of other activities has to be recognized. In some countries an explosion of oil wealth would appear to have destroyed or at least severely eroded the harsh survival incentives to farm at those productivity levels which traditional technology and opportunity allowed (see Chapter 12). This has also been accompanied by the down-grading of agriculture, relative to other sectors, as a profitable field for investment. Saudi Arabia provides the most obvious example here, resulting particularly in the rural depopulation of what was once one of the country's most important agricultural areas, Assarah (Mughram 1973).

Such forces, moreover, ignore national boundaries so that the deleterious effects on agricultural production of what might be termed 'the Dick Whittington syndrome' – the lure of perceived opportunities elsewhere – are not confined within changing domestic situations. Within the Arab world, a common language and a sufficient degree of cultural homogeneity has further encouraged international as well as inter-regional movements out of rural areas, easily demonstrable in many countries as different as Lebanon, the Yemens and Tunisia (Sinclair and Birks 1979). The residual non-mobile populations are in turn behaviourally affected not only by remittances and injections of cash by migrants, but also by the novel and inchoate aspirations to which the former become exposed (Bowen-Jones 1970). The operation of such spread and backwash effects on, for example, technological innovation, consumerism and investment, is of course neither random nor accidental and is susceptible to spatial analysis; some examples appear elsewhere in this volume. Of relevance here is the fact that much contemporary development thinking is vitiated by the over-simple acceptance of uniformitarian theories of social response to change and the lack of appreciation not only of the spatial and functional variety of process operation but also of the fact that much of this variety is rooted in fundamental differences in resource-use opportunity. Some examples from classic studies in rural Turkey serve to illustrate this point.

'Its pride and independent spirit are declining and sense of inferiority increasing' (Stirling 1965) is a judgement on a Turkish village that has been echoed in Turkey by Yasa (1957) and by others elsewhere. Valid in themselves, such asseverations of the decay of rural life have become too easily utilized as the foundation elements in general radical theories which seek to find the causes of such decay in general and/or external malevolence. In fact, what indisputably has been observed in Turkey is firmly rooted in the incompatibility between hypothetically legitimate aspirations for material improvement and the reality of the limited household incomes which can be derived, under any socio-political order or feasible technology, from farms smaller than 7 ha, e.g. 68 per cent in Hasanoğlan (Yasa 1957) and 40 per cent in Sakultutan (Stirling 1965) on the Anatolian steppe. Thus, when we endeavour to ascribe causes to the disturbing features of Table 1.3 (Bilgin 1970) it would be as well to consider more fully the causes of rural disillusionment with change. These, with all that may be implied in terms of stagnant or declining farm production, are in fact born of frustration with binding ecological limitations which are spatially specific rather than the resultant of forces which are apparently eradicable simply by socio-political action.

The recent tendency, for a variety of reasons, has been to concentrate attention on institutional obstacles to agricultural development. These often can be identified as important factors; it must be accepted, for example, that the land-tenure systems which dominated the rural scene in Iraq and Syria up

Table 1.3 Evaluation of past and future trends in eastern Anatolia (% of respondents)

	Market-ignoring villages	*Market-seeking villages*
Evaluation of the past		
Nothing has changed	58.4	38.9
Changed for the better	6.9	9.7
Changed for the worse	9.7	38.9
Don't know	25.0	12.5
	100.0	100.0
Evaluation of the future		
Would change for the better	8.3	13.9
Would change for the worse	12.5	34.7
Would be the same	48.6	29.2
Don't know	30.6	22.2
	100.0	100.0

Source: Bilgin (1970)

to the middle 1950s were, at the least, not conducive to high productivity (Dowson 1931). It is equally true that a deliberate and planned concentration on industrialization, particularly in highly centralized situations, for example in Algeria (Lawless 1976), can lead to the absolute neglect of agriculture.

Many other examples could be adduced to demonstrate the depressive effects on agriculture of such social and economic forces, but if we examine their operation in real space and time we come to realize that the resultants are not merely the product of historical accident and that alternative more desirable scenarios are not easily created. We can illustrate this truth and take the argument further by reference to land reform, a field in which too many false hopes have been raised by reality-remote theory.

In Egypt, greater equity and security has been achieved in landholding and a larger proportion of hard-earned income has remained in the hands of the fellah as the result of land reforms carried out in 1952, 1961 and 1969. Nevertheless, the sad truth is that land reform *per se* has done nothing and can do nothing significantly to alter the rural state. Half the labour force is employed in agriculture, typically on 2 ha farms (21 ha is the maximum legally permitted size). The small size of landholdings is a function of very limited land and water availability in conditions of total dependence on irrigation and of rapid population growth. The labour-absorptive capability of cities and of industry is already overstretched and so a continued increase in the demand for farmland is ineluctable. At the most sanguine estimates, the increased supply of cultivable land in Sinai or New Valley schemes could do no more than hold the present balance. All that is left is to increase productivity and the efficiency of use of those land and water resources already utilized, and these offer but limited hope. Mechanization is largely irrelevant to small farms, fertilizer application rates and crop-yields per hectare can rise but slowly from traditionally high base-levels (in contrast to most Middle-Eastern countries). Multi-cropping based on perennial irrigation already gives a cropped area 30 per cent larger than the cultivable area and cannot be extended much more. A further switch to high-value, export-orientated agricultural crops is technically feasible but would increase the already considerable dependence on a small number of affluent markets in which competition with other specialist suppliers (e.g. Israel) is tough. In this situation, the determinants of rural development are inescapably associated with specific man–land ratios: institutional or redistributional policies are frankly not relevant to the fundamental dynamics of rural poverty.

Since it may be held that Egypt is merely a case of exceptionally adverse man–land ratio, Iraq, long considered to have large agricultural resources under-utilized by a proportionately small population, is worthy of brief examination.

Modern intervention by rulers in the general field of land title and tenure dates from 1858, culminating in the Agrarian Reform Law of 1958 with Amendments in 1964 and 1970, all increasingly designed to 'circumvent the vested interests of the landlords' (Gabbay 1978), to improve the deplorable lot of the fellah and reduce the dependence of the country on external forces. Unfortunately, except for some material amelioration in farmers' living standards, policies have not been very successful to date, primarily because the complexity of realistic agricultural planning has not been appreciated.

Of the approximately 48 million donums (12 million ha) estimated total of cultivable land, the areas under crop in rainfed and irrigated systems each occupy between 6 and 7 million donums, approximately half of the utilized area in each case as the result of the predominance of an alternate-year fallow system. The average holding size is now about 36 donums (9 ha), effectively reduced by one-half by that same fallow system; where perennial (tree-crop and vegetables) irrigated farming is practised, holdings are smaller. Raising production, productivity and incomes clearly requires that more land be cultivated more efficiently under irrigation and that rainfed farming technology be improved, but this is far from easy.

Iraqi irrigation relies almost entirely, directly or indirectly, on the flow of water in the Tigris and Euphrates river systems. It already consumes more than one-third of such surface flows, and Syria and Turkey, the upstream riparian states, effectively control supply. Moreover, within the 11.6 million donums described as having irrigation networks (Iraq Ministry of Irrigation 1972), 64 per cent of the area is inhibitingly salinized with electrical conductivities of more than 4 mmhos/cm including 43 per cent very heavily salinized (more than 8 mmhos/cm). Significantly and stably to extend the cropped area under present conditions of water use is therefore not feasible, and to improve these conditions requires the effective application of considerable technological skill on a very large scale by small-scale farmers. So too with the rainfed areas, the main suppliers of wheat and barley staples, in which traditional adaptations to variable and low precipitation, cold winters and hot summers have to be replaced with all that is implied by 'dry-farming', with the appropriate skills and 'upstream' and 'downstream' inputs and organization.

In both agricultural systems in Iraq, actual as opposed to nominal rural development, as elsewhere in our region, then requires a new rather than a conventional radicalism, one which is prepared to jettison terms such as 'intermediate' or 'revolutionary' for the simple 'appropriate'. Everywhere in the Middle East, not only are the basic resources of land and water limited in quantity, adequately identified in distribution, and possessing critical physical sensitivities – this can be said of many areas of the world – but, to an overwhelming extent, are and for long have been occupied, appraised and worked by myriads of successful survivors. In such situations, the demands on technical, financial, organizational and simply human capability made by apparently simple policies of institutional change such as the transfer of land title, the establishment of cooperatives, agricultural credit, central purchasing agencies and many others, become too great to be met effectively and urgently. This, in Iraq, as almost everywhere in the Middle East, explains very largely why, for example, sequestrated land is never completely re-allocated to farmers, why increases in agricultural productivity are minimal, why the effectiveness of new official agencies is so low, why the flight from land is so strong and why national dependence, even in this primary sector, grows rather than declines.

In the countries of the Middle East, therefore, we find a great variety of agricultural situations and potentials. Only one, Sudan, possesses considerable tracts of under-utilized land and even here ecological and human sensitivities are such as to make development very demanding of skills as well as of capital. Elsewhere, the requirements are even more stringent, since they involve the appropriate transformation of a great diversity of existing holistic environmental and human interrelationships. There are no simple, universally applicable answers.

Diversification and industrialization

The Middle East is not and, historically, never has been merely rural in character. Manufacturing, service industries and commerce have for long been associated with a host of urban centres such as Cairo and Damascus, Istanbul and Isfahan. Nevertheless, as indicated in Table 1.2, columns C and E, industry in general and manufacturing in particular remain today remarkably undeveloped. In this overview attention is confined to an identification of some of the main reasons why neither traditional manufacturing experience nor development policies have resulted as yet in a 'take-off' in manufacturing comparable, for example, with that in Japan at the turn of the century.

In general, Middle-Eastern approaches to growth in manufacturing have depended on three main articles of faith: first, that industrialization is desirable; secondly, that industrialization can only rapidly proceed through public sector involvement; thirdly, that the main problem is associated with two dissimilar structural disequilibria – scarcity of capital resources in some countries, acute labour shortage in others (El Mallakh *et al.* 1977). However, if we briefly explore a little further we are returned once more to the facts of regional heterogeneity and the need for identifying what is nationally appropriate.

In Table 1.4, the data indicate the low manufacturing status attained in 1975 by a range of Middle-Eastern countries. In terms of total value added in

Table 1.4 Indicators of manufacturing, trade balances and external debt

	A *1975*	*B* *1975*	*C* *1975*			*D* *1976*	*E* *1976*	*F* *1977*	*G* *1977*	
			Value added as % all production							
	Value added in manufacturing (million 1970 US $)	*Column A as estimated per capita population $*	*Machinery & transport equipment*	*Chemicals*	*Other manufactures excl. textiles & clothing*	*Manufactured exports as % of all merchandise exports*	*Manufactured imports as % of all merchandise imports*	*Current account balance excl. foreign debt service million $*	*Foreign public debt, outstanding and disbursed*	
									million $	*% of GNP*
Egypt	1,651	44	11	13	25	27	59	−529	8,099	69.2
Morocco	551	30	5	7	32	16	62	−1,743	3,469	36.0
Syria	341	44	—	3	15	10	69	−137	1,528	20.7
Algeria	913	54	11	6	36	1	77	−1,935	8,165	42.5
Turkey	3,426	82	*	*	30	24	68	−3,155	4,323	9.5
Iraq	503	43	*	3	38	**	81	+1,209	761	4.0
Iran	3,240	93	26	6	31	1	86	+5,371	6,198	7.5
S. Arabia	453	60	*	*	*	**	84	+12,791	*	*
Libya	154	59	*	*	*	**	79	+2,905	*	*
Kuwait	200(est.)	180	*	*	*	3	*	+5,483	*	*
Spain	15,234	420	18	9	43	69	44	−2,055		
Japan	91,071	811	36	11	38	96	18	+10,911		

Source: World Bank, *World Development Report* 1979
* Not available
** Insignificant

manufacturing (column A), only two countries, Turkey and Iran, could claim some slight degree of world importance, comparable for example with Finland or the Korean Republic; Egypt lying somewhat lower, approximately equivalent to Pakistan and Israel. The more striking though somewhat crude indicator of column B further reinforces this point. Within the manufacturing sector as a whole, the key areas of mechanical and chemical engineering appear very undeveloped, except for Iran, possibly Egypt and the growing capability of Algeria.

Dependence on imported manufactures, especially of machinery and transport equipment (column E), is remarkably uniform but, when combined with the dependence on imported foodstuffs noted earlier, clearly imposes great strain on national balances of payment (column F). Unhappily, as all developing countries with time-ambitious industrializing programmes have found, such programmes require the importation of capital equipment for manufacturing processes, for industrial raw material production and a range of physical infrastructure requirements, particularly in communications. When these bills for capital imports have to be met, together with those for manufactured consumer goods (the demand for which increases if the development dynamo works), and those resulting from growing food deficits, then, inevitably, debt dependence grows (column G).

These broad trends are common to most developing countries but in the context of the Middle East two particular sets of complications become significant, the first a variant on the theme of 'absorptive capacity', the second and interrelated group having to do with human attitudes and behaviour.

Much has been written about countries' absorptive capacity for industrial investment but it is still legitimate to maintain that too little attention has been paid both to the multi-variable complexity of this problem and to the critical significance of some regional and country-specific factors. Here one can but illustrate the situation.

A common starting point for Middle-Eastern analysis is the differentiation between the oil exporters and the rest. In the case of the former 'increased oil revenues have overcome the poor countries' traditional and often most serious bottleneck, the shortage of capital' (Al-Hamad 1977). Almost invariably attention is then drawn to the survival of other bottlenecks, structural and functional in type, ranging from deficiencies in labour, skills and transport to inflationary pressures; it is usually further assumed that these can be broken by a set of more or less uniform procedures which can be taken by governments. However, if we take analysis a little deeper then we uncover a greater degree of heterogeneity at different levels.

Morgan (1979) differentiates two groups of oil-exporting countries, six with 'relatively low short-run absorptive capacity': Saudi Arabia, Kuwait, the UAE, Qatar, Iraq and Libya, and three with high absorptive capacity:

Algeria, Iran and Oman. It is surely no accident that in the group of six the three Gulf Emirates have an almost total deficiency in the non-oil and gas material resources base for manufacturing, Saudi Arabia and Libya are severely limited in the range and volume of their endowments; Iraq's slightly stronger position rests mainly on its relatively greater wealth in agricultural raw materials and water. Conversely, the higher absorptive capacity of Algeria and Iran follows in large measure from their far superior endowment in minerals and other raw materials; Oman is weaker in this respect. It could be maintained that this is to exaggerate the significance of the domestic material resource base but if we consider it in context the reverse is true.

In all nine countries the concentration in the hands of governments of the largest element in revenue and foreign exchange, export earnings from oil and gas, has encouraged the dominance of public sector investment and expenditure whether in countries wedded to principles of free enterprise such as Saudi Arabia and Oman or to state ownership and control as in Algeria and Iraq. In the industrially nascent countries various temptations then appear. The first is for the state to invest large blocks of the only readily available resource – capital – in those few basic industrial sectors where material resource advantage is greatest. This is precisely what has happened, capital-intensive industrialization utilizing domestic hydrocarbon resources, hydrocarbon feedstock for petrochemicals, and the supplementation, to a greater or lesser extent, of other material requirements by imports; necessarily also importing capital equipment and production systems from a small number of industrially advanced countries (which are also the main markets for oil and gas) (Turner and Bedore 1980). The greater the dependence on oil and gas to supply energy, raw materials and hard currency export revenue, the less is the freedom of manoeuvre and the more critical becomes the balance of factors in the equation. Thus, whilst Kuwait can claim some success in its industrial programme, with an annual rate of growth (value added) in manufacturing of 28.7 per cent between 1971 and 1977, between 1974 and 1977 the only industrial sector to show an absolute and proportional decline was the chemical industry, from 37.2 million KD to 26.8 KD, because of a sharp fall in 1976 and 1977 in the marketable price of ammonium sulphate and sulphuric acid (IBK 1979), partly as the result of over-production of these commodities by oil-based economies. In Libya, the Misurata steel-making project, with a planned capacity of 7 million tons a year at a total capital cost of some $5 billion, and based on domestic energy and high-cost domestic ore, is only rational today in the context of a desperate desire to diversify an economy with a limited resource base in which almost all activities are nationalized.

The central committee of the Algerian ruling National Liberation Front Party, in its reappraisal of development policy in January 1980, illustrated the dangers of the second temptation besetting relatively affluent oil and gas states even with fairly high absorptive capacities, that of over-relaxing economic discipline in the mixing of social, political and economic objectives for state enterprises. The committee planned to slow the rate of industrial investment because of an excessive reliance on foreign technical and financial assistance (see Table 1.3) and at the same time to restructure over-large, unwieldy state organizations which, whilst not necessarily inefficient, have lost any true accountability because their objectives are not well defined. One of the catalysts in the Iranian revolution of 1979 was undoubtedly the fact that a confusion of aims in increasingly complex and ambitious development programmes led to a low level of return on social as well as economic inputs and to frustration in a growing proletariat.

Such a situation is most likely to develop in the socio-political circumstances of emergent countries and, in price terms as well as in industrial mix, where the resource base is severely skewed. In all the relevant Middle-Eastern countries an internal conflict of price interests appears between state agencies concerned with the production, distribution and processing of, for example, oil and gas, depending on which is selling and which is buying. Blandly to dismiss the problem as one to be solved in 'the national interest' is to avoid rather than resolve the issue, as many developed countries have found in analogous situations; for those countries, such as the United Arab Emirates, to which oil and gas resources offer the only foreseeable basis for development, early and correct decisions are critical. 'In the long term [*sic*], 30 or 40 years from now, the Arab countries could find themselves hungry for the energy which now forms the mainstay of their economic life' (El-Saadi 1980).

The last temptation is to accept that successful, large-scale industrial investment policies are difficult to implement rapidly in the Middle East, and to turn to expenditure in socially or politically desirable fields which are consuming rather than productive in type – housing, artificially created employment, military expenditure, subsidies, etc. The dangers here have been earlier exemplified in Egypt:

> the control of the prices of commodities produced by the public sector means . . . a reduction in the size of the profits needed for re-investment. Since the rent sector has been already confiscated and redistributed for consumption, the confiscation of the profits sector as well leaves some countries (e.g. Egypt) with no domestic source of saving and investment. The position is further aggravated by channelling most of the foreign aid received to consumption either directly or indirectly (through budget support used to cover up losses of the public sector and the subsidization of basic commodities). (Union of Arab Chambers of Commerce 1979)

In this case the situation was of course exacerbated by Nasserite socio-economic policy and a low national income. However the same basic elements, the control of prices of commodities produced in the public sector, the channelling of externally derived revenue to consumption directly and indirectly, may also be observed in the totally different national situation in Saudi Arabia. The effect on investment in both cases has been remarkably similar and the only significant difference is that Saudi Arabia is uniquely capable, over time, of buying its way out of the predicament while Egypt has to rely on external assistance.

Clearly, a great number of related issues cannot be considered in this essay, for example the extent to which sets of determinants of absorptive capacity for specific types of industrial investment regionally recur and the influence which this may have on the regional rationalization of investment policies. Nevertheless, it is equally clear that the choice of what is appropriate to a range of relevant national and regional scenarios is much more critically constrained than contemporary development theory allows.

When, finally, one briefly considers attitudinal and behavioural aspects of industrialization and economic diversification the same themes of unity and diversity reappear.

'The major capital stock of an industrially advanced country is not its physical equipment; it is the body of knowledge amassed from tested findings and the training of the population to use this knowledge' (Kuznetz 1955). If we extend this dictum to the countries of the Middle East, we find ourselves concentrating neither on the transfer of technology nor on formal education and training requirements but rather on the nature and variety of past experience. As noted earlier, the peoples of the Middle East have extensive histories of initiative as well as participation in secondary and tertiary economic activity. But, 'In the Middle East, millennia of experience in complex commerce and finance . . . have been characterized by a socio-economic ability continuously to adapt and adjust a survival-orientated, non-contractual economy and society' (Bowen-Jones 1972). We are not examining conventional 'pre-industrial' societies but rather states which have inherited certain types of 'tested findings' and structural responses which communities have found acceptable and practicable over the ages. Of especial significance here is the continuing strength of primary and personal rather than contractual and purely economic relations.

It is tempting to relegate such phenomena to a 'traditional' past and to insist on the presence of change but, I believe, dangerously mistaken. The industrial and craft guilds of Egypt disappeared as legal entities earlier this century but the numerically preponderant small industrial establishments of Cairo and Alexandria still maintain a highly personalized family-based ethos. So too, the continued growth of small industries in Turkey (Manisali 1978), and even more so in Lebanon, demonstrates the strength of survival-experienced small production systems in conditions of instability. The old special relationship between merchant dhow-owner and nakhoudar dhow-captain in the Gulf may no longer be important, but if one looks behind the sectoral breakdown of private sector industrial projects in Saudi Arabia (SIDF 1399) and Kuwait (IBK 1979) one finds, little-changed, the individuals and networks of combined entrepreneur-industrialist-financier, whose characteristics are fundamentally different from those of salaried industrialists. These experience-based networks reinforce and are reinforced by the tendencies, long observable in developing countries, for private sector investment initiative to be concentrated in certain types of immediate market-orientated industry, for example the simpler forms of fabrication and of construction material, clothing and footwear, etc. Such industries offer more certain returns than do the 'industrializing industries' but, even more important, require little radical change in the managerial, entrepreneurial and financial systems already established and understood. Conversely, it is worth considering whether, in specific cases and geographical circumstances, the ways in which the public sector is invoked to fill an otherwise underplayed modernizing role may, in fact, create a type of rentier economy in which behavioural transformation does not occur. 'The transitional path from one form of rentier economy to another . . . is highly unstable' (Abdel-Fadil 1979) is a dictum not only relevant to oil-rich communities with large foreign assets. Half a century of modernization in Iran did virtually nothing to transform the powerful and long-established bazaar systems in Tehran and provincial Iranian cities (Kano 1978).

Again, it can be maintained that neither the general nor the spatially particular manifestations of idiosyncratic and varied approaches to non-rural change are merely accidental or perverse. The thesis advanced here is that within this region we call the Middle East the strengths – and weaknesses – of established systems indicate very powerfully the spatial heterogeneity of opportunities afforded to and constraints upon the complex sets of specific components of development.

The countries of the Middle East, in different ways, have achieved a great deal in the field of development but it sometimes appears that in their haste to utilize that scarcest of all resources, time, they are tempted to ignore the full implications of geographical realities. The Middle East is a thick palimpsest of heterogeneous cultural appraisals of material potentials, of unique two-way challenges and responses between men and their environments. We cannot simply rewrite the human roles without reference to the setting.

This essay is not an exercise in geographical determinism, not a restatement of 'small is beautiful', and certainly not an argument for 'whatever is, is best'. Rather, it serves to illustrate some of the reasons why it is necessary to utilize, not only one but all types of 'engineering', social and technical, economic and scientific, in the search for what is appropriate, for what variously may be attained somewhere between what is feasible and what is desired.

References

ABDEL-FADIL, M. (1979) 'The pure oil-rentier states', *Oil and Arab Cooperation*, 5.3, Kuwait, OAPEC.

AL-HAMAD, A. (1977) *Arabian Markets and Arabian Development*, Kuwait, Kuwait Fund for Arab Economic Development.

AL KUWARI, A. K. (1978) *Oil Revenues of the Gulf Emirates*, London, Bowker for Durham University Middle East Centre.

AMIN, S. A. (1974) *The Modernization of Poverty*, Leiden, E. J. Brill.

BILGIN, B. (1970) 'Non-economic factors and their influence on economic development in the Middle East with special reference to Turkey', unpublished PhD thesis, University of Durham.

BOWEN-JONES, H. (1970) 'Land and labour productivity in the Middle East', *Proc. I.G.U. Agricultural Typology Conference*, Verona, IGU.

—— (1972) 'Urbanization and economic development', in Wright, W. D. C. and Stewart, D. H. (eds) *The Exploding City*, Edinburgh University Press, 94–105.

DOWSON, Sir ERNEST (1931) *An Inquiry into Land Tenure and Related Questions*, Letchworth.

EL MALLAKH, R., KADHIM, M., and POULSON, B. (1977) *Capital Investment in the Middle East*, New York, Praeger.

EL-SAADI, A. (1980) 'Aspects of Arab cooperation and coordination in the energy field', *OAPEC Bulletin*, 6.1, Kuwait.

FISHER, W. B. (1950) *The Middle East*, 1st edn, London, Methuen.

FOOD RESEARCH INSTITUTE (1956) *Food and Inflation in the Middle East 1940–45*, California, Stanford University Press.

GABBAY, R. (1978) *Communism and Agrarian Reform in Iraq*, London, Croom Helm.

GADDAFI, M. (1973) *The Revolution of First September: the Fourth Anniversary*, Tripoli, Libyan Ministry of Information and Culture.

IBK (1979) *Annual Report 1979, Industrial Bank of Kuwait*, Kuwait.

IRAQ MINISTRY OF IRRIGATION (1972) *General Scheme of Water Resources and Land Development in Iraq*, Baghdad.

KANO, H. (1978) 'City development and occupational change in Iran', *The Developing Economies*, 16.3, Tokyo, Institute of Developing Economies.

KING ABDALLAH OF JORDAN (1978) *My Memoirs Completed (Al-Takmilah)*, London, Longman.

KUZNETZ, S. (1955) *Processes and Problems of Industrialization in Underdeveloped Countries*, New York, UNO.

LAWLESS, R. I. (1976) 'Industrialization in the Maghreb – progress, problems and prospects', *Maghreb Review*, No. 2, 6–9 and 18.

MANISALI, E. (1978) *The Place of Small Industry in the Turkish Economy*, Istanbul, Faculty of Economics, Istanbul University.

MORGAN, D. (1979) 'Fiscal policy in oil-exporting countries, 1972–78', *Finance & Development*, 16.4, Washington, IMF & World Bank.

MUGHRAM, A. A. (1973) 'Assarah, Saudi Arabia', unpublished PhD thesis, University of Durham.

OMAN (1976) *The Development Plan 1976–1980*, Muscat, The Development Council.

SAUDI ARABIA (1975) *The Second Development Plan*, Riyadh, Ministry of Planning.

SAYIGH, Y. A. (1978a) *The Economies of the Arab World*, London, Croom Helm.

—— (1978b) *The Determinants of Arab Economic Development*, London, Croom Helm.

SIDF (1399 A. H.) *Saudi Industrial Development Fund Annual Report 1398/1399 and Review 1394/95 – 1398/99*, Riyadh.

SINCLAIR, C. A. and BIRKS, J. S. (1979) 'International migration in the Arab Middle East', *Third World Quarterly*, 1.2.

STEVENS, P. (1977) 'Planning in the Middle East', *An-Nahar Arab Report & Memo*, vol. 1, No. 8.

STIRLING, P. (1965) *Turkish Village*, London, Weidenfeld & Nicolson.

TIBAWI, A. L. (1969) *A Modern History of Syria*, London, Macmillan.

TURNER, L. and BEDORE, J. M. (1980) *Middle East Industrialisation*, Farnborough, Gower for RIIA.

UNIDO (1976) *Comparative Study of Development Plans of Arab States*, New York, United Nations Industrial Development Organization.

UNION OF ARAB CHAMBERS OF COMMERCE (1979) *Arab Economic Report*, No. 8. Beirut, General Union of Chambers of Commerce, Industry and Agriculture for Arab Countries.

WORLD BANK (1979) *World Development Report*, Washington.

YASA, I. (1957) *Hasanoğlan*, Ankara, Yeni Matbaa.

CHAPTER TWO

Renewable natural resources in the Middle East

J. A. ALLAN

The Middle East is poorly endowed with renewable natural resources other than solar energy. Water resources are concentrated in a few well-favoured areas, and in the absence of humid climate even the limited potential of the region's soils cannot be harnessed for agricultural production. The renewable natural resources of the region are not only limited, however, they are often marginal, deteriorating or liable to deteriorate if they are not managed skilfully.

Surveys of renewable natural resources become desirable if decisions are to be made about their management. The need to make such decisions or at least to investigate alternative management strategies arises when:

1. There is a need to improve the productivity of a resource mix, as measured in agricultural terms or in terms of water utilization for industrial and agricultural purposes.
2. There is a need to alleviate a deteriorating resource position. The deterioration may be a quantitative one, for example a declining groundwater body, or an eroding soil resource; or it may be qualitative decline through the pollution of a water resource or the salinization of an alluvial soil.
3. There is a change in the balance in the factors of production and new management options become possible. Many countries in the Middle East have experienced an extreme version of a change in this balance through the availability of oil revenues, especially in the recent past. Governments, as the main agencies of investment, need information in order to evaluate to what extent the newly won oil capital can be substituted for deficiencies in land (soil and water) resources or for shortages of skilled and unskilled labour.

In this review we shall discuss the disposition of renewable natural resources and then examine briefly the pattern of survey activity, first in a temporal sense, and next with respect to national and regional variation. There will be a concluding discussion on the special measures which have been taken to manage limited resources in parts of the Middle East, and the revised perspectives with which limitations, investments and returns must be viewed in the light of such experience. It will be suggested that the 'arid zone inferiority complex' (Amiran and Wilson 1973: 27) is not a satisfactory starting point for resource survey and evaluation. The perception of the most limited resource in a region changes according to the levels of management, the scale of inputs and the quality of product-marketing able to be brought to bear. Such development may be discontinuous in terms of space, in that one or other resource may be limiting, but the concentration of agricultural activity in space as well as by the season is not an unfamiliar device even in humid let alone arid environments. The implication for resources surveyors in the Middle East is, however, that it will rarely be appropriate to limit surveys to high rainfall areas. In some circumstances the prevalence of clear skies and therefore the availability of a timely warm growing season means that indifferent soils, watered by expensive water, make agricultural production viable. Current agricultural development experience in the Middle East does much to confirm the hypothesis of the environmental determinist, but the region is becoming rich in examples of unconventional resource use and it is time, therefore, that renewable natural resources be unconventionally appraised.

The disposition of renewable natural resources in the Middle East

Imperfect, but helpful, indicators of the current status and development of renewable natural resources in the Middle East are data relating to irrigated and dryland agriculture (Tables 2.1 and 2.2). Only the northern parts of Turkey, the coastal area of the Caspian in Iran, and parts of North Yemen are well watered in agricultural terms. The rest of the region is subject to deficiencies of moisture, and except in the relatively favoured coastal areas of Syria, Lebanon and Israel the annual rainfall is insufficient for crop production without supplementary irrigation.

Rainfed farming, even in the cooler winter season when evapotranspiration is not so excessive as in summer, depends on a well-distributed season of rainfall of upwards of 200 mm. Table 2.1 shows that only two countries of the region have more than 200 mm per year in all of their national areas. All the rest have about half or more than half of their areas deficient in rainfall for cultivation. Nine countries (Bahrain, Kuwait, Qatar, Saudi Arabia, the United Arab Emirates, South Yemen, Algeria, Egypt and Libya) have less than 10 per cent of their land with rainfall sufficient for rainfed farming.

Table 2.1 Climatic and terrain constraints on land use in the Middle East and North Africa

	Area with more than 200 mm mean annual rain		*Estimated percentage of the area with 200 mm rainfall in which terrain limits arable use*
Asia	*thousand ha*	*% of national area*	
Bahrain	0	0	—
Iran	72,260	44	80
Iraq	12,190	28	70
Israel	907	45	55
Jordan	1,400	15	20
Kuwait	0	0	—
Lebanon	1,017	100	70
Oman	4,460	12	99
Qatar	0	0	—
Saudi Arabia	5,000	3	80
Syria	8,950	49	40
Turkey	76,745	100	65
UAE	0	0	—
Yemen North	10,179	52	90
Yemen South	0	0	—
Africa			
Algeria	18,300	8	65
Egypt	0	0	—
Libya	6,268	4	62
Morocco	22,730	58	65
Sudan	140,650	59	75
Tunisia	4,830	31	20

Source: Author's estimates

Rainfed farming is not by any means possible everywhere within the rainfed area. Table 2.1 also shows that a very high proportion of the areas within the 200 mm isohyet comprises terrain unsuitable for cultivation. This is a predictable situation since much of the land which attracts the higher levels of precipitation is the upland and broken terrain of Turkey, Iran, Lebanon, Yemen, Algeria and Morocco. A similar, though less extreme position obtains in Syria and Libya where it is mainly the upland areas which receive the reliable rainfall, but these are frequently too broken for other than grazing use. For example the terrain factor reduces the area of land suitable for cultivation to less than 10 per cent in Iran, to 30 per cent in Lebanon, to 5 per cent in North Yemen, and from 50 per cent to 15 per cent in Sudan.

Table 2.2, which shows estimates of the arable land areas of Middle-Eastern and North African countries together with the area used for dryland farming,

Table 2.2 Agricultural indicators of the limited character of renewable natural resources in the countries of the Middle East and North Africa as managed in the mid-1970s

	Land area (thousand ha)	*Arable land* %	*Irrigated area* %	*Dryland farming area* %
Asia				
Bahrain	62	*	*	*
Iran	163,600	10.0	3.1	6.9
Iraq	43,542	15.6	6.6	9.0
Israel	2,005	20.8	8.7	14.1
Jordan	9,100	14.3	0.7	13.6
Kuwait	2,072	0.1	0	0.1
Lebanon	1,017	33.9	6.7	27.2
Oman	21,242	0.1	*	*
Qatar	2,201	*	*	*
Saudi Arabia	154,600	0.1	*	*
Syria	18,268	32.3	3.3	29.0
Turkey	76,745	36.7	2.5	34.2
UAE	8,360	*	*	*
Yemen North	19,500	0.1	*	*
Yemen South	28,768	0.1	*	*
	(511,082)			
Africa				
Algeria	220,486	3.1	0.2	2.9
Egypt	100,000	2.9	2.9	0
Libya	175,954	1.4	0.1	1.3
Morocco	38,990	19.1	16.9	2.2
Sudan	237,632	3.0	0.6	2.4
Tunisia	15,583	2.9	0.1	2.8
	(788,645)			

Source: World Bank Atlas 1977, FAO Production Yearbooks, and author's estimates
* Not available

reflects the rainfall and terrain conditions discussed in the preceding paragraph. The estimate of usable arable land for dryland farming is at best only just over 10 per cent for the Asian countries of the region and 7 per cent for the countries of northern Africa. The position would be much worse in the Asian countries if Turkey were to be excluded from the calculations. Surface runoff and groundwater extend the region's usable land into areas of deficient rainfall, especially in the Nile and the Tigris–Euphrates valleys, but still only bring the usable area to just over 11 per cent for the Asian and to under 8 per cent for the African parts of the region. The importance of the river lowlands is not only in increasing the area suitable for arable farming; much more important is the qualitative matter of the high potential productivity of the irrigated river lowlands. This high potential is frequently impaired by mismanagement, but the river lowlands have for the past five millennia produced and will in the future produce a disproportionate amount of the agricultural production of the region.

The renewable natural resources of concern to agriculture are soil and water. These are the resources which can be enhanced or degraded according to the method of management. Shortage and unreliability of rainfall are the most severe problems of the region with respect to agricultural production. Deficiencies in rainfall also affect runoff and the recharge of groundwater which are the sources of supplementary moisture for irrigated farming. The surface runoff resources of the region have been managed for many centuries, and river works have been constructed on all the major and on some of the minor rivers. There are few further possibilities of a major engineering nature for manipulating river water in the Middle East; however there are still many desirable engineering activities of a less spectacular nature than the large dams so far given attention. These minor works which should be implemented are to do with efficient water distribution and soil drainage.

Groundwater may be renewable or non-renewable. In this chapter we are concerned with the former. As rainfall is low and unreliable in the region, the recharge of aquifers is similarly impaired. Geological conditions are favourable to the accumulation of groundwater in many of the coastal areas with Secondary and Tertiary sediments which mantle the ancient basement underlying most of northern Africa and South-West Asia. These sediments have proved to be very reliable reservoirs for water accumulated through the last 30,000 years. Some aquifers lie in the Nubian sandstone of northern Africa and equivalent sediments of Arabia. These are fossil resources and no current recharge has as yet been proven.

The renewable groundwater of the region is associated with areas of current rather than ancient precipitation. On occasions the groundwater catchment may be much more extensive than the area of intensive use through favourable geological circumstances, for example along the North African coast of Libya where gently dipping relatively undisturbed sediments slope northwards to the coast permitting the aquifers to transmit water to the main areas of agriculture. Unfortunately, the aquifers of the Middle East and North Africa are nowhere sufficient for the demands being made upon them. Withdrawal gradually increased annually up to the 1950s. Since then the availability of pumps and the competence to purchase them in large quantities in the oil-rich countries has led to dramatic rates of decline in groundwater storage. For example, in the most favoured area of Libya, the Jefara plain, withdrawal is on average four times that of recharge and in places eight times. Even in Israel, where water resources are monitored very carefully, the tendency is to manage them in a way which leads to excesses of withdrawal rather than a conservation or enhancement of storage.

An overall assessment of the region's soil resources is made difficult by the extreme climate which causes more than 80 per cent of the area south of Iran and Turkey to be desert, as well as 30 per cent of Iran. That soils in these desert areas will often produce reasonable yields provided water can be led to them has been proved for millennia in the major river valleys, and very recently in the central Sahara of Libya, as well as in remote areas of Saudi Arabia, where fossil water has been pumped to irrigate field and orchard crops.

The main environmental factors affecting the soils of the rainfed tracts of the region are the seasonal character of the rainfall, the high summer temperatures, the underlying lithology and the general sparseness of natural vegetation, a condition aggravated by over-use. The development of soil profiles is affected by the downward leaching associated with winter conditions and a potential upward capillary movement of water and associated soluble salts in summer. The development of soil is impaired by the relatively slow rate of weathering associated with low precipitation, and the organic content is also generally low because of the specialized character of the natural and semi-natural vegetation which has adjusted its production of biomass to the irregular and restricted rainfall. Limestone underlies substantial tracts of northern Africa and South-West Asia, especially the upland areas of Turkey, the Levant and North Africa where are found the patches of terra rossa soils (on the hard limestones) and rendzina soils (on the soft limestones). Elsewhere, soils with a high silica content are common, and it is on these mobile unconsolidated soils that much of the recent extension of irrigated agriculture has taken place. Soils used for dryland farming, or those which are only partially irrigated, are subject to a severe erosion hazard. For half the year at least the upper part of the profile is dry and liable to wind erosion. The generally poor vegetation cover means that sudden storms lead to severe erosion, and when the wadis flow they are always highly charged with suspended material.

The soils of the river lowlands have developed from the alluvial processes which transport and sort silt and clay fractions, sometimes brought vast distances by rivers such as the Nile. The presence of the rivers leads to seasonally, and often to perennially, high levels of soil-moisture content and high water tables. In the latitudes of the Middle East the temperature regime is such that all major river valleys are likely to have naturally occurring saline–alkali soils. These develop through the upward movement of moisture and associated soluble soils caused by capillary action during periods of prolonged high temperatures. Water tables within one or two metres or less of the surface provide an uninterrupted source of moisture for such processes which can lead to toxic accumulations of salts. Where high water tables naturally occur, there are associated tracts of saline–alkali soils. Elsewhere, extensive areas have been made unusable by the introduction of irrigation schemes with inadequate drainage provision. A high proportion of the irrigated land of Iraq is severely affected, the irrigated land of the Nile is to some extent affected and some of the new schemes, such as at Nubariyah, have caused very rapid rises in the groundwater which has in turn led to the degradation of the reclaimed land.

In short, the soil and water resources of the Middle East are limited, and the expensive processes of resources survey and evaluation should be deployed judiciously to maximize the returns on survey investments. For this reason, it has recently become commonplace to approach the evaluation of renewable natural resources through an integrated approach incorporating studies of environmental, economic and social variables relevant to the use of an area for a purpose defined by an appropriate politically responsive agency.

Before discussing the past and current trends in resources studies in the Middle East, it is appropriate to mention in general terms the approaches open to those managing renewable resources. A 'production approach' can be taken in which soil and water are used to produce crops and livestock. This 'production approach' can be favourable or not according to the time frame within which it is judged in a particular case. A production system may allow the sustained use of a region or it may lead to environmental degradation. A second approach is that of 'conservation', familiarly applied with respect to soil, water and vegetation resources, but it may also have a sociological or cultural connotation, as implicit in the 'folk ecology approach'. In the latter case, traditional institutions and land-management systems such as nomadism might be preserved by government policy. Happily, extreme policies of gross resource exploitation or inflexible conservation of environmental and institutional circumstances are rarely consciously pursued, although imperfect or misleading evaluations have on occasions led to development projects which have caused severe resource deterioration or irrevocable deterioration in living conditions. It will be shown that development goals have varied in response to the changing emphasis given to environmental and economic features of development, which in the recent past have yielded to complex compromises of the goals of welfare, export earnings and resource conservation.

The pattern of resource survey activity in the Middle East

The history of resources surveys in the Middle East

The Middle East has a long tradition in resource-evaluation studies. The rich soil and water resources of the Nile drew the attention of the Condominium government in Egypt, and a group of experienced irrigation and hydraulic engineers from India reviewed the resources and recommended a number of hydraulic works to regulate water supply and improve drainage (Hurst 1957). A number of studies reviewed by Hurst also dealt with Sudan and attributed a high value to the soils of the Gezira area between the White and the Blue Niles, and led to the creation of the settlement scheme which has been the model, sometimes misguidedly, for much subsequent agricultural development elsewhere in Sudan.

Interest of the same sort followed in the valley of the Tigris–Euphrates, when with the arrangement of the political frontiers after the First World War the Turkish influence over the Levant and Mesopotamia was replaced by a British and French presence. In the major river valleys the experience of engineers gained in India and Egypt was deployed to study the very evident problems of poor drainage and associated saline–alkali soils. And this was the beginning of a continuous interest by British irrigation engineers and other scientists in the management of the water resources of Iraq.

Elsewhere, colonization by French and Italian agencies in the late nineteenth and early twentieth centuries led to the best surveys to date of soil, water and vegetation resources of northern Africa. On the basis of these surveys, irrigated and dryland farming were extended and in most of the countries of northern Africa the cropped area was more than doubled.

This early phase of resources surveys was understandably concentrated on the regions of high potential for agriculture, and especially in the major river valleys. After the Second World War, however, greater attention began to be paid to other regions of the semi-arid and arid Middle East, and much of the work was reviewed and to some extent given a focus by the UNESCO Arid Zone Research symposia and publications. Hydrology, the utilization of saline waters and other water-related studies were certainly included, but the topics discussed were very wide ranging, with emphasis in climate and on the ecology of plant and animal production and land use (UNESCO 1960, 1963 and 1964). Another useful co-ordination of natural resources information came through the collation of surveys for the Soil Map of the World (UNESCO 1974). Two sheets of this series cover northern Africa and the Middle East, and these are currently being used by FAO to assess the crop potential of the region.

The early concentration of survey activity in the environmentally well-favoured areas was consistent with the limited survey resources able to be deployed in the region, and also consistent with the currently available agricultural technology. Further, as water was the limiting resource in arid-zone agricultural development, many countries had to wait for the extensive drilling activity funded by oil exploration to reveal the extent of the groundwater resources beneath the deserts.

Two factors have brought about a change in the emphasis in resources surveys since the Second World War. First, a number of arid countries, mainly with low populations, have joined the oil-rich and have been able to embrace ambitious development schemes involving extensive resources surveys. Secondly, the technologies for water management, and especially for the economical use of water, have caused a reappraisal of remote areas. These changes took place in circumstances of relative political instability and consequently of inexperienced national leadership. This inexperience extended to the evaluation and management of renewable natural resources. There are examples of re-surveys of areas previously evaluated unfavourably because more optimistic consultants have come forward with promises of resources which would realize the economic aims implicit in national development plans.

Increased activity in resource surveys took place according to the availability of oil-generated finance. Iran and Kuwait were the first economies to be transformed, followed by Iraq, all in the 1950s. The 1960s saw Algeria, Libya and the Gulf States join the oil-rich, and the profile of resources survey activities has only slightly lagged behind the revenue-generating capacity of the respective oil economies. A useful review of the soil and water research information fairly readily accessible was compiled by Clawson *et al.* (1971). The most interesting feature of this review is not the wealth of information assembled, but the gaps and deficiencies in resources data which it revealed. Despite the stimulus given to resources studies through the increased competence to commission surveys, governments were in the 1970s grossly deficient in data for resource management. It is not surprising, therefore, that the pace of resources surveys increased markedly in the oil-rich countries after 1973, when the impact of oil price increases began to affect disbursements to resource surveys. It is likely that the number of specialists from outside the region engaged in studies of soil, water and vegetation resources more than trebled between 1973 and 1976, and allocations from the national budgets of the oil economies to natural resources research increased by similar proportions. Saudi Arabia, for example, with its huge area had little water resources information available for planning purposes in 1970 (Beaumont 1977) although this was being remedied by country-wide surveys. Elsewhere expensive resources surveys were being commissioned to assess for a second and even third time, but more precisely than before, the agricultural and range potential of huge tracts of arid and semi-arid terrain. The cost of such surveys was so high that their commissioning was very significant, and by 1978 it revealed to increasingly experienced governments that the temptation to utilize resources on the basis of inadequate surveys had proved to be an extravagance which even the oil-rich could not afford. It is nevertheless understandable that politicians, in a hurry to demonstrate effective deployment of investment deriving from oil revenues, should be attracted to experimentation with the environment prior to detailed evaluation, because such evaluation might reach 20 per cent of the development costs of the scheme being proposed. Such high-level costs were doubly unattractive in the inflationary times of the past decade.

The high cost of resources surveys which the oil-rich can sustain are quite unacceptable in countries without the advantages and distortions of oil wealth. Moreover, a more rational approach to resource inventory has been demanded in semi-arid areas where potential productive returns can never justify expensive investments in research or other inputs, a demand also made very loudly by resource managers in other arid regions such as Australia (Robertson and Stoner 1970: 3–4). Flexibility and economy in surveys began to be seen as essential, rather than the rigid adoption of general specifications for soil survey and land classification. Most important, it was recognized that sampling procedures should be adopted.

At the same time a comprehensive approach to land evaluation had become the conventional wisdom by the late 1960s, despite the difficulties which had been experienced in integrating the myriad of disciplines needed to survey all the environmental, economic and sociological factors pertinent to the evaluation of a region. By the mid-1970s, the integrated approach was seen by many to be just as relevant to the whole development process. In Sudan, however, the methodologies for the acquisition of survey data for planning purposes have been tempered in the harsh circumstances of huge area-covering requirements, seasonally dynamic traditional methods of land use, a very low environmental potential and scarce capital for resource surveys. The large area suggested aerial methods of survey, the need for economy urged the use of sampling, and the desirability of a comprehensive approach led to the development of the aerial transect surveys of ecological zones; these zones are defined by preliminary work on aerial imagery, often on satellite imagery, and occasionally on existing photography or by aerial reconnaissance (Watson and Tippett *et al.* 1975). Those who have developed these economical methods do not pretend that ground survey is not also necessary, and in Sudan complementary detailed surveys and case studies were implemented on the ground. The relevance of the approach is very clear for Middle-Eastern and North African environments, where productivity is likely to be low and real returns can only justify an inexpensive monitoring system. One of its drawbacks is its sophistication in terms of the statistical and remote sensing skills required to implement the surveys, with the main constraint being the availability of experienced ecologists who can also pilot aircraft to carry out the inventory work. However, it seems to be proven that without the judicious use of such survey methodologies it will not be possible to provide data for the integrated rural development of the huge semi-arid tracts, such as those of Sudan, which must remain for environmental reasons areas of low-intensity use.

The disposition of survey activity

The extent and purpose of resources surveys in the Middle East and North Africa has varied according to government motivation and financial competence, as well as with respect to the survey methodologies in currency. A comprehensive regional assessment of the surveys carried out to date is difficult because of the confidentiality of many of the data gathered. Also current activity is extremely dynamic and survey methodologies are in a state of flux, both in the sense of the comprehensive, integrated and economical approach discussed in the previous section and also through the use of new technologies associated with satellite sensing.

The regions of traditional high-intensity use remain the best known. The irrigated lowlands of the Nile and Tigris–Euphrates have the most detailed soil surveys. Such regions are being subjected to major changes, however, and efforts are being made to monitor the impact of such river works as the High Dam on the Nile completed in 1970 and the Euphrates Dam completed in 1973. A number of Egyptian studies have looked at recent changes in the Nile regime and consequent soil and water quality. Outside agencies have been invited to examine the recent changes in the Nile, some to enhance the monitoring procedures such as the study mounted by the University of Michigan and the Egyptian National Research Centre (Gischler 1976: 36), others to report on specific aspects of water management and in cases of such severe soil deterioration, as at Nubariyah, on the progressive salinization of recently reclaimed areas. Parallel studies have been completed in Iraq and Syria, although these are, if anything, more difficult of access than the Nile studies (FAO 1966; Ubell 1971).

Outside the areas irrigated by surface flow, the land-use options are more limited. In some areas access to groundwater makes irrigated farming possible, and increasing emphasis has been given to the quantification of groundwater reservoirs and their recharge, if any. The early research carried out in the north-eastern Sahara was summarized by Murray (1952), who also suggested some controversial long-distance groundwater movements from the Nile to the Western Desert. More recently the water resources of the Nubian formations were evaluated on the basis of new work discussed at a meeting in Cairo (Pallas and Fadel 1974) and in publications elsewhere. The understanding and quantification of the groundwater resources of the Arabian peninsula were shown to be deficient by Burdon (1972). Extensive surveys have been completed in Jordan and in the Maghreb, where a very thorough and extensive assessment has been made of the region's aquifers by UNESCO (1972).

It is to be expected that soil surveys will not cover all areas; they have generally only been initiated in areas where groundwater or other water resources had been located. It is predictable, therefore, that Turkey should have the most extensive and most detailed soils maps in the Middle East. A high proportion of the agriculturally productive parts of Turkey has been surveyed at a scale of 1:25,000. Iraq, less well blessed with areas of high soil moisture, has a less comprehensive coverage of soils maps, although the position was well reviewed by Buringh (1960) and since then a substantial additional area has been surveyed at a detailed level with special emphasis given to soils hazardous to agriculture. Jordan has special problems with respect to its soils resources in that so much of its area is only able to sustain dryland farming and the demands made on the limited areas with sufficient soil moisture for dry farming have put them under great pressure for many decades. It is not surprising, therefore, that surveys in Jordan emphasize soil conservation (Fisher *et al.* 1966). The Arabian peninsula, like Jordan, has few areas of rich agricultural potential. The best-endowed region, North Yemen, and the only part of the peninsula with sufficient rainfall for agriculture, has recently opened its borders to resources scientists and the first studies are just appearing. Elsewhere reconnaissance surveys, such as those in the Gulf carried out in the late 1960s (Bowen-Jones *et al.* 1967), have been supplemented by detailed soils studies in regions of proposed agricultural development. A similar position obtains in Iran, although Iran was the first to attempt a national-level resources assessment based on a variety of data sources including remote sensing.

Even before the recognition of the need for an integrated approach to resources surveys, it was usual to include studies of land use including forestry and range management practices. Agriculturists, foresters and ecologists have surveyed and evaluated the soil and vegetation resources of the region for agriculture, grazing and forestry. Existing land-use studies are frequently commissioned to determine current land-use management practices and traditional expectations of the environment. The resulting studies have been diverse, and a useful attempt was made at the UNESCO Heraklion Symposium in 1962 to classify the climatic and hydrological variables which bear on land use in the northern part of the region (UNESCO 1964); the same study reviewed the status of land-use survey in most of the countries of northern Africa and the eastern Mediterranean.

There have been a number of classic studies which have taken advantage of aerial survey methods. The first, that of Lebon (1965) in Sudan, exploited small-scale photography of the huge and otherwise unwieldy national study area. The second was a survey of Cyprus (Hunting Technical Services 1969) where the complexity of two-season cropping was monitored by using timely photographic imagery for winter and summer activity.

Most areas of the Middle East and northern Africa are only suitable for utilization at the lowest level of intensity, namely grazing. All the countries of the region have published floras, and the status of such studies together with an invaluable bibliography are available in FAO's study of ecological management of arid and semi-arid rangelands in Africa and the Near and Middle East (Kernick 1978). Another wide-ranging study is the two-volume compendium of Zohary's (1973) own extensive but by no means comprehensive surveys of geobotanical resources. His fieldwork extended over thirty years and led to publications in many areas such as Sinai and concerning the region in general (Zohary 1954). An early flora of Syria, Palestine and Sinai appeared in 1933 and has now been superseded and made comprehensive in the national flora of the respective countries as well as in some special publications such as the atlas of range plants in Syria. Iraq has had a great deal of international agency attention, and published surveys of vegetation are numerous; here Kernick (1978) is the best source of information. The management and conservation of the vegetation of semi-arid areas has led to numerous studies directed towards the evaluation of vegetation for livestock and timber production and for the reclamation of otherwise unusable areas. Overviews of the region provided by Tisdale (1967) and Rossiter (1966) are supplemented by more detailed descriptions of range conditions in particular countries (Kernick 1978).

Changing perceptions of renewable natural resources: new survey technologies and new environmental management possibilities

The improvement in the economic circumstances of the region together with advances in technology have made it necessary to reappraise the resource base of the Middle East and North Africa. The unconventional appraisal of resources referred to by Amiran (1978: 127) is based on a wide range of experience in intensive vegetable raising (Kloner 1967), in the use of low-quality water (Evanari 1971) and the creation of controlled environments for horticulture. Extremes of lateral thinking have even led to discussions about the possibility of using transported icebergs as a source of fresh water. It is no longer relevant to take as the starting point a traditional view of the constraints of arid and semi-arid areas. Similarly it is no longer possible to take a traditional approach to the completion of resources surveys. Semi-arid areas which are normally cloud free are very susceptible to survey by remote sensing techniques from satellite platforms. These methods have a place as do field methods which yield data gathered according to appropriate sampling frames. There will always be a temptation to survey in greater detail in the oil-rich countries, such as Saudi Arabia and Libya, than in countries with more limited resources for survey purposes. Experience so far indicates, however, that very significant advantages in survey methodology have been made in countries which have to watch their survey budgets most carefully, as in the examples from Sudan already discussed, in Israel and Jordan, as well as in Egypt.

The most striking experience of those engaged in the survey and development of renewable natural resources in the Middle East is that the crucial factor in the successful implementation of surveys and subsequent development is the availability of a trained and highly motivated population. This factor is taken into account in the integrated approach to land evaluation, but surveys carried out by outside agencies with the necessary scientific and socio-economic skills take the process of development only so far. Sustained production from agricultural development schemes cannot depend on expatriate management and technicians in the fashion of the oil industry. Rates of return from the two industries are very different. In the rush to 'develop' natural resources, Middle-Eastern governments have sometimes been deceived by the flexibility and mobility of the skilled and unskilled labour forces from both Arab and non-Arab sources into the belief that development can be purchased. These labour movements reflect the great regional shortage of critical skills as well as of adequate institutions for mobilizing human effort (UNESCO 1968: 35). Nevertheless, it is only when a society has a broad foundation of trained manpower, as well as financial resources, that optimum and truly relevant surveys leading to effective development can be implemented.

References

AMIRAN, D. H. K. (1978) 'Geographical aspects of national planning in Israel; the management of limited resources', *Transactions of Institute of British Geographers*, 3, 115–28.

AMIRAN, D. H. K. and WILSON, A. W. (1973) *Coastal Deserts: their Natural and Human Environments*, Tuscon, Arizona.

BEAUMONT, P. (1977) 'Water and development in Saudi Arabia', *Geographical Journal*, 143, 42–60.

BOV, N. L. and GUEST, E. (1968) 'Flora of Iraq', *Gramineae*, 9, 588.

BOWEN JONES, H. *et al.* (1967) *Survey of Soils and Agricultural Potential in the Trucial States*, Department of Geography, University of Durham.

BURDON, D. J. (1972) 'Groundwater resources of Saudi Arabia', in Ayorty, M. K. (ed.) *Groundwater Resources in Arab Countries*, Cairo, ALECSO.

BURINGH, P. (1960) *Soils and Soil Conditions in Iraq*, Baghdad, Ministry of Agriculture.

CLAWSON, M. *et al.* (1971) *The Agricultural Potential of the Middle East*, New York, Elsevier.

EVANARI, M. (1971) *The Negev: The Challenge of a Desert*, Cambridge, Massachusetts, Harvard University Press.

FAO (1966) *High Dam Soil Survey, General Report*, FAO/SF: 16/UAR, Rome.

FISHER, W. B. *et al.* (1966) *Soil Survey of Wadi Ziqlab, Jordan*, Department of Geography, University of Durham.

GISCHLER, C. E. (1976) *Present and Future Trends in Water Resources Development in the Arab States*, Cairo, UNESCO/ROSTAS.

HUNTING TECHNICAL SERVICES (1969) *Land Use in Cyprus 1968. A Study Using Aerial Photographs*, Ministry of the Interior, Government of Cyprus.

HURST, H. E. (1957) *The Nile, a General Account of the River and the Utilization of its Waters*, rev. edn, London, Constable.

KERNICK, M. D. (1978) *Indigenous Arid and Semi-Arid Forage Plants of North Africa and the Near and Middle East*, EMASAE, Phase II, vol. 4, Rome, FAO.

KLONER, U. (1967) *Vegetable Growing under Plastic Cover in Israel*, Production and Extension Services, Israel, Ministry of Agriculture.

LEBON, J. H. G. (1965) *Land Use in Sudan*, World Land Use Survey, Berkhamsted, UK, Geographical Publications.

MURRAY, G. W. (1952) *The Artesian Water of Egypt*, Publications of the Survey Department of Egypt, Report 52.

PALLAS, P. and FADEL, M. (1974) *Note on Kufra Agricultural Project*, General Water Authority, Tripoli, Libya, Paper at a meeting on Nubian Sandstone Aquifers, 1974, Cairo.

ROBERTSON, V. C. and STONER, R. F. (1970) 'Land use surveying: a case for reducing the costs', in Cox, I. H. (ed.) *New Possibilities and Techniques for Land Use and Related Surveys*, Berkhamsted, UK, Geographical Publications.

ROSSITER, R. C. (1966) 'Ecology of Mediterranean annual type pasture', *Advances in Agronomy*, 18, 1–56.

TISDALE, E. W. (1967) *A Study of Dryland Conditions and Problems in Portions of South-West Asia, East and North Africa and the Eastern Mediterranean*, Drylands Research Institute, University of California, Riverside, Report No. 3.

UBELL, K. (1971) 'Iraq water resources', *Nature and Resources*, UNESCO, 7, 3–9.

UNESCO (1960) *A History of Land Use in Arid Regions*, Arid Zone Research Series, Paris, UNESCO, vol. 17.

—— (1963) *Nomades et Nomadisme au Sahara*, Arid Zone Research Series, Paris, UNESCO, vol. 19.

—— (1964) *Land Use in Semi Arid Mediterranean Climates*, Arid Zone Research Series, Paris, UNESCO, vol. 24.

—— (1968) 'Some aspects of the development of human resources in various countries of the Middle East', in *Studies on Selected Development Problems in Various Countries in the Middle East*, New York, United Nations Economic and Social Office in Beirut, 35–46.

—— (1972) *Etude des Ressources en Eau du Sahara Septentrional*, Paris, UNESCO.

—— (1974) *Soil Map of the World*, vol. 6, 'Africa including the Middle East', Paris.

WATSON, R. M. and TIPPETT, C. *et al.* (1975) *Sudan National Livestock Census and*

Resource Inventory, Khartoum, for the Ministry of Agriculture, Food and Natural Resources.

ZOHARY, M. (1954) 'Hydro-economical types in the vegetation of Near East deserts', in *Biology of Deserts*, London, Institute of Biology, 55–67.

—— (1973) *Geobotanical Foundations of the Middle East*, 2 vols, Stuttgart, Gustav Fischer Verlag.

CHAPTER THREE

Water resources and their management in the Middle East

PETER BEAUMONT

The Middle East is composed of a mosaic of often widely differing environments. In the northern part of the region, through Turkey and Iran, there are ranges of high mountains with crestlines up to 4000 m in height enclosing the high plateau regions of Anatolia and central Iran. In winter temperatures fall well below freezing, and snowfall often occurs associated with eastward-moving low-pressure systems. In contrast, the summers are nearly everywhere hot and dry. Further south in the Arabian peninsula and North Africa the mountains give way to a landscape of plains and plateaux, and here one finds the sand seas of the Rub al Khali and the Western Desert of Egypt. All of this region is characterized by low annual precipitation totals and hot summer temperatures; biological productivity is almost everywhere low.

The most important features in terms of pressure on resources in the region today have been the massive growth in population and, in particular, the increase in size of urban centres which has occurred during this century; these phenomena are examined elsewhere in this volume. In addition, significant changes in resource demand have followed changes in life-style. In the 1940s the populations of all the countries in the Middle East depended largely on agriculture and related activities for their livelihood. Society tended to be organized at the village or tribal level and many parts of the region were mainly self-sufficient with regard to basic foodstuffs. Urban centres existed, many of ancient foundation, such as Damascus, Jerusalem and Isfahan, in which market activities and small-scale industries flourished. Nowhere, however, did there exist the industrial city which has become one of the characteristic features of western Europe and North America in the late nineteenth and early twentieth centuries.

Equally important was the essentially symbiotic relationship which existed between the urban centres and their surrounding hinterlands in the more arid parts of the region. The towns provided a range of services and goods, while the agricultural areas around them provided foodstuffs, fibres and other materials (Beaumont, Blake and Wagstaff 1976). Until the time of the Second World War most rural/urban communities operated almost entirely on solar, animal and human energy and were dependent on those land and water resources which could be thus exploited, within relatively limited territories. As we shall see, all this has changed radically.

Surface water resources

Throughout the Middle East the areas of water surplus (precipitation minus evaporation) are surprisingly small and confined entirely to the northern part of the region (Thornthwaite, Mather and Carter 1958) (Fig. 3.1). Only in parts of north-eastern Turkey and north-western Iran are there major areas where the water surplus exceeds 1200 mm per annum; smaller areas of high water surplus occur along the highland regions of southern Turkey, the higher parts of the Elburz mountains in Iran, along the coastal strip of Syria and the Lebanon and the Black Sea coast of Turkey.

The isoline of 100 mm per annum of water surplus picks out the major uplands of South-West Asia and in a general way reflects the pattern of precipitation. However, the high evapotranspiration rates experienced by the Middle East mean that the water surpluses are only a fraction of the total amount of precipitation falling on the region (Perrin de Brichambaut and Wallen 1963).

It is these areas of water surplus which permit river systems to exist in the region and which are also responsible for recharging the groundwater resources. However, water surpluses generated in the northern part of the region are often transported very great distances into areas of severe water deficiency by river systems and groundwater aquifers. A classic case of such movement of water resources in South-West Asia is provided by the Tigris and Euphrates rivers, which transport the water surpluses of central and eastern Turkey to the intensely arid regions of southern Iraq (Al-Khashab 1958; Ubell 1971; Beaumont 1978b).

The major river systems of South-West Asia all rise in the areas of high water surplus. Not surprisingly, given the juxtaposition of highland areas and the major seas, many of the rivers of the region are short in length and possess only small drainage basins. Despite their small size, runoff rates are often very high. This is clearly seen in Turkey, where the small rivers draining into the Black Sea and Mediterranean Sea possess annual runoff rates in excess of 450,000 m^3 per sq km and sometimes even greater than 700,000 m^3 per sq km (General Directorate of State Hydraulic Works, Turkey 1968). In Iran, rivers draining the Elburz and Zagros mountains can also attain high annual runoff values of more than 250,000 m^3 per sq km (Beaumont 1973a), whilst further

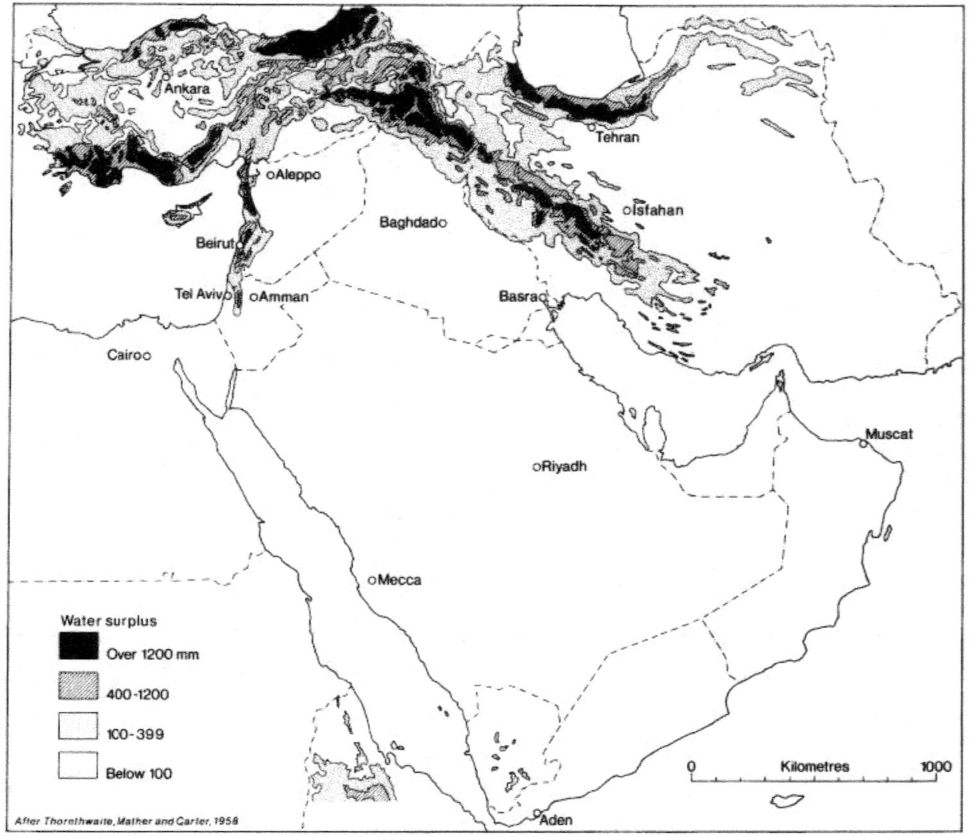

Fig. 3.1 Areas of water surplus in the Middle East.

south, the Karun River produces runoff at rates in excess of 300,000 m^3 per sq km in the upper part of its basin.

Of the three largest rivers of the region in terms of their total discharges, the Nile, Tigris and Euphrates (Table 3.1), the Nile is fed by water surpluses outside our region, in Ethiopia and East Africa, and has a discharge regime at Aswan in upper Egypt which is quite different from the other rivers of the Middle East (Khalil 1975). Along the Nile, peak discharges are attained during the period August–October as the result of monsoonal rains falling over the highlands of Ethiopia. During September the mean discharge is in excess of 8000 m^3/second; a truly staggering volume, and mostly contributed by the Blue Nile. Minimum-flow conditions on the Nile occur during April and May and at this time almost all of the water is brought by the White Nile from the plateau region around Lake Victoria (Hammerton 1972).

On most of the other rivers of the Middle East maximum discharge most frequently occurs during spring or early summer, despite the fact that precipitation within the region normally attains its highest values between November and February. The reason for the delay in the discharge peak is that most of the winter precipitation which occurs in the highlands of the Middle East falls as snow. The result is that spring snow-melt produces floods in all the major river systems draining the uplands of the region, giving characteristic March/April/May discharge peaks (Table 3.1). Minimum-flow conditions on most rivers are recorded at the end of the long, dry summer period in the months of August, September and October. At this time almost all of the water flowing in the rivers is being discharged from groundwater sources, as precipitation amounts are negligible.

The rivers of the region, therefore, have regimes which tend to be dominated by surface water runoff between March and May inclusive; by groundwater discharge between June and October, and then by a combination of the two for the rest of the year.

Like the Nile in Egypt, many of the rivers of the Middle East which drain into interior basins reveal a marked decrease in discharge in a downstream direction. Although this is believed to be a commonly occurring feature within the region, it is often difficult to prove quantitatively owing to the absence of gauging stations along individual rivers. Along the Zayandeh River in Iran, on which the oasis of Isfahan is located, there are four gauging stations which permit the downstream decrease in volume to be clearly seen (Beaumont in press) (Fig. 3.2). As with the Nile, the reasons for this are complex, being partly natural through evaporation and percolation and partly man-induced as a result of the diversion of water for irrigation. In many parts of lower Egypt the excessive use of irrigation water has disturbed the natural hydrological balance and caused the water table to rise substantially (Schulze and de Ridder 1974).

Table 3.1 Discharges of selected major rivers in the Middle East (m^3/second)

River	*J*	*F*	*M*	*A*	*M*	*J*	*J*	*A*	*S*	*O*	*N*	*D*	*Year*	*Record*
Nile, Aswan, Egypt	1,110	1,020	834	819	698	1,340	1,910	6,570	8,180	5,200	2,270	1,400	2,650	1912–62
Tigris, Fatha, Iraq	1,080	1,640	2,290	3,210	2,940	1,540	737	443	356	363	539	727	1,320	1931–66
Euphrates, Hit, Iraq	677	812	1,140	2,140	2,380	1,250	545	322	271	328	453	570	906	1932–66
Karun, Ahwaz, Iran	423	639	644	1,046	900	680	343	260	217	189	218	361	492	1955/6–1964/5
Sakarya, Botbasi, Turkey	242	328	350	354	273	162	101	73.8	92.1	115	123	213	202	1961–6
Ceyhan, Ceyhan, Turkey	265	324	421	446	326	153	66.9	47.6	48.3	61.2	70.9	265	199	1954–66
Kizilimak, Inozu, Turkey	174	234	308	260	190	172	89.5	62.0	87.1	98.9	97.7	132	159	1962–6
Karkheh, Hamidiyeh, Iran	117	195	166	266	342	120	43.7	28.4	20.7	24.4	46.9	91.5	121	1955/6–1964/5
Buyuk Marderes, Soke, Turkey	167	190	168	124	83.7	53.7	28.1	23.8	31.6	38.9	67.7	103	90	1951–66
Ghezel Ozan, Ostur, Iran	39.1	51.5	79.6	220	254	66.6	9.8	1.9	5.0	11.1	28.0	36.4	66.9	1955/6–1964/5
Zohreh, Deh Molla, Iran	74.1	105.3	77.6	91.7	67.7	41.5	40.2	15.4	15.6	17.7	25.9	94.9	53.7	1955/6–1964/5
Zarineh, Sara Chamish, Iran	17.2	34.5	75.7	166.9	157.3	29.6	10.0	4.2	2.9	5.9	9.0	17.0	44.2	1955/6–1964/5
Jordan, King Hussein Bridge, Jordan	44.4	57.8	44.8	34.3	27.4	24.2	23.9	23.3	24.9	26.2	29.5	31.5	32.5	1932/3–1958/9
Litani, Khardale, Lebanon	30.6	49.8	43.5	31.7	19.0	9.8	6.5	5.4	5.8	6.6	9.3	15.6	19.5	1947–69

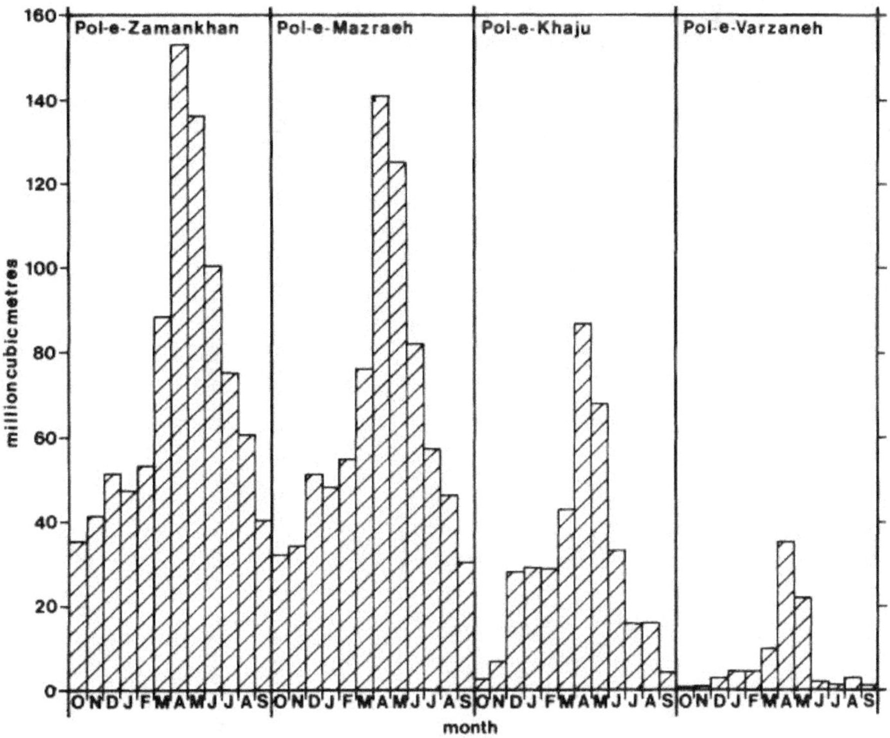

Fig. 3.2 Downstream variations in discharge along the Zayandeh River, Iran.

Until recently, even with those few rivers across which barrages had traditionally been built, it was the natural unregulated river regime which dictated the seasonal rhythm of water use in irrigation and therefore, also, many other periodic activities. Today the hydrology of many rivers of the Middle East has been considerably modified through the construction of large dams and associated reservoirs. Indeed, by the mid-1970s in this region 156 large dams (more than 15 m in height) had been built, were under construction, or were in an advanced planning stage (International Commission on Large Dams 1973; Beaumont 1978a). Of these dams, 94 were located in Turkey, and their reservoirs are able to store 43.6 per cent of the average annual discharge of all the rivers in that country.

Nowhere are the effects of human action better seen than on the Nile following the building of the Aswan High Dam. Since the final closing of the dam, the Nile downstream has been reduced to the role of a large irrigation canal, with the discharge regime now controlled entirely by the sluice gates on the dam. The result is that the pattern of water flow into Lake Nasser is quite different from the outflow into the Nile below the dam (Fig. 3.3). Maximum

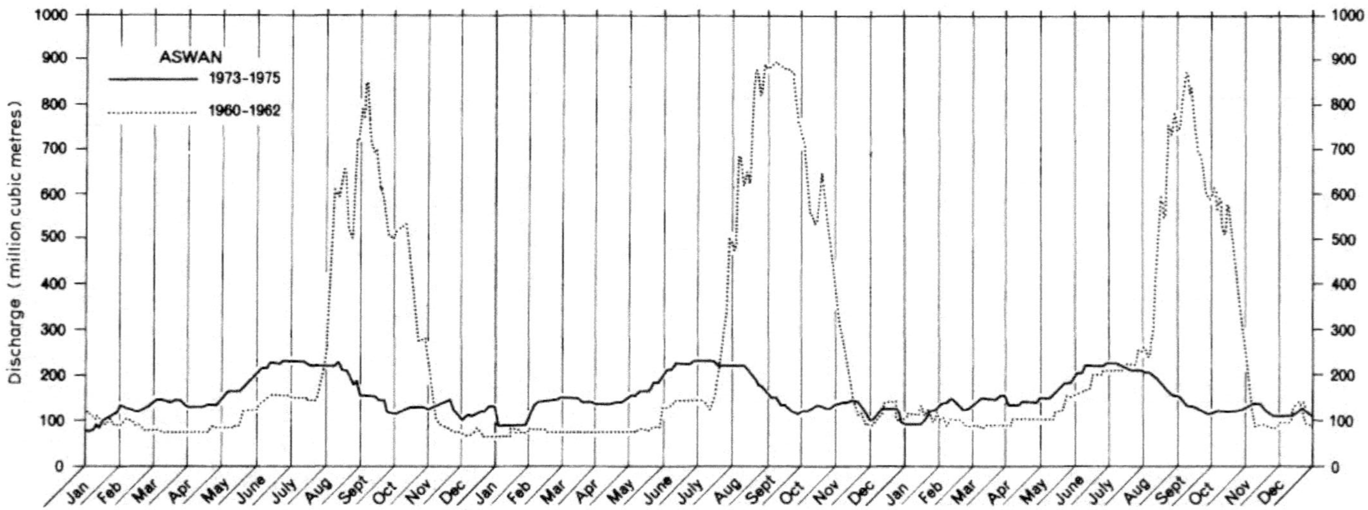

Fig. 3.3 Daily discharge regime of the River Nile before and after the construction of the Aswan High Dam.

discharge downstream from Aswan now takes place during the period of highest irrigation demand.

The building of the Aswan High Dam has also led to very high evaporation losses from Lake Nasser. The actual area of the lake has been the subject of some dispute, but it seems likely to be about 6000 sq km. Using the lower of the two available open-water evaporation estimates, of 2.2 m/annum, over this area Gischler (1979) gives an annual water loss of 13,200 million m^3. The likely evaporation losses from the lake are therefore slightly over 15 per cent of total river discharge. Considerable but unknown amounts of water are also lost by percolation through the bed and banks of the reservoir (Wafer and Labib 1973) – part of which provides a useful groundwater recharge for the Nubian Sandstone aquifer.

Another major problem facing many of the large dams of the Middle East has been the very high rates of sedimentation which have occurred in the reservoirs behind them. The high rates of erosion in the watersheds result from the generally arid conditions, the low-density vegetation cover and the widespread outcrop of rock types which break down easily into small particle sizes (Berry, Brophy and Naqash 1970; Philip 1968). This has meant that raindrop impact and the overland flow of water are highly erosive. Each storm, therefore, can carry away large volumes of sediment. In many areas the naturally high rates of erosion have been increased still further as the result of man's pastoral and cultivation activities. Research on the reservoir behind the Sepheed River Dam in Iran, for example, has revealed that 403 million m^3 of sediment accumulated during nine and a half years after the construction of the dam (Vali Khodjeini and Mohamed 1975). For the basin as a whole above the dam this produced the very high annual specific degradation rate of 1040 tonnes per sq km.

These high rates have meant that the life expectancy of many dams has often been greatly reduced. For example, the Dez Dam in Iran had an initial life expectancy of more than 100 years but this has been revised downwards to less than 50 years as a result of sediment accumulation. Many other dams in the region have suffered similar fates. In turn this leads to the expected benefits of a particular project being less than the costs of establishing the scheme in the first place, and so produces a major drain on a country's resources.

Taking into account both the direct and indirect effects, there can be no doubt that man has produced major changes to the surface water hydrology of the Middle East, to such an extent that it is only of academic interest to separate natural from man-induced conditions. Despite the growing impact of man, surprisingly little information is available dealing with water quality variations along rivers. Indeed it is only along the River Nile that detailed analyses of water quality have been available for a considerable time (Talling 1957; Mustafa 1973).

Groundwater resources

Groundwater conditions also vary considerably but, in general, it is possible to distinguish between shallow alluvial aquifers along river valleys and beneath alluvial fans and deep rock aquifers, usually of sandstone or limestone. In any particular region both types may be present. The shallow alluvial aquifers are generally unconfined, small in area, and have water tables which respond rapidly to local precipitation conditions (Bergstrom and Aten 1964). In contrast, the deep rock aquifers are often confined, sometimes of tremendous areal extent, and contain water which in part can be many thousands of years in age.

The natural recharge of the shallow alluvial aquifers takes place along broad gravel-floored valleys in the highland zones or, in the plains, on the upper parts of alluvial fans where the mainstream of a river breaks up into a number of distributaries. In these areas water percolation into the underlying gravels can be very rapid, and it is not unusual for all the waters of a river to disappear while crossing an alluvial fan zone. Indeed, in some cases water only manages to cross this zone and flow into the central part of a basin during extreme flood events. Natural recharge of the small alluvial aquifers of the Middle East usually occurs during the spring and early summer season at the time of maximum river discharge.

The major rock aquifers are usually much greater in extent than their alluvial counterparts, covering many thousands of square kilometres. Natural recharge of these aquifers occurs in upland or foothill regions where surface outcrops of an aquifer are found. The actual mode of recharge is still to some extent uncertain. It seems likely that recharge is achieved by the concentration of storm runoff into river channels followed by percolation through the bed into the underlying aquifer. In this way water draining from a considerable area can be recharged into a rock unit which may have only a limited surface outcrop area.

With these large aquifer systems there is dispute as to the degree to which natural recharge is occurring at the present day in extremely arid areas such as Egypt and Saudi Arabia. Part of the problem is that we still have little idea as to how much runoff is generated in these regions during the rare, but often locally intense, rainfall events. At Tehran, with an average annual precipitation of only 216 mm, a daily fall of 35 mm can be expected to occur with a return period of ten years (Gordon and Lockwood 1970). It is also not known with any certainty how much of this runoff is recharged into the underlying aquifer systems.

Recent research has suggested that the contemporary natural recharge of groundwater in very arid zones may be much greater than was previously believed. Work in Saudi Arabia, in the area of the Dahna sand dunes to the

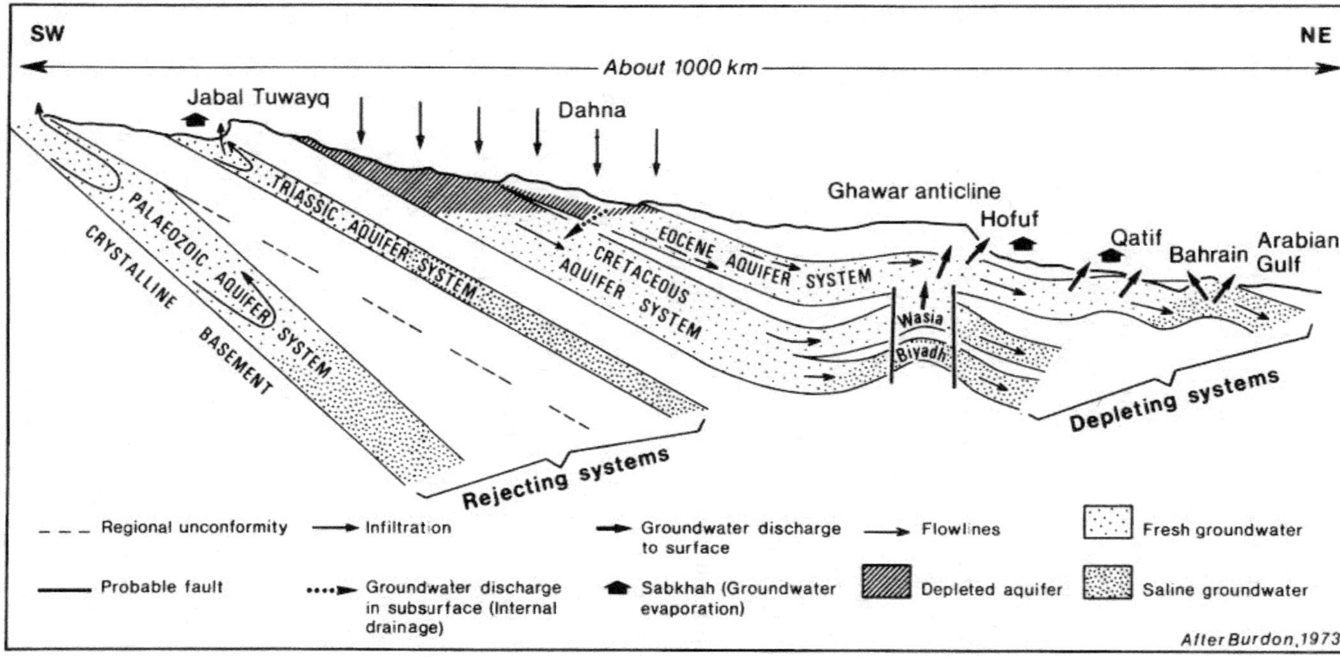

Fig. 3.4 Major aquifer systems of Saudi Arabia.

east of Riyadh, has revealed that even when annual precipitation totals are as low as 80 mm, about one-quarter of this amount can percolate downwards and recharge the groundwater systems (Dincer, Al-Mugrin and Zimmerman 1974). In dune areas it appears possible that groundwater recharge may occur even where annual precipitation is as low as 50 mm, provided that the sand is coarse grained.

There are no perennial streams in the whole of the Arabian peninsula, although perennial sections of rivers are found in parts of the Asir mountains of Saudi Arabia and the Yemen, and the Jabal Akhdar of Oman, where groundwater is forced to the surface to cross solid rock outcrops. Given this general lack of surface waters it is not surprising that attention has always been focused on groundwater reserves (Beaumont 1977a and b). Two major groundwater provinces can be recognized in the Arabian peninsula. In the western uplands ancient igneous and metamorphic rocks outcrop to form a rigid Basement Complex. Permeability and porosity of these rocks is low and so the only significant groundwater reserves are found in alluvial materials flooring the major wadi systems. Water quality in these small aquifers can vary considerably. Recently the artificial recharge of wadi aquifer systems has been carried out by the construction of small dams and the controlled release of floodwaters from them (Shaikh 1971). The rate of water discharge is designed to be such that the water can percolate into the alluvial deposits downstream from the dam.

Overlying the Basement Complex and forming surface outcrops throughout the eastern and southern parts of the peninsula are sedimentary formations attaining thicknesses of up to 500 m. Eleven major aquifers have been identified in the sequence ranging in age from the Cambrian to the Quaternary, within which four major aquifer systems have been described in the Arabian peninsula (Burdon 1973) (Fig. 3.4). The two oldest systems, the Palaeozoic and the Triassic, contain mainly sandstone aquifers with the Basement Complex forming a major aquiclude beneath the Cambrian rocks. Both appear to function as closed and confined aquifer systems, suggesting that natural discharge from them is minimal. Analysis of water samples from the Suq aquifer, of Cambrian age, revealed ages of 28,000 years BP (Otkun 1971). This suggests that most of the water in the aquifer is fossil in nature and that major recharge took place during the wetter stages of the Quaternary, probably between about 16,000 and 35,000 years ago (Thatcher, Rubin and Brown 1961).

The Wasia and Biyadh formations are the main water-bearing strata of the Cretaceous aquifer system. This, too, functions largely as a confined system, though some subsurface transfer to overlying aquifers does appear to take place. Natural discharge from the Cretaceous aquifer system seems to be greater than recharge at the present day as water-table levels are declining, even where no significant abstractions of water are occurring.

The Eocene aquifer system, which is largely composed of carbonate units, is also confined. During the Tertiary period the rocks were subjected to karstic conditions which have greatly improved the water-holding and transmission capabilities of the aquifer system (Burdon and Al-Sharhan 1968). Discharge from it takes place through terrestrial and submarine springs as well as by evaporation from coastal sabkhahs along the Gulf coast (Pike 1970). As groundwater levels are falling in the Eocene aquifer system, it would appear that recharge is less than discharge.

In Egypt, the only major water resources, apart from the Nile, are found in the Nubian aquifer system. The Nubian series, consisting of sands, sandstones, clays and shales, cover extensive areas of north-east Africa and attain a maximum thickness of more than 3500 m. Although a number of different aquifer units are present within the Nubian series it would appear that it functions as a single multi-layered artesian system covering about 2.5 million sq km (Hammad 1970; Himida 1970). Natural discharge from the aquifer system occurs into a series of depressions extending across the northern part of the region, the largest of these being the Qattara depression and the Siwa, Farafra, Bahariya and Dakhla oases. Most of the water which has been recharged into the system is believed to have originated as runoff from the highlands of north-east Chad, the western uplands of the Sudan and the Ennedi and Tibesti plateaux. Some recharge may occur directly from the Nile in the area to the south of Khartoum (Fig. 3.5). Seepage through the bed of Lake Nasser may also contribute to the total recharge of the aquifer system (Wafer and Labib 1973). The water-level contours of the Nubian aquifer system indicate that the dominant direction of water movement is towards the north-east; this tends to push the fresh/saline water interface, now situated about 200 km inland from the Mediterranean coast of Libya and Egypt, further towards the coast.

For the eastern Mediterranean littoral a large number of published works have appeared dealing with the groundwater resources of Israel and Jordan. A characteristic feature of the hydrology of Israel in particular has been the way in which it has been changed as a result of man's activities. These include overpumping of aquifers and artificial recharge in the coastal regions, the draining of the Huleh marshes and the transfer of large quantities of water from the catchment of the River Jordan (Schick and Inbar 1972). One example of human impact can be found in the case of the major aquifer of the Israeli coastal plain. Of Plio-Pleistocene age and consisting mainly of alluvial sands and gravels, the aquifer is between 7 km and 20 km in width, reaching depths of 180 m near the coast (Aberbach and Sellinger 1967; Ronen, Kanfi and Rebhun 1975). Natural recharge is the result of percolation of between one-

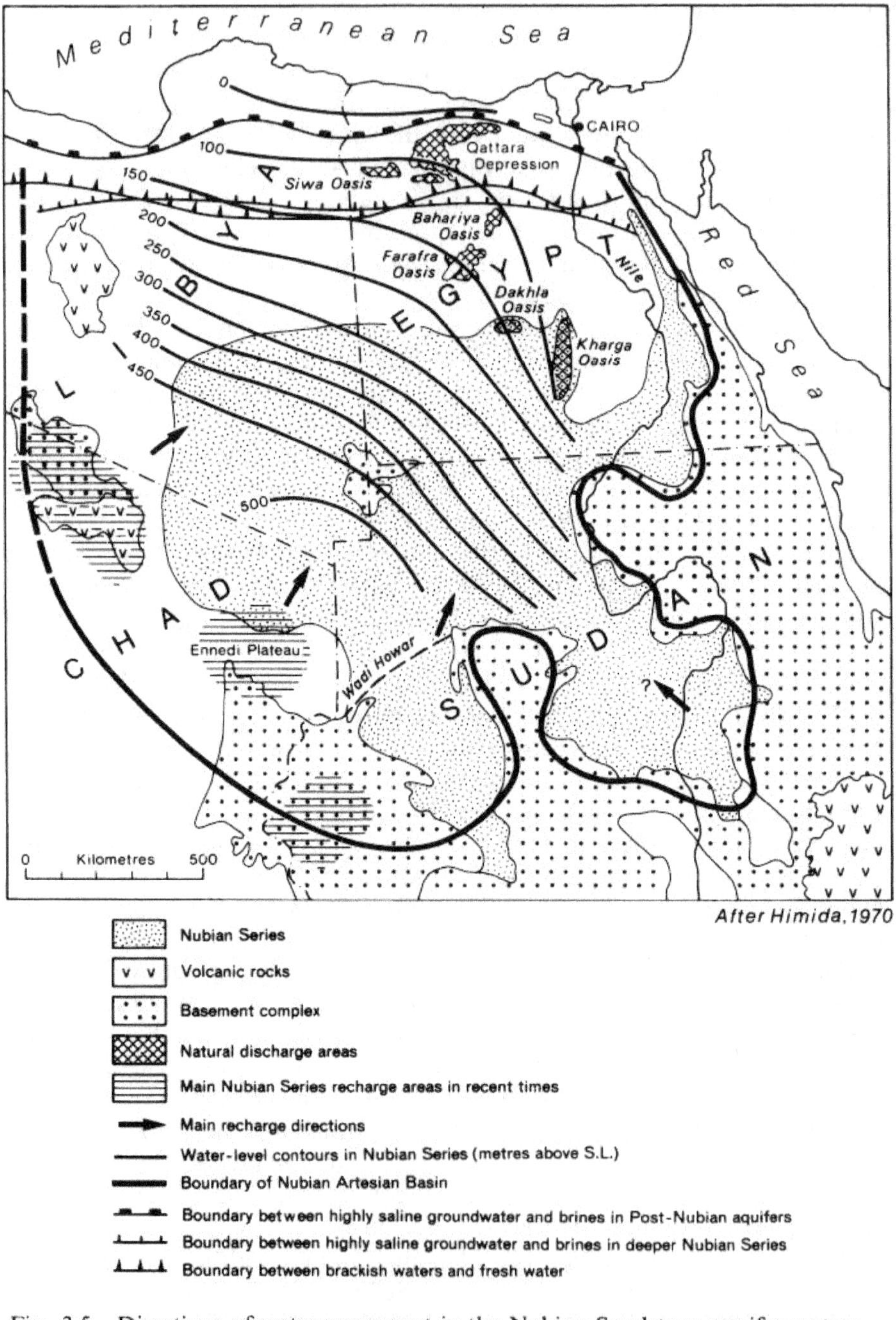

Fig. 3.5 Directions of water movement in the Nubian Sandstone aquifer system.

third and one-half of the precipitation falling on the uplands to the east. The major discharge from the aquifer is believed to take the form of submarine flow into the Mediterranean Sea. As the result of overpumping to supply irrigation needs during the 1950s and early 1960s, the water table dropped to levels of 2 m to 4 m below sea level at distances up to 3.5 km from the coast. This caused a reversal of the hydraulic gradient and the penetration of sea water into the aquifer for a distance of up to 2.5 km inland. To combat the sea water intrusion problem in the Tel Aviv area a freshwater mound was injected into the aquifer from 1964 onwards by the use of wells. More recently further artificial recharge of this aquifer system has been achieved by the diversion of floodwaters from the coastal wadis into sand dune zones where infiltration rates are high.

Winter precipitation falling on the uplands of Jordan, Israel and the West Bank is also the main source of recharge of a major aquifer in mid-Cretaceous limestones and dolomites. Two major discharge points from the aquifer are through the Tanninim springs in the north with an annual average discharge of 100 million m^3 and the Yarkon springs in the south with an average discharge of 220 million m^3 (Harpaz and Schwarz 1967; Schneider 1964). Here also the hydrological balance of the limestone aquifer has been disturbed by pumping from wells, most of which have been drilled between 1950 and 1965. As a result of this overpumping, which reached a value of 280 million m^3 in 1963, the discharge from the Tanninim and Yarkon springs has markedly declined.

During the 1960s considerable attention was focused in Jordan on elucidating the properties of the main aquifer systems of the East Bank (Mitchell 1967). As a result of this work, seven major aquifers have been recognized in rocks of Mesozoic or younger age, of which the Amman–Wadi Sir aquifer system is of by far the greatest economic importance. It outcrops in the high rainfall zone of the western highlands overlooking the Dead Sea lowlands and is also present beneath younger sediments in the plateau region. Water recharge to this system is thought to average about 336 million m^3/annum, with most of this being derived directly from percolating precipitation and runoff, but C_{14} dates for the groundwater have revealed ages of up to 30,000 years BP, suggesting that at least some of the water was recharged into the aquifer during the latter part of the Pleistocene period. Almost two-thirds of the total, 199 million m^3, is believed to be discharged from springs or seepage points along the rift escarpment (UNDP/FAO 1970). The remainder is discharged either eastwards or westwards as groundwater flow.

In Israel the salt content of groundwater appears to be largely the result of the deposition of airborne salts derived from marine sources. These salts are carried eastwards by the prevailing winds and then precipitated by rain, especially in the coastal areas, but also well inland. Estimates for the whole of Israel suggest that about 100,000 tonnes of NaCl are deposited from the air each year (Yaalon 1963). This compares with a maximum figure of 500 tonnes per year released by weathering and about 30,000 tonnes as the result of industrial production. The importance of irrigation's effect on the salt content of groundwater should not, however, be underestimated anywhere in our region, as drainage waters can have salt contents ten times or more than those in the original waters. Some of these drainage waters will undoubtedly percolate into the soil to become groundwater, and if the groundwater bodies are small the overall effects on water quality may be considerable.

In Iran the most important aquifer systems are the alluvial deposits of Tertiary and Quaternary age in the northern, eastern and central parts of the country, and the carbonate rocks of Jurassic, Cretaceous and Eocene age in western and south-western Iran (Issar 1969). Elsewhere, other aquifers may be of local importance. The most widespread aquiclude in the country is a sequence of marls and evaporites of Tertiary age. As a result of Tertiary and Pleistocene tectonic movements, it is possible to identify five major groundwater provinces in Iran.

In Azerbaijan the most important aquifer consists of a series of alluvial tuffs of Plio-Pleistocene age, which occur to the south of Tabriz (Fookes and Knill 1969). In the Zagros mountains a thick sequence of calcareous rocks, mainly of Oligocene, Miocene, Eocene and Cretaceous age, outcrop in a series of anticlines, some of which form important aquifers. The water in the limestone aquifers appears to form part of a regional aquifer system connecting up different anticlinal structures. Large springs draining this major aquifer system feed the headwaters of the Karun and Dez rivers.

The Elburz mountains are relatively unimportant from the groundwater point of view, largely as a result of the complex geological structure of the area and the widespread occurrence of impermeable or semi-impermeable rocks. To the north of the mountains are a series of sedimentary plains fringing the Caspian Sea consisting of layers of sands, gravels, silts and clays, and so confined aquifer systems commonly occur.

The main aquifer system on the central plateau of Iran is made up of thick Pleistocene alluvial deposits which attain thicknesses of 400 m (Vita-Finzi 1968). Recharge of this aquifer system occurs almost exclusively as the result of floodwaters leaving the Elburz and Zagros mountains and spreading out over the fringing alluvial fans. Recharge is greatly helped by the numerous unlined irrigation canals which exist in the region (Beaumont 1974). The actual patterns of water movement are complex and it seems likely that water is often transferred from one aquifer to another by subsurface flow. Marked water-quality variations can also occur over small areas (Beaumont 1973b; Ighanian 1967).

Water balance and competition for resources

Owing to a lack of data, it is still not possible to calculate an accurate water balance for the Middle East as a whole. However, approximate water balances can be drawn up for certain countries or river basins, which permit one to discover just how much water is available for beneficial purposes. In Turkey, for example, it has been estimated that the mean precipitation provides $518 \times 10^9\ m^3$ of water each year (Fig. 3.6), of which about $351 \times 10^9\ m^3$, or 68 per cent of the total, is lost through evapotranspiration or deep percolation. The majority of this evapotranspiration occurs from non-agricultural lands, but about 15 per cent of the total ($76 \times 10^9\ m^3$) is from areas on which crops are grown. The total amount of surface runoff is, therefore, of the order of $167 \times 10^9\ m^3$ each year, about one-third of the mean annual precipitation to provide for man's water needs – for irrigation, for industry and for domestic supplies. In the mid-1960s it was estimated that about $80 \times 10^9\ m^3$ of water each year were being abstracted for beneficial uses and that, after use, about $46.3 \times 10^9\ m^3$ were being returned to the river systems, though sometimes with a much lower water quality. Almost all the consumptive use of water was accounted for by irrigation.

These data can be used in conjunction with population statistics to illustrate the varying pressures which are being put on Turkey's water resources. Surface runoff is not all available for human use since a proportion of this water will flow directly into the sea during flood events. To capture some of these floodwaters a number of dams have been constructed in Turkey. Even so, one has to realize that dams rarely control the total discharge from a catchment and consequently a large proportion of water will flow directly into the oceans before man is able to use it.

In 1900, the population of Turkey was about 14 million (Dewdney 1972), so the per capita water availability was about 11,929 m^3/annum. By 1977, with continued population growth to 41.9 million, this figure had fallen to 3986 m^3; by the year 2000, when the population is expected to reach about 72.4 million, it will be only 2307 m^3. What should be remembered is that these figures represent the likely maximum values for water availability and that in reality much lower figures may prevail. At the mid-1960s when the water balance was calculated, about $80 \times 10^9\ m^3$ of water were being utilized per annum by a population of 31.4 million. This represents a per capita water-use figure of 2548 m^3/annum. It will be noted that this is higher than the predicted water availability value for the year 2000.

The water-resource position in Turkey is further complicated by the fact that a large proportion of eastern Turkey forms the headwaters of the Euphrates, one of the largest rivers of the Middle East. The average volume of water leaving Turkey via the Euphrates each year is $28.4 \times 10^9\ m^3$ (Garbrecht

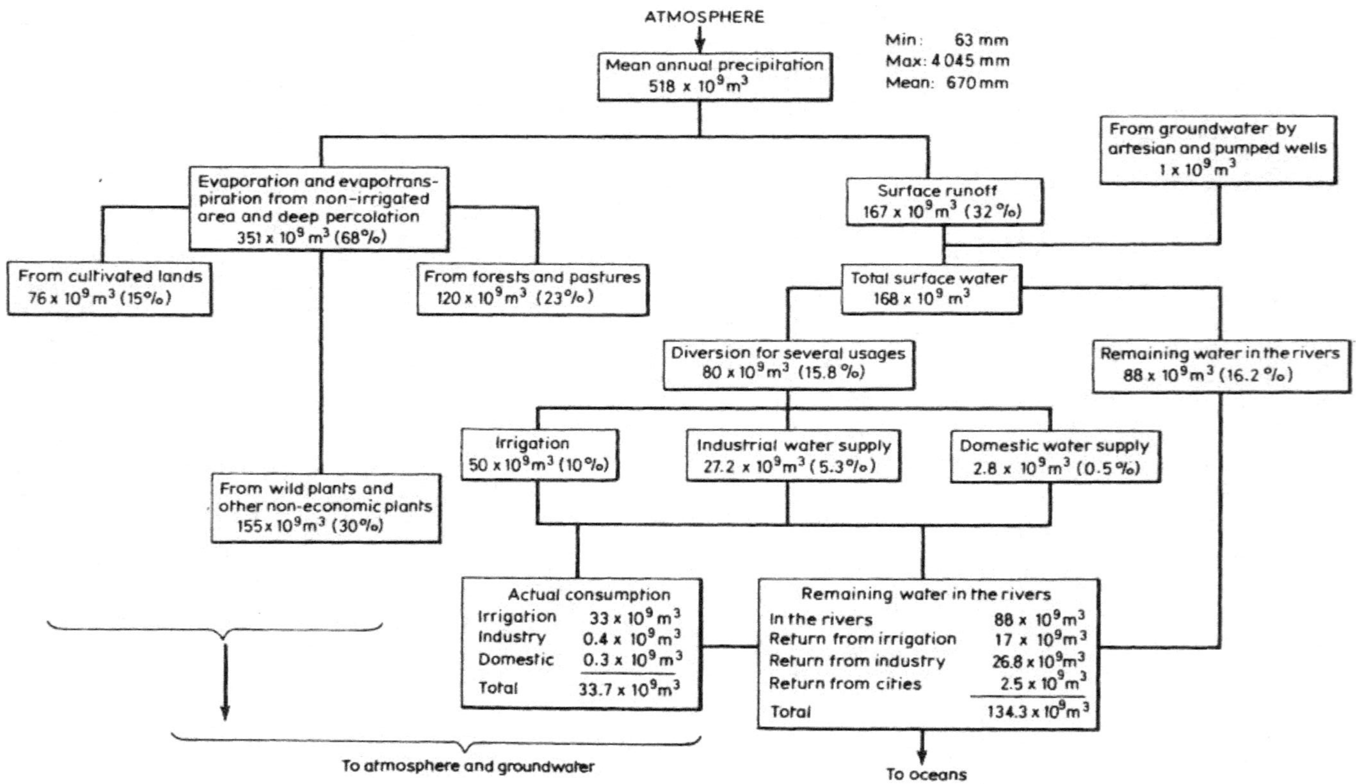

Fig. 3.6 Annual water balance for Turkey.

1968), representing about 17 per cent of the total runoff of the country, and about 88 per cent of the total flow of the Euphrates in Iraq. Downstream, Syria contributes a further 12 per cent, while virtually no water is generated within Iraq. It is, therefore, obvious that Turkey will only be able to utilize the waters of the Euphrates for consumptive uses at the expense of the countries downstream.

Until the last decade or so Iraq has been the only country with land within the basin of the Euphrates which has drawn extensively on the available water resources. Most of the water which is abstracted by Iraq is used for irrigation and so losses through evapotranspiration are high. Irrigation systems in Iraq have been in operation, though not always with the same intensity, for many thousands of years. Over the last few decades the actual volume of water abstracted by Iraq from the Euphrates has increased quite markedly from 8640 million m^3/annum in the 1940s to 16,370 million m^3 in the 1960s (Ubell 1971). This latter figure represents 45 per cent of the mean flow of the river at Hit, Iraq. Currently Iraq is also constructing a new dam at Haditha so that it will be able to store and use an even greater proportion of the water resources of the river.

Within the last decade both Syria and Turkey have begun to implement plans for the utilization of large volumes of water from the Euphrates. In Syria, the Tabqa Dam is to supply water to a new urban/industrial complex, as well as to new irrigated land. In Turkey the Keban Dam is to act as a major storage facility with which to regulate the flow of the Euphrates, whilst further downstream the Karakaya Dam will form part of a long-term water management project to develop irrigation in south-east Turkey.

When these schemes in Syria and Turkey are in full operation by the end of the twentieth century, they will have a major impact on the overall hydrological balance of the Euphrates. Estimates suggest that the demand for water may be so great that Iraq, the country furthest downstream, might have great difficulty in supplying her water needs (Beaumont 1978b). The predicted maximum water use when all the current development projects have been completed is between 28,000 and 36,500 million m^3/annum (Table 3.2). The mean annual flow of the Euphrates at Hit in Iraq is only 31,800 million m^3. It seems inevitable, therefore, that very severe problems will arise with regard to water use during periods of low flow conditions lasting over a number of years. What the above figures cannot take into account is the effect which the irrigation schemes will have on the overall water quality of the river. This is largely unpredictable, as it depends to a large degree on the amount of salt in storage within the upper layers of the soils which are being irrigated. There seems little doubt though that, unless a comprehensive strategy for the long-term development of the River Euphrates is agreed upon by Turkey, Syria

Table 3.2 Annual potential water use along the Euphrates River

Mean annual discharge at Hit, Iraq	*Million m³/annum* *31,820*
Turkey	
Evaporation from reservoir above Kaban Dam	476 (max)
Potential evaporation from reservoirs above Karakaya, Gölköy and Karabaka dams	607 (max)
Potential water withdrawal for irrigation	3,500–7,000
Syria	
Evaporation from reservoir above Tabqa Dam	630 (max)
Potential water withdrawal for irrigation	5,000–10,000
Iraq	
Current water use (1960–9)	17,213
Potential evaporation from reservoir above Haditha Dam	602 (max)
Predicted water use as a result of schemes actually built, under construction or planned	Minimum = 28,028 Maximum = 36,538

Source: Beaumont 1978b

and Iraq, there are likely to be serious disputes concerning water rights and water usage.

Irrigation

In the Middle East irrigated agriculture continues to be by far the largest user of water. The actual volumes of water used for irrigation are almost impossible to come by, but the size of the irrigated areas today in each of the countries does give some indication of the relative magnitude of water use. Iran contains by far the largest irrigated area followed in turn by Iraq, Egypt and Turkey. In all the other countries irrigation is of only minor significance in absolute terms, though relatively speaking, as in the case of Kuwait, it can have tremendous significance for the country as a whole.

The ways in which irrigation waters have been utilized reveal large regional variations. Until the Second World War traditional methods predominated almost everywhere, but since then rapid changes have occurred. The simplest form of irrigation involved the use of natural flooding, as on the banks of the Nile, or the diversion of surface flows and water transport by canal. Some ancient systems, for example the Nahrawan Canal constructed during the sixth century AD to irrigate the lower part of the Diyala plains and over 300 km in length and in places more than 50 m wide (Adams 1965), were extremely extensive.

Throughout the drier parts of the Middle East groundwater has played a vital role in sustaining human life. In eastern Arabia, southern Egypt and Libya classic small desert oases are found where groundwater seeps out at the surface to form springs or pools, but elsewhere man has had to devise methods to recover water from depths often in excess of 50 m. The simplest technique is the well shaft sunk to a few metres below the water table but the limited energy available to bring water to the surface traditionally imposed many constraints on its usefulness.

In Iran and other parts of the Middle East an ingenious method has long been used to extract groundwater and overcome the problem of energy shortage. The method consists of the construction of a sloping tunnel dug from the ground surface into alluvial material until it intersects an aquifer (Bybordi 1974). Water flows into the tunnel below the water table and then flows along by gravity until it reaches the ground surface (Fig. 3.7). These structures, known as *qanats* in Iran, are also common in Afghanistan, in which two countries they may have originated and whence they may have spread to parts of Arabia and North Africa (English 1968). The vast majority of qanats in Iran are between 1 km and 5 km in length and between 10 m and 50 m in depth (Beaumont 1968, 1971), with an average water discharge of between 1 m^3 and 80 m^3/hour, though sometimes values in excess of 300 m^3/hour have been recorded.

The greatest change to Middle-Eastern irrigation systems has been brought about by the widespread introduction of electric and diesel water pumps. These have done away with the need for human and animal labour, and during a time of cheap energy supply have meant that water could be pumped economically from considerable depths. As a result, many new wells have been sunk in areas not formerly cultivated, so permitting an extenstion of the agricultural area. Even in formerly cultivated areas, pumps have been installed on wells originally worked by animals and in some qanat-watered areas.

The use of the water pump has not been without severe environmental cost. In many areas the overpumping of aquifers has caused the water table to fall sharply, often leading to the drying up of qanat systems and so causing serious social and economic problems (Beaumont, Blake and Wagstaff 1976). So great did these difficulties become in Iran that the government passed a bill nationalizing all water resources in 1968 (Ministry of Information, Tehran 1970), and requiring that licences be issued for new wells.

A feature of many countries of the Middle East since the mid-1960s has been the implementation of large-scale irrigation projects, but many of these schemes have not lived up to the high design expectations. With most

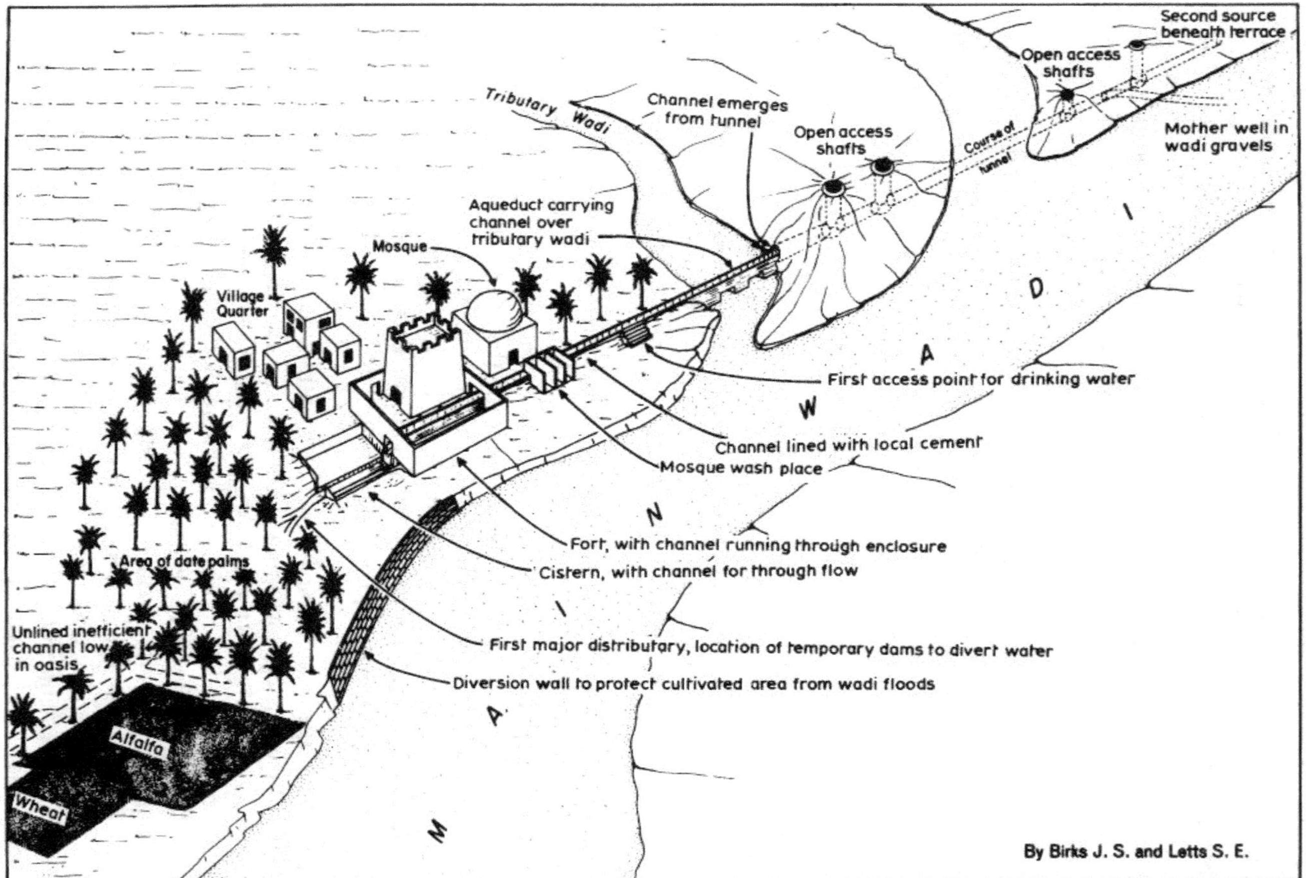

Fig. 3.7 Cross-section and plan of a falaj or qanat in Oman.

irrigation projects there is usually an initial period following construction when all parts of the system function properly, so permitting satisfactory crop yields to be produced. However, even during this time the groundwater table is rising in response to the new hydrological conditions. In many cases, soon afterwards the water table rises so close to the ground surface that salination of the soil becomes a major problem, especially if the irrigation project does not possess a satisfactory drainage network (Elgabaly 1977). The net result is that after a period of successful operation, lasting normally 2–20 years, the crop yields begin to fall as soil salination and waterlogging become ever more serious problems. At this stage, reclamation of the land becomes a very expensive and sometimes impracticable proposition, as can be seen today in parts of the Nile delta (Schulze and de Ridder 1974).

The Kufrah Agricultural Project is another example of schemes which have not fully lived up to expectations. This is a 10,000 ha project in south-east Libya geared to support 50,000 sheep on irrigated alfalfa (Allan 1976). Water was supplied from 102 deep wells, drilled in the early 1970s, and the water was distributed through self-propelled rotary boom sprinklers watering over 100 ha in every 60 hours. Hydrological modelling suggested that over a 40-year period of use the water table would fall by 20 to 36 m if the extraction rate was 4.3 m^3/second. However, within one year of operation, at an extraction rate of only 4.0 m^3/second, water levels had fallen by between 5 and 15 m. This figure suggested that the assumptions on which the programme had been designed might not have been correct (Gischler 1979). If it indeed proves that less water is available or that extraction costs are more expensive than was previously thought, then parts of the project might have to be re-evaluated. Similar problems have also arisen in the New Valley Project of Egypt.

Of all the countries in the Middle East, Israel has undoubtedly the most modern and mechanized irrigation systems (Wiener 1972). These have been introduced mainly to reduce the consumption of water for agricultural purposes, so that today almost all irrigation water is applied by sprinkler or drip/trickle techniques. Although these systems are efficient in terms of the water they use, they are costly to operate in energy terms. Pressure usually has to be maintained by pumping, and the pipe systems have to be renewed normally at least once a decade. Despite these sophisticated applications systems, irrigation is still by far the largest consumer of water in Israel and this position does not seem likely to change substantially in the future (Table 3.3).

Even more energy intensive are the controlled environment methods used in Israel for the production of specialized high-value crops such as fruit and flowers. Although these proved economical propositions in the cheap energy era of the 1960s and early 1970s, it is difficult to believe that they will be able to continue successfully for very much longer without heavy government subsidies (Stanhill 1974).

Table 3.3 Current and predicted water use in Israel (million m³/annum)

Use	*1971*	*1975*	*1980*	*1985*
Agriculture	1,170	1,210	1,250	1,250
Domestic	250	300	350	430
Industry	55	100	140	220
Losses	50	50	50	60
	1,525	1,660	1,790	1,960

Source: Jacobs, M. and Litwin, Y. (1973) 'A survey of water resources development, utilization and management in Israel', *First World Congress on Water Resources, International Water Resources Association*, Chicago

As a result of the high energy and water demands which are generated by modern irrigated agriculture, the question must be posed as to what is the future for this type of crop production. Some countries in the Middle East, especially the very arid ones, have really no option but to continue with irrigation if they are to produce certain foodstuffs within their boundaries. Other larger countries with more varied natural environments, such as Turkey, Iran, Iraq and Syria, may well have some choice as to how their agriculture should develop. The choice is between putting large sums of money into new irrigation projects in areas which have never been cultivated, or at least not for a considerable period, or else investing in rainfed agricultural areas in such a way as to minimize the effects of drought conditions when they occur. Another choice concerns the size of a proposed project. Over the last decade or so very large irrigation projects have gained most support from the major aid-awarding bodies. They have also been popular with many governments as national status symbols. One of the difficulties which has become apparent with these large schemes is that it is possible for a single malfunction in a key part of the system to jeopardize the operation of the whole project. In future, it seems sensible to introduce smaller irrigation projects, which are more in keeping with the size of indigenous rural settlements and which ought to be handled more easily by prevailing levels of education and technology.

Urban and industrial water needs

The most spectacular changes in water use in the Middle East since the Second World War have been associated with the growth of urban centres (see Chapter 9). This urbanization has concentrated large numbers of people in relatively small areas, while the increasing standards of living of urban dwellers has meant that per capita water consumption has increased rapidly as well. The net result has been to put severe strains on the water-resource base of many regions, besides causing other environmental problems as well (Beaumont 1976; Beaumont 1980).

The response to these pressures has generally been to build water-resource projects which will permit the import of water from outside the immediate region. The case of Tehran provides a good example of this approach. Originally water for the city was supplied from groundwater by qanats. However, by 1920, when the population had reached 200,000, the indigenous water supplies proved insufficient and a scheme to abstract water from the River Karaj, some 40 km to the west, was implemented in the 1930s. Soon after the Second World War this scheme proved inadequate and a major dam was built on the Karaj River. This was a multi-purpose project, but one of its prime aims was to supply Tehran annually with 184 million m^3 of water. During the 1960s even more water was required to supply the growing demands of the capital, and a new barrage was completed on the River Jaj, to the east of Tehran, which was capable of supplying the city with a further 80 million m^3 per annum. The population of Tehran had grown to about 3 million by 1970, so that it became obvious that still more water was needed. This time it was decided to construct a dam on the River Lar, which flows into the Caspian Sea, and to divert water through a 20 km tunnel under the Elburz mountains into a reservoir behind the River Jaj Dam, whence about 180 million m^3 a year could be supplied to Tehran by the early 1980s (Marwick and Germond 1975). Now all the relatively easily available water resources are being fully utilized and in the future Tehran will have to go even further afield to supply its needs. This will inevitably mean very high costs in terms of the infrastructure required to supply the water.

In contrast to the Tehran situation, Kuwait City has attempted to solve its water problems by the desalination of sea water, as available water supplies within the country were so restricted (Beaumont 1977b). The first government desalination plant was commissioned in 1958 with a daily capacity of 9100 m^3. From this time onwards the production of desalinated water has grown rapidly to a figure in excess of 280,000 m^3/day in 1975. In Kuwait a dual water-supply system is in operation with fresh water being utilized for drinking, washing and cooking, while brackish water is used for toilet flushing and other uses where water quality is not of prime importance.

The rapid rise in freshwater production in Kuwait has been necessary not only to supply the rapidly growing population but also to meet the marked rise in per capita water consumption associated with increased living standards. Per capita potable water consumption increased threefold between 1957 and 1975, while per capita brackish water consumption went up by more than four times (Table 3.4). By the mid-1970s total daily water consumption (potable and brackish) per capita had reached a figure in excess of 280 litres. On the basis of current trends, estimates for the year 2000 suggest that total daily water demand per capita may well be in excess of 1500 litres (Beaumont 1977b). Already it would seem that Kuwait is concerned about the proportion of its available resources which it has committed to water production. Likely control measures, which could be introduced, include rationing on a volumetric basis or else the use of the price mechanism to limit water consumption. To date, neither of these measures has been rigorously applied with the intention of cutting down water demand.

Qatar, too, has a long history of desalinated water production going back to 1954. The first plant had an output of 0.099 million m^3/annum, but by 1977 23.7 million m^3 of water was being produced per annum (Gischler 1979). To make more water available, the desalinated water is blended with groundwater before distribution to the consumer takes place. By 1994 it is estimated that annual water demands in Qatar will have risen to 100 million m^3. It is current government policy to ensure that this demand is met by desalinated water production. Over the years the groundwater resources of Qatar have been severely depleted by overpumping, so that by the late 1970s it was believed that annual water extraction was more than 30 million m^3 in excess of natural groundwater recharge in northern Qatar alone. Since it was feared that the intrusion of sea water or the upwelling of brackish groundwater could seriously impair the quality of the superficial lens of fresh groundwater, an unusual and highly expensive scheme has been proposed (Gischler 1979). This calls for the annual production of 60 million m^3 of desalinated water, which would then be fed through a number of injecting wells to recharge the fresh groundwater lens. It is believed that if this project could be initiated by the

Table 3.4 Potable and brackish water use in Kuwait

Year	*Potable water production ($m^3 \times 10^3$ per year)*	*Potable water consumption (litres per capita per day)*	*Brackish water production ($m^3 \times 10^3$ per year)*	*Brackish water consumption (litres per capita per day)*	*Total water consumption: potable and brackish (litres per capita per day)*	*Population: census (thousands)*
1957	3,136.6	41.50	2,377.6	31.55	73.05	206.5
1961	7,432.7	63.33	8,573.6	73.05	136.38	321.6
1965	12,560.6	73.65	19,475.1	114.20	187.85	467.3
1970	32,840.3	121.80	35,276.9	130.84	252.64	738.7
1975	54,893.0	151.70	47,219.3	130.49	282.19	991.4

Source: Beaumont 1977b

mid-1980s the depletion of reserves could be stopped and at the same time an increased rate of groundwater extraction permitted.

An even more impressive commitment to desalinated water production than in either Kuwait or Qatar is found in Saudi Arabia. Here the desalination of sea water did not begin until 1970, but the growth of production has been quite remarkable, reaching a figure of 100,000 m^3/day by 1977 (Beaumont 1977a). The Second Development Plan of Saudi Arabia called for an increase in desalinated water production to about 500,000 m^3/day by 1980, to be followed by further growth to 1.22 million m^3/day (Saudi Arabia Central Planning Organization 1975). It does, however, seem that these ambitious targets are likely to be scaled down in view of escalating costs and other technical difficulties. The desalination plants in Saudi Arabia are located on the Red Sea and the Gulf coasts. When all the major plants are completed, approximately two-thirds of the installed capacity will be situated on the Gulf coast. It is hoped that many of the plants, on both coasts, will be interconnected by pipeline networks. Estimates for the six largest urban centres in Saudi Arabia – Riyadh, Jeddah, Makkah, Medinah, Ta'if and Dammam – suggest that between 1974 and 1980 water demand is likely to increase by more than 150 per cent to about 550,000 m^3/day. A new demand for water, of the order of 100,000 m^3/day, is likely to be generated by the industrial centres at Jubail and Yonbu.

All the large desalination plants in Saudi Arabia, Kuwait and Qatar have been built in association with electricity stations. The aim of this has been to utilize the waste energy from the generating stations to heat the incoming water for the desalination plants and so minimize energy costs. Costs have also been kept down by low-interest grants made available by the government for plant purchase. As yet, relatively little is known about the life expectancy of these plants before major replacements become necessary, but few workers suggest that they are likely to operate satisfactorily for more than about 25 years.

The urban and industrial complexes of the Middle East are now producing other problems through the disposal of waste waters. It is somewhat ironic in a region characterized by water shortage that all the large urban centres should have opted for water-borne sewage systems in the last few years. It is only since the beginning of the 1960s that many of the large cities of the region have obtained integrated sewage networks, and almost all of these have been constructed by western-based companies. A fundamental point about the water-borne sewage system is that for really successful operation the treated effluent must be discharged into a water body where further dilution can occur. In a Middle-Eastern context many river systems only flow during the winter period, so that dilution of the discharged effluent cannot always be guaranteed and severe pollution may result. An example of such widespread pollution can be seen along the Nile downstream from Cairo, where nutrient levels and biodegradable organic matter contents are high as the result of the vast quantities of sewage effluent which are discharged from the growing city (Saad 1973).

In some nations where water-borne sewage systems are used the governments are looking into the possibility of utilizing their treated effluents (Shuval and Gruener 1973). The greatest worries are caused by disease potential and the presence of a wide range of chemicals, some of which may have toxic effects. While it is true that conventional sewage treatment methods do appear to kill most bacteria and many viruses which are present, it is believed that some of the more resistant viruses may survive. The net result has been that, to date, it has been felt best to confine the re-use of sewage waters to agricultural and industrial activities (Shuval 1977). In 1976, about 240 million m^3 of household and industrial effluent were used for irrigation in Israel. Plans to use purified effluents for the recharge of aquifers have also been considered, but not implemented as yet owing to a lack of data on the likely effects of such procedures (Laster and Brovender 1976).

It has been suggested that in rural areas sewage disposal may be profitably achieved through the use of composting toilets. These are installed in individual households and so do away with the need for an expensive sewerage infrastructure. Water use per household is reduced, and up to 30 kg of fertilizer per person is produced each year (Ortega and Lefebvre undated). Experiments with this type of toilet have been carried out in the United Arab Emirates and it seems likely that its use will increase in popularity in the future.

Pollution is not only coming from urban areas. The increasing use of fertilizers and pesticides in modern agriculture is leading to the contamination of surface water bodies and aquifer systems (Kahanovitch and Lahav 1974). Nowhere are these problems better seen than in Lake Tiberias, which forms the main freshwater body of Israel and has been subjected to such growing pressures from tourist and agricultural activities that they have given rise to concern about the long-term deterioration of water quality and have led to the establishment of an authority to minimize the impact of new developments on water quality (Laster and Brovender 1976).

Conclusion

The Middle East is a region where the management of water resources has a history going back at least 10,000 years in settlement sites such as Jericho (Kenyon 1969). Throughout most of this period man's chief aim has been the provision of a sufficient volume of water for agricultural needs, with only minor amounts of water being required for domestic purposes. Since the middle of the twentieth century this situation has changed dramatically. Although agriculture continues to be the largest single user of water, the growth of domestic and industrial water consumption has increased sharply. Today the quality dimension of water is of increasing importance, both in urban water supply and also in the impact which polluted water can have on both surface water and groundwater reserves.

Another feature of the region in recent years has been the way in which water-management decisions have tended to move from the village or small town to the national level. As pressure on water resources has increased, almost all countries have established central water-planning divisions, which have attempted to oversee the management of water resources. These organizations are usually part of larger ministries. For example, in Turkey the main agency dealing with water is the General Directorate of State Hydraulic Works, which is part of the Ministry of Energy and Natural Resources.

The Islamic legal framework within which most water management is carried out is also of considerable importance. As Islam was founded in the deserts of Arabia, it is not surprising that it should contain a large body of law dealing with water resources. With elements of 'appropriate rights' built into it, Islamic law is quite well adapted to coping with the range of problems which are experienced in the region at the present day (Caponera 1973; Wilkinson 1978).

What is still lacking in most countries of the Middle East is a conceptual overview of the role of water in the process of economic development. To date water seems to have been regarded by most governments not only as a prerequisite for development, but it seems to have been assumed that if water is provided development will occur. The industrial complexes now being constructed on the Gulf coast are based on this idea. Very few studies seem to have been carried out to discover what other conditions are important, and little has been done to assess the 'multiplier' effect of water-resource development.

Wiener (1977) has suggested a more sophisticated approach to water-resources planning, involving four 'geometries': water-resources geometry; water-requirements geometry; pollution geometry and socio-political institution transformation geometry. Of these four, only the pollution geometry can be regarded as being relatively new and of being able to impose constraints on the institutional decisions which might be reached. The water-resources geometry has remained virtually unchanged until the last 25 years or so when desalination has produced new freshwater for intensely arid areas, though the actual quantities of water produced by these methods has remained small.

The institutions involved in water-resource management have tended to become larger, though methods of operation remain largely unchanged. With the exception of Israel, very few nations have utilized many methods to manage water demand. All too often predicted future water requirements have tended to be accepted as fixed quantities which have to be met at all costs. The net result has been that large capital sums have often been committed to major projects when smaller schemes may well have sufficed. In fairness, it must be admitted that demand management is not very advanced even in the western industrial nations, despite its apparent simplicity. The price mechanism represents an easily applied way by which water demand can be reduced. This could be effectively employed to limit demand in the more affluent urban areas, where significant quantities of water are used for the watering of gardens and for swimming pools. As yet, little attempt has been made to influence the actual methods of using water, such as the introduction of sewage systems with smaller water requirements or irrigation networks utilizing less water.

In the future it seems likely that the pressure on available water resources will continue to grow and so lead to the construction of ever more costly water projects. Such schemes will use larger and larger amounts of a country's energy and capital resources. Eventually a situation will be reached when the benefits which can be obtained will be less than the costs of obtaining them. Hopefully, this will lead to a thorough appraisal of the current uses of water and a reassessment as to whether the water could be used in a more productive manner. Whether a satisfactory water-planning methodology exists which will permit this to happen is something that only time will tell.

References

ABERBACH, S. H. and SELLINGER, A. (1967) 'Review of artificial groundwater recharge in the coastal plain of Israel', *Bulletin of the International Association of Scientific Hydrology*, 12, 65–77.

ADAMS, R. M. (1965) *Land behind Baghdad*, Chicago, The University of Chicago Press.

AL-KHASHAB, W. H. (1958) *The Water Budget of the Tigris–Euphrates Basin*, University of Chicago, Department of Geography, Research Paper No. 54.

ALLAN, J. A. (1976) 'The Kufrah agricultural schemes', *Geographical Journal*, 142, 48–56.

BEAUMONT, P. (1968) 'Qanats on the Varamin Plain, Iran', *Transactions of the Institute of British Geographers*, 45, 169–79.

—— (1971) 'Qanat systems in Iran', *Bulletin of the International Association of Scientific Hydrology*, 16, 39–50.

—— (1973a) *River Regimes in Iran*, Occasional Publications (New Series) No. 1, Department of Geography, University of Durham.

—— (1973b) 'A traditional method of groundwater extraction in the Middle East', *Groundwater*, 11, 23–30.

—— (1974) 'Water resource development in Iran', *Geographical Journal*, 140, 418–31.

—— (1976) 'Environmental management problems – the Middle East', *Built Environment Quarterly*, 2, 104–12.

—— (1977a) 'Water and development in Saudi Arabia', *Geographical Journal*, 143, 42–60.

—— (1977b) 'Water in Kuwait', *Geography*, 62, 187–97.

—— (1978a) 'Man's impact on river systems – a world-wide view', *Area*, 10, 38–41.

—— (1978b) 'The Euphrates River – an international problem of water resources development', *Environmental Conservation*, 5, 35–43.

—— (1980) 'Urban water problems in the Middle East', in Blake, G. H. and Lawless, R. I. (eds) *The Changing Middle Eastern City*, London, Croom Helm.

—— (in press) 'Hydrology and water resources of the Iranian plateau', *Encyclopaedia Persica*, New York, Columbia University.

—— BLAKE, G. H. and WAGSTAFF, J. M. (1976) *The Middle East – a Geographical Study*, London, John Wiley & Sons.

BERGSTROM, R. E. and ATEN, R. E., (1964) 'Natural recharge and localisation of fresh groundwater in Kuwait', *Journal of Hydrology*, 2, 213–31.

BERRY, R. W., BROPHY, G. P. and NAQASH, D. (1970) 'Mineralogy of the suspended sediment in the Tigris, Euphrates and Shatt-al-Arab Rivers of Iraq and the recent history of the Mesopotamian Plain', *Journal of Sedimentary Petrology*, 40, 131–9.

BURDON, D. J. (1973) *Groundwater Resources of Saudi Arabia*, Groundwater Resources in Arab Countries, ALESCO Science Monography No. 2.

BURDON, D. J. and AL-SHARHAN (1968) 'The problem of the Palaeokarst Dammam limestone aquifer in Kuwait', *Journal of Hydrology*, 6, 385–404.

BYBORDI, M. (1974) 'Ghanats of Iran: drainage of a sloping aquifer', *Journal of the Irrigation & Drainage Division, ASCE*, 100 (IRC) Proc. Paper 10785, 245–53.

CAPONERA, D. A. (1973) *Water Law in Muslim Countries*, FAO Irrigation & Drainage Paper, No. 20/1.

DEWDNEY, J. C. (1972) 'Turkey: recent population trends', in Clarke, J. I. and Fisher, W. B. (eds) *Populations of the Middle East and North Africa*, London, University of London Press, Ltd, 40–67.

DINCER, T., AL-MUGRIN, A. and ZIMMERMAN, U. (1974) 'Study of the infiltration and recharge through the sand dunes in arid zones with special reference to the stable isotopes and thermonuclear tritium', *Journal of Hydrology*, 23, 79–109.

ELGABALY, M. M. (1977) 'Water in arid agriculture: salinity and waterlogging in the Near East region', *Ambio*, 6, 36–9.

ENGLISH, P. W., (1968) 'The origin and spread of qanats in the Old World', *Proceedings of the American Philosophical Society*, 112, 170–81.

FOOKES, P. G. and KNILL, J. L. (1969) 'The application of engineering geology in the regional development of northern and central Iran', *Engineering Geology*, 3, 81–120.

GARBRECHT, G. (1968) 'Gediz und Euphrat als Beispiele grossraumiger Wasserwirtschaftsplanungen', *Wasser und Boden*, 20 Jahrgang, Heft 2, 42–6.

GENERAL DIRECTORATE OF STATE HYDRAULIC WORKS, TURKEY (1968) *Statistical Bulletin with Maps 1968*, Ankara, Turkey.

GISCHLER, C. (1979) *Water Resources in the Arab Middle East and North Africa*, Wisbech, Middle East & North African Studies Press.

GORDON, A. H. and LOCKWOOD, J. G. (1970) 'Maximum one-day falls of precipitation in Tehran', *Weather*, 25, 2–8.

HAMMAD, H. Y. (1970) *Ground Water Potentialities in the African Sahara and the Nile Valley*, Beirut, Beirut Arab University.

HAMMERTON, D. (1972) 'The Nile River – a case study', in Oglesby, R. T., Carlson, C. A. and McCann, J. A. (eds) *River Ecology and Man*, New York, Academic Press, 171–214.

HARPAZ, Y. and SCHWARZ, J. (1967) 'Operating a limestone aquifer as a reservoir for a water supply system', *Bulletin of the International Association of Scientific Hydrology*, 12, 78–90.

HIMIDA, I. H. (1970) 'The Nubian Artesian Basin, its regional hydrogeological aspects and palaeohydrological reconstruction', *Journal of Hydrology, New Zealand*, 9, 89–116.

IGHANIAN, R. (1967) 'Geochemistry of ground waters in Iran', in ECAFE/UN, *Methods and Techniques of Ground Water Investigation and Development*, Water Resource Series No. 33, New York, UN, 147–50.

INTERNATIONAL COMMISSION ON LARGE DAMS (1973) *World Register of Dams*, Paris, 943–9.

ISSAR, A. (1969) 'The groundwater provinces of Iran', *Bulletin of the International Association of Scientific Hydrology*, 14, 87–99.

KAHANOVITCH, Y. and LAHAV, N. (1974) 'Occurrence of pesticides in selected water sources in Israel', *Environmental Science & Technology*, 8, 762–5.

KENYON, K. (1969) 'The origins of the Neolithic', *Advancement of Science*, 26, 144–60.

KHALIL, M. B. (1975) 'River regime with special reference to River Nile', *Journal of the Hydraulics Division ASCE*, 101 (NY1) Proc. Paper 11072, 135–53.

LASTER, R. and BROVENDER, S. (1976) *An Overview of Israel's Environmental Problems and Programs for their Solution*, Selected Papers on the Environment in Israel, No. 4, Environmental Protection Service, Jerusalem, Ministry of the Interior.

MARWICK, R. and GERMOND, J. P. (1975) 'The River Lar multi-purpose project in Iran', *Water Power and Dam Construction*, 27, 133–41.

MINISTRY OF INFORMATION, TEHRAN, IRAN (1970) *Nationalisation of Water Resources of Iran*, Tehran.

MITCHELL, J. E. (1967) 'Planning of water development in the Hashemite Kingdom of Jordan', in *Water for Peace – Vol. 7 – Planning and Developing Water Programs*, Washington, US Gov. Printing Office, 282–96.

MUSTAFA, M. A. (1973) 'Appraisal of the water quality of the Blue and White Niles for irrigation use', *African Soils / Sols Africains*, 18, 113–24.

ORTEGA, A. and LEFEBVRE, B. (Undated) *Water Saving Devices for Sanitation*, United Arab Emirates, UNDP and Ministry of Public Works & Housing.

OTKUN, G. (1971) 'Palaeozoic sandstone aquifers in Saudi Arabia', Presented to the International Association of Hydrogeologists, Tokyo Conference, March 1971 (typescript).

PERRIN DE BRICHAMBAUT, G. and WALLEN, C. C. (1963) *A Study of Agroclimatology in Semi-Arid and Arid Zones of the Near East*, World Meteorological Organization, Technical Note 56, Geneva.

PHILIP, G. (1968) 'Mineralogy of the recent sediments of Tigris and Euphrates rivers and some of the older detrital deposits', *Journal of Sedimentary Petrology*, 38, 35–44.

PIKE, J. G. (1970) 'Evaporation of groundwater from coastal playas (Sabkhah) in the Arabian Gulf', *Journal of Hydrology*, 11, 79–88.

RONEN, D., KANFI, Y. and REBHUN, M. (1975) 'Trace metals in Israel groundwater 1. The coastal plain aquifers', *Proc. 6th Scientific Conference of the Israel Ecological Society, Tel Aviv*, Dept. of Zoology, Tel Aviv University, 331–46.

SAAD, M. A. H. (1973) 'Distribution of phosphates in Lake Mariut, a heavily polluted lake in Egypt', *Water, Air & Soil Pollution*, 2, 515–22.

SAUDI ARABIA CENTRAL PLANNING ORGANIZATION (1975) *Development Plan 1395–1400 (1975–1980)*, Riyadh, Saudi Arabia.

SCHICK, A. P. and INBAR, M. (1972) 'Some effects of man on the hydrological cycle of the Upper Jordan–Lake Kinneret watershed', in *Run-off Regimen & Water Balance II*, Freiberg, Institut der Albert-Ludwigs-Universität, 27–43.

SCHNEIDER, R. (1964) *Cenomanian Turonian Aquifer of Central Israel, its Development and Possible Use as a Storage Reservoir*, Washington D. C., US Geological Survey, Water Supply Paper, No. 1608-F.

SCHULZE, F. E. and DE RIDDER, N. A. (1974) 'The rising water table in the West Nubarya area of Egypt', *Nature & Resources*, 10, 12–18.

SHAIKH, A. H. (1971) 'The Marjmaah Dam, Saudi Arabia', *Water & Water Engineering*, 75, 70.

SHUVAL, H. I. (ed.) (1977) *Water Renovation and Reuse*, New York, Academic Press.

SHUVAL, H. I. and GRUENER, N. (1973) 'Considerations in renovating waste water for domestic use', *Environmental Science & Technology*, 7, 600–4.

STANHILL, G. (1974) 'Energy and agriculture: a national case study', *Agro-Ecosystems*, 1, 205–17.

TALLING, J. F. (1957) 'The longitudinal succession of water characteristics in the White Nile', *Hydrobiologia*, 11, 73–89.

THATCHER, L., RUBIN, M. and BROWN, G. F. (1961) 'Dating desert groundwater', *Science*, 234, 105–6.

THORNTHWAITE, C. W., MATHER, J. R. and CARTER, D. B. (1958) *Three Water Balance Maps of Southwest Asia*, Publs. Clim. Drexel Inst. Technol., vol. II.

UBELL, K. (1971) 'Iraq's water resources', *Nature and Resources*, 7, 3–9.

UNDP/FAO (1970) *Investigation of the Sandstone Aquifers of East Jordan – Jordan – The Hydrogeology of the Mesozoic-Cainozoic Aquifers of the Western Highlands and Plateau of East Jordan*, Rome.

VALI KHODJEINI, A. and MOHAMED, A. (1975) 'Étude du débit solide et de la sédimentation du barrage de Shah-Bonou Farah', *Bulletin of the International Association of Scientific Hydrology*, 20, 223–31.

VITA-FINZI, C. (1968) 'Late Quaternary alluvial chronology of Iran', *Geologische Rundschau*, 58, 951–73.

WAFER, T. A. and LABIB, A. H. (1973) 'Seepage from Lake Nasser', In Ackerman,

W. C., White, G. F., Worthington, E. B. and Young, J. L. (eds) *Man-Made Lakes: their Problems and Environmental Effects*, American Geophysical Union, Geophysical Monograph, 17, 287–91.

WIENER, A. (1972) 'Comprehensive water resources development. Case history: Israel', in Weiner, A. *The Role of Water and Development*, New York, McGraw-Hill, 401–11.

——(1977) 'Coping with water deficiency in arid and semi-arid countries through high efficiency water management', *Ambio*, 6, 77–82.

WILKINSON, J. C. (1978) 'Islamic water law with special reference to oasis settlement', *Journal of Arid Environments*, 1, 87–96.

YAALON, D. H. (1963) 'On the origin and accumulation of salts in groundwater and in soils of Israel', *Bulletin of the Research Council of Israel*, II G, 105–31.

CHAPTER FOUR

Irrigation in the Arab countries of the Middle East

JOHN H. STEVENS

Civilization in the Middle East has always been synonymous with irrigation. The empires of the past have had as their bases the irrigated areas of the Fertile Crescent and the Nile, and the technology of the irrigated agriculturist was of an extremely high order. Agriculture, though, has always been a duality with, on the one hand, the sedentary farmers practising irrigated agriculture along the rivers and in the scattered inland oases, and on the other, the nomadic livestock herders. The latter were reliant upon the sedentary farmers for considerable quantities of foodstuffs, notably dates, and, in some cases, owned date gardens in the oases.

With the advent of modern states, the old traditions are changing. Attempts are made to integrate the nomadic pastoralist more closely with the sedentary farmers while the whole pattern of irrigated agriculture is changing. There are still the traditional irrigated areas, but modern technology has opened up new areas as well as changing the methods employed. Most of the traditional areas still practise flood irrigation, though with modifications, and their agricultural potential is based on fertile alluvial soils. In contrast, the new, irrigated areas have often been established on sandy soils, generally calcareous, which in the past would never have been considered for anything more than extensive pastoralism. However, establishment of irrigation in such localities has impinged on the traditional movements of the pastoralists – even the earliest of the major modern schemes, the Gezira in Sudan, was not without problems resulting from this.

This paper will analyse some of the problems arising from modern irrigation by way of examining some case studies. Obviously, such an approach cannot attempt to be definitive but it does enable certain salient, and common, problems to be discussed in more detail. Furthermore, it also allows attention to be focused on future developments in the Middle East. The examples have been chosen to reflect the geographical diversity of the area.

Traditional irrigated areas – surface irrigation

After centuries of being the basis of economic life, the traditional irrigated alluvial areas are increasingly facing problems brought about by changing pressures on both land and water resources. In Egypt, these problems have been caused by the unrecognized side effects of the construction of the Aswan High Dam. In the past the fertility of the Nile valley and delta soils was replenished by the deposition of silt from the annual flooding. Now this deposition has ceased and fertilizers are necessary, while the continual availability of water has led to salinity and a raising of the water table.

These latter problems are also endemic in Iraq. Buringh (1960) considered that most of the groundwater in central and southern Iraq is now saline, while Haradan (1970) suggests that salinization in southern Iraq started some 7000 years ago. The problem has become acute in recent decades as increasing pressure has been exerted on the irrigated lands to provide more food for the urban centres. In the 1960s the average annual rate of population growth was about 3.4 per cent but at the same time the contribution of the agricultural sector to the national income was steadily declining.

The soils of the Tigris–Euphrates lowlands are basically fine-textured alluvial soils but they contain unusually high contents of gypsum (up to 5 per cent) and calcium carbonate – $CaCO_3$ (20–30 per cent). Whilst there are about 3.6 million ha of traditional cultivable land in the Tigris and Euphrates basins, now only about 1.7 million ha are cultivated. The remainder are either too saline for agriculture or are waterlogged. The traditional canal system was designed to provide about 85 per cent of the water for winter crops and 10 per cent for summer crops, and the canals were operated in rotation, except for late April to mid-June. Irrigation has always been by surface methods, and while the irrigation water is of low salinity (at Baghdad the conductivity of irrigation water is 0.387 mmhos/cm at 25 °C), irrigation and more particularly over-irrigation for hundreds of years has caused a build-up in soil salinity. Even with irrigation water of this low salinity, it has been estimated that one ton of salt is added to each hectare of cultivable land each year.

Unfortunately, the alluvial soils of the Tigris–Euphrates also have high water tables. Water tables have been rising owing to over-irrigation and resulting in a range of allied problems. With deep, light-textured soils, which occur in desert areas, over-irrigation can be beneficial at times, since it washes salts out of the topsoil. However, the salts that have accumulated in the alluvial soils of Iraq have been washed into the groundwater. At Baghdad, for instance, the conductivity of the groundwater is 6.0 mmhos/cm at 25 °C. This has risen in many places sufficiently close to the surface for capillary action to add to the salts in the topsoil. Part of the problem has been that in the summer months over 50 per cent of the cultivated area is in fallow and secondary salination occurs.

The problems thus facing modern irrigation in Iraq stem from the association of saline soils and saline groundwater. Attempts to introduce more efficient irrigation methods have had to include costly drainage schemes to make such projects effective. Since 1952 some 2.5 million ha of new irrigation have been developed in the Tigris and Euphrates basins, whilst reclamation of soils in the traditional areas is going ahead. Better understanding of crop water requirements, reduction in irrigation water applied and drainage, as well as increased use of fertilizers, have had a profound effect on yields of crops in the traditional areas. For instance, while the average yield of wheat in Iraq is only about 750 kg/ha, where reclamation of traditional areas has taken place yields have risen to as high as 3400 kg/ha, though more usually in the range 1500–2500 kg/ha.

Reclamation of desert areas

Flood irrigation

Apart from irrigation changes in the traditional farmed areas in Egypt, expansion of agricultural areas has taken place where there has been reclamation of sandy desert soils. These soils are also highly calcareous with up to 50 per cent $CaCO_3$. Over 2 million ha have so far been reclaimed, though not without problems. Criteria that have to be considered include leaching requirements to remove soluble salts from the topsoil, depth of reclamation, drainage requirements, cropping and management practices. In the initial stages of planning, many of these areas were to be reclaimed by Nile waters and the fertility of the reclaimed soils was to be achieved by the addition of Nile silts. The construction of the High Dam altered this.

The criteria affecting reclamation are interrelated. Continuous irrigation is necessary for the quick leaching of soluble salts, whilst regular intermittent leaching is necessary to prevent resalinization. To undertake such a programme, drainage is essential as the initial leaching requires the application of about 25 cm of water to dissolve the soluble salts and a further 40 cm to leach the salts out. The areas reclaimed, mainly to the west of the Nile delta, consist of two ridges, a depression and a plateau. The ridges tend to have shallow soils but where deeper soils occur (greater than 50 cm) in the depressions and on the plateau, calcium-cemented horizons occur in the soil profile where there is impedence of internal water movement. These calcium-cemented horizons have caused local perched water tables and subsequent salinization.

The depth of reclamation, i.e. removal of salts, is very much dependent upon the proposed cropping, but about 1.5 m is generally regarded as necessary for field crops, and as much as 2 m of reclaimed soil are required for tree crops. The speed at which these soils can be reclaimed varies considerably, but soils with good permeability can show desalinization of the top 90 cm within a year.

These calcareous soils apparently have good physical properties but, as soon as irrigation commences, chemical changes occur. The carbonates are converted to bicarbonates and precipitation of the latter on drying causes the formation of a hard surface crust (Elgabaly 1973). Introduction of organic matter can modify this condition – barley, berseem and alfalfa are helpful in this respect.

The commonly practised form of irrigation is surface flooding. In the Nile valley the traditional practices were to use border strips or small basins. Neither was applicable without modification. Without the provision of a comprehensive training programme, the Egyptian farmer did not easily accept the application of contour irrigation or farming to reduce the difficulties and expense of land levelling. Subsequently, many of these reclamation areas have been made over to government-sponsored companies but not before salinization, waterlogging, etc., became apparent.

Sprinkler irrigation

Theoretically, sprinkler irrigation should be the most efficient form of irrigation in the Middle East. Many of the agricultural soils are of sandy texture which lend themselves to this form of irrigation, while the irrigation efficiency is very much higher than with flood irrigation. However, the area devoted to sprinkler irrigation is still relatively small and scattered. The reasons for this relatively slow acceptance are many and complex but include initial high capital cost, especially if subsurface water sources are used, excessive salinity of some irrigation water, lack of trained personnel and consequently high maintenance costs, lack of suitability of equipment for particular environments and the virtual absence of any basic research. Expansion of areas under sprinkler irrigation will have to take place if water resources are to be conserved. Conservation of water resources is, perhaps, of more significance than that of oil, as water extraction is proceeding far faster than replenishment while, in addition, further reserves are polluted. Currently, 75 per cent of all water usage in Saudi Arabia is utilized in irrigating only 1 per cent of the country.

Centre-pivot irrigators which cause the distinctive circular patterns so common in the mid-west and west of the United States, are gradually being introduced into the Middle East. Water and fertilizer or sprays are introduced from the central point and the arm of the irrigator makes a complete rotation at set intervals. Libya has been in the forefront of their introduction at Kufrah oasis. Whilst there were difficulties in their early introduction, mainly due to a lack of understanding of the conditions under which they were to operate, the new large Magnousa development is making use of this method of irrigation. The capital cost of such equipment alone amounts to about $4000 per ha installed (but excluding well drilling) and each centre pivot can irrigate up to 40 ha. A particular feature of this type of irrigation is that land levelling is kept to a minimum, thus conserving what little natural structure and organic matter there is in the topsoil. The major drawback is that this is a relatively sophisticated form of irrigation technology and needs proper maintenance and back-up. Furthermore, design of such schemes needs to take much care in incorporating environmental detail such as salinity of the water, frequency and direction of winds. The large area that can be cultivated allows mechanized farming to take place, as at Magnousa, where the first crops of cereals have yielded about 90 bushels per acre with good management.

Travelling irrigators have been introduced into probably more Middle-Eastern countries than any other form of modern irrigation, with the exception of conventional sprinklers. However, the results from this type of irrigation have been very mixed and in some cases disastrous. Quite apart from problems of trained labour and maintenance, the main difficulties have been brought about by the large size of water droplets formed by the emittance of the irrigation water at pressures of up to about 80 per sq in. at the rain gun. This high pressure is fundamental in the design of such equipment but can be extremely detrimental on the weakly structured desert soils. Erosion and removal of fine seeds, such as those of alfalfa, result. A variant of the travelling irrigator, on which the rain gun has been replaced by a boom with conventional sprinklers, has also been introduced but development is not as far advanced as that of the rain-gun irrigator. Experiments such as those at Al Hasa oasis, Saudi Arabia, or field operations at Salalah, Oman, may provide the operating data necessary for a new generation of irrigators at moderate cost ($400 per ha excluding water-drilling costs) coupled with operation flexibility and droplets of conventional sprinkler size.

Conventional sprinkler systems are in operation in all Middle-Eastern countries but schemes of any size are few. Most countries in the Middle East suffer from a shortage of labourers in the agricultural sector and this absence of a large labour force precludes the extensive use of hand-move sprinkler systems. This is compounded by the fact that to overcome the high capital cost of such systems (up to $1000 per ha excluding pumps and water drilling) it is often necessary to shift laterals at night. Solid set systems, where the mains and laterals are buried to allow farming operations and the sprinklers are permanently in position, are far too expensive (up to $3500 per ha excluding pumps and water drilling) except for small areas of very high-value cash crops.

Despite these drawbacks, sprinkler irrigation offers the best hope for expansion of irrigated agriculture in the Middle East. This form of irrigation is particularly well adapted to light, sandy soils, and the distribution of irrigation water can be well controlled. Irrigation efficiencies of 60–75 per cent can be obtained compared with 20–30 per cent for surface irrigation. However, its success will only be apparent if design and management are correct. Lack of basic data on crop water requirements hampered early schemes, but matching irrigation applications to crop requirements and water-holding capacity of the soils is now less of a problem. The use of slightly saline water, though, still precludes the use of this form of irrigation on crops such as tomatoes whose skin can be blemished. Even where slightly saline water is used on field crops, there is the need for periodic leachings to remove accumulated salts from the rooting zone of plants. However, the ultimate success of sprinkler schemes depends upon a trained labour force; far too often schemes have not been as successful as they might have been owing to the lack of one.

Trickle irrigation

Drip or trickle irrigation is the newest irrigation method to be developed. In theory water is delivered to plants in exact quantities by means of emitters inserted into plastic or polyethylene tubes. The water is only delivered to the plant in quantities that the latter requires and fertilizer is also applied through the system. The system can be automatically controlled and, because water is only delivered to the crop plant, weed growth is minimized, thus reducing labour costs. Furthermore, because exact quantities of water are applied it is possible to use more saline water than with sprinkler irrigation, making allowances for the leaching requirement. It has an extremely high theoretical efficiency of over 90 per cent but also has a number of limitations such as blockage of emitters, as well as a high degree of skill in design, installation and operation together with high unit costs ($2000 per ha installed).

Trickle or drip irrigation is particularly suited to tree crops and it is with forestry and orchards as well as shelterbelts that it has found its widest acceptance. Its use in forestry projects in Abu Dhabi is well described by Wood *et al.* (1975). Water was abstracted from boreholes and the young trees, mainly *Acacia spp., Prosopis spicigera* and *Zizyphus spina-christi* which were raised in local nurseries, were planted once the irrigation system had been installed, but not in the hottest periods of the year. Correct irrigation and proper forestry methods have allowed a relatively high success rate in the establishment of trees, though as many as 40 per cent of the trees may need replacement within the first twelve months. Forestry plantations were first established in Abu Dhabi in the late 1960s and the area has been expanding ever since. Trickle irrigation has been extremely effective and growth rates of trees planted at densities of 200 per ha have exceeded 1.5 m per year, depending on the species.

Hydroponics

This is a totally new approach to crop production and is based on a high degree of technical expertise. The basic problems in crop cultivation in the Middle East stem from a harsh natural environment – poor soils, poor-quality irrigation water, high solar insolation and great ranges of temperature. Control of the environment is all important and the use of hydroponics in temperate zones has shown that extremely high outputs can be achieved albeit with the use of high inputs.

Early attempts at hydroponic cultivation in the Middle East took place in the 1960s and early 1970s in Kuwait and Abu Dhabi. The newest approach to hydroponics is the nutrient film method in which the roots are placed in plastic or polythene troughs and a thin film of nutrient solution is allowed to trickle down the channel. No soil or peat is necessary and, because the roots are enclosed, disease control is also made much easier. Countries such as Saudi Arabia are showing great interest in this method and it obviously has major advantages, though the expertise has to be introduced as part of the package.

Future developments of irrigated agriculture

The bulk of the irrigated land will continue to be the traditional irrigated areas along the main rivers of the region, but as knowledge of the groundwater resources increases, then there will be further reclamation of desert areas. The short reviews of the different types of irrigation have identified the strengths and weaknesses of the systems, but above all there must be a change to making more effective use of the irrigation water. Land is not a major constraint, though the situation has been made worse by malpractices in irrigation. If the premise is accepted that the Arab countries must grow more of their own food, a point emphasized in the development plans of most countries, then this can only be achieved by the introduction of new technology incorporating greater irrigation efficiency. This latter is also identified with increased crop yields as well as better-quality produce, a feature that is increasingly being demanded by the local markets.

However, to achieve a change in approach to irrigation, there must be a commensurate amount of effort and time expended on education and technology transfer. The cultivator of the traditional lands around Baghdad cannot be expected to take up overnight the new technology and associated mechanization and there has to be a long period of assimilation. Thus, while improvements can and must be made to the traditional areas, there are proportionately better prospects in the newly reclaimed areas.

Where irrigation is starting with reclamation of virgin areas, the whole development can be properly designed. This is not just a question of irrigation layout and mechanization, but also of management structure and financing. There are two approaches to the technology transfer problem: either the operators can be imported as part of the package or the project can develop more slowly as pilot schemes are established and training given to local operators. The former approach is that currently being adopted in Libya where the new developments at Magnousa are being undertaken on a turnkey basis by Americans, including the provision of the operators. On the other hand, the Egyptians are now pursuing a more cautious approach on their newly developed lands, first testing irrigation equipment and assessing its various capabilities under different operational conditions and then training local operators. Obviously, the former approach is more valid in countries of low population density and where the workforce is at a premium, whilst the latter approach is more pertinent in the more densely populated territories.

The technology for this improvement in irrigation cannot be considered in isolation since it involves totally new approaches to farming. Without improved seeds, there may still not be the expected response to irrigation developments, and fertilizer and spray usage are essential. Appropriate mechanization is synonymous with improved irrigation, but perhaps the biggest changes have still to be made in crop storage, processing and marketing. New management structures have to be established to operate extensive areas (in comparison with the small plots cultivated traditionally), whilst there have often to be changes in rural credit facilities.

One of the biggest gaps at present is in identifying the people who are to benefit from the project. It has already been commented that new projects often impinge on traditional tribal grazing lands and all too often the pastoralist is forgotten. There is an increasing tendency to try to integrate the nomadic lifestyle with that of sedentary irrigated farmers – an example of this can be seen in the thinking behind the Seheit scheme in Sudan. The object of the scheme is to grow fodder (using surface irrigation owing to the heavy clay soils) and fatten livestock for slaughter, the livestock being purchased from the pastoralists. How effective such a scheme can be, remains to be seen. Other opportunities for improving the lot of the pastoralist already exist, but are unexploited. Residues from irrigated areas, such as the Guneid sugar scheme, again in Sudan, can be used to enable the pastoralist to take some benefit from what is essentially an alien form of farming in his environment.

Thus, the future for irrigation is complex and is undoubtedly going to cause social strains. The technology is available, as are the products, but their application needs to be treated with consideration. Given the resources available, the benefits of improving, perhaps only temporarily, the traditional areas of irrigation by expensive drainage schemes must be balanced against the development of new lands, again at high cost.

References

BURINGH, P. (1960) *Soils and Soil Conditions in Iraq*, Iraq Ministry of Agriculture.

ELGABALY, M. (1973) 'Calcareous soils', *Soils Bulletin*, 21, 123–8, Rome, FAO.

HARADAN, A. A. (1970) *Archaeological Dating of Soil Salinity in the Lower Palaeopedology*, Amsterdam, Elsevier.

WOOD, P. J., WILLENS, A. F. and WILLENS, G. A. (1975) 'An irrigated plantation project in Abu Dhabi', *Comm. For. Rev.*, 54.2, 139–46.

CHAPTER FIVE

The impact of economic development on pastoral nomadism in the Middle East: an inevitable eclipse?

J. STACE BIRKS

Drought in the Sahel of Africa has recently drawn attention away from the major underlying force eroding pastoral nomadism with increasing rapidity – modern economic development. It is because of oil-stimulated economic growth and its urgent pace that the pastoral nomadic societies of the Middle East continue to decline.

To state this is not to accept that traditional, successful pastoral nomadism only survived because of lack of contact with a broader world. The network of external links, both social and economic, integral to successful survival of non-sedentary peoples is not questioned: traditionally, sedentary and non-sedentary groups were economically interdependent (English 1973). What is asserted is that during the present century in general, and the past decade in particular, pressures from modern economic expansion upon nomadic societies have been so strong, and so deleterious to their contact with settled groups, as to militate against healthy nomadic society; nomads have become much the weaker of two unequal partners.

Earlier this century, colonial contact with nomadic society was not so deleterious. Despite some attempts to settle nomads, peaceful conditions strengthened many facets of non-sedentary society and economy.

In reality, this apparent survival of pastoral nomadism as a traditional part of a dual economy is probably best thought of as a slow decline, for numbers of nomads have been falling since the late nineteenth century. However, the present rapid breakdown of pastoral societies in the Middle East suggests that recent oil-stimulated economic development is having a new and acute impact upon the traditional economies of the area.

Some definitions

Little is gained in a paper of this length from a necessarily brief attempt to sort out the involved terminology which nomadic studies, despite their relative paucity, have thrown up. It is a major task to rationalize the plethora of names used to describe pastoral groups, and varied interpretations of words borrowed from Arabic, Fufulde and other languages used by nomads (Johnson 1969). Even that word most cherished by western commentators to describe nomads – *bedu*, and its many derivations – is not used consistently in Arabic to mean 'nomad', but has a wide range of meanings according to the context in which it is used.

Here 'nomadic' will be used synonymously with 'non-sedentary'. It is used to describe groups falling within a spectrum of mobility which includes, at one extreme, a pastoralist with no permanent dwelling following no agricultural pursuits and, at the other extreme, animal herders who move only short distances on a seasonal or irregular basis and who derive much of their livelihood from farming. Thus 'semi-nomadic' groups are included as nomads. To be included as nomads, as defined here, the mobility of peoples must be, or have been, associated with livestock – moves such as those of villagers and townspeople from winter to summer quarters are, therefore, omitted.

The numbers of non-sedentary peoples in the Arab world

The axiom that numbers of non-sedentary people in the Middle East have been declining for most of this century is in some countries difficult to demonstrate empirically, despite a probable rapid increase in the rate of their decline during the 1970s (George 1972).

In some states census figures do not relate to the non-sedentary component of population, largely because of the logistics; theoretical difficulties of definition pale when compared with the challenges of enumeration. Some governments deliberately suppress numbers of nomadic groups, not admitting their existence. Nomads are widely and probably increasingly considered anachronistic by Arab governments, an embarrassment rather than an asset. This is sometimes engendered by consideration of nomadic populations as a security risk. Their persistent tribal, rather than national, loyalties and their reputations as fighting men are notable features. Where empirical data are available, they rapidly become of historical rather than contemporary interest in the Middle East, where change proceeds especially quickly. Moreover, the value of these empirical data is further reduced by definitional inexactitude and inconsistency because of the difficulty in distinguishing the boundary between settled and non-sedentary peoples.

Unfortunately, the findings of detailed case studies of nomadic pastoralists are not often extrapolated to a national or even a regional scale. Their utility is thus limited, an aspect of 'the curiously inchoate nature' of nomadic studies, little remedied since Dyson-Hudson's observations of 1972.

In view of these shortcomings of available data, it is unfortunate that Middle-Eastern governments appear increasingly hostile towards attempts to study nomads. As noted by Swift (1978) in a broader context, they see in such enquiry an unwarranted meddling, usually patronizing, and based upon romantic western notions about 'noble savages'. This is felt by oil-rich states in the Arab world, who are aware enough of the decline of pastoral society to enshrine the paraphernalia of nomadic life in museums, and also by the poor countries, to whom the lingering continuance of a traditional pastoral sector is a source of embarrassment as an indicator of underdevelopment.

Today there are probably fewer than 2.5 million nomads in the Middle East. This excludes Sudan, where the 1973 census records some 1.6 million non-sedentary peoples (although a truer value might approach 4 million). Probably only Iran has approaching one million nomads. Even Saudi Arabia, the country associated most closely with the bedu in the popular mind, has now probably fewer than 550,000 nomads within its territories.

Grossly misleading statements pertaining to non-sedentary populations continue to be made. Fyfe (1977) recently asserted that 30 per cent of the population of the United Arab Emirates are bedouin. In fact, the proportion of the economically active population of the Emirates engaged in agricultural pursuits is as low as 3.2 per cent (United Arab Emirates 1976). Of these 10,000 people, only a few are actively engaged in livestock production. Fyfe seems to be using the term 'bedouin' to imply dispersed population, regardless of whether they are really geographically mobile or from where they derive their livelihood. In fact, most of these households derive their income from the modern sector, particularly employment in government services.

Government settlement of nomadic populations

Government efforts to induce the settlement of nomads in the Middle East have not approached the scale of those in the Soviet Union, where nearly one million Kazakhs were reorganized to eliminate their 'political, economic and cultural backwardness' (Tursunbayev and Potapov 1959). Significant instances of induced settlement in the Middle East include the schemes of the 1920s to settle some of the Basseri of Fars in Iran. These were successful in so far as they cut down transhumance movements, but also resulted in massive losses of livestock (Barth 1962).

Probably the best-known settlement scheme to induce sedentarization of nomads in the Middle East is the al-Hijar scheme of Saudi Arabia, which was instituted in 1912. The scheme represented a deliberate policy of the breaking down of nomadic tribal loyalties, which was thought to be best achieved by sedentarization. The al-Hijar scheme had settled some 250,000 nomads by the 1930s when interest in the programme waned.

More recently, the government of Kuwait has made great efforts to settle their bedouin population, building areas of low-cost housing, and providing free services as at Jahra and near Fahahil; but this has by no means stopped all the nomadic movements.

Despite the continued desire of most governments to sedentarize their nomadic populations, such schemes to induce settlement have fallen from favour. Today it is spontaneous settling (though often encouraged subtly by government incentives) that accounts for an increasing decline of pastoral nomadism.

Spontaneous settlement of nomadic populations

Extensive spontaneous settlement might be expected; there is an argument for considering nomadic society to be transitional, in constant evolution towards a settled existence. Nomads can thus always be considered to be seeking opportunities to settle (Wilkinson 1977). At the same time, new nomadic groups are being created; surplus agricultural populations can be thought of as being 'expelled' into nomadic life or, alternatively, new nomadic groups have been demonstrated to be created by choice. Haaland (1970), for example, details the transition of sedentary Fur into nomads. Also in Sudan, erstwhile settled Fulani families have been observed to adopt a nomadic lifestyle (Birks 1976).

Thus, nomadic society might be considered, at a global level of assessment, to be in a state of dynamic equilibrium, sedentarization occurring contemporaneously with the generation of new nomadic populations elsewhere. In the case of Saudi Arabia, new nomadic groups have originated in the south-west of the peninsula, dispersions of pastoralists radiating from this hearth or core region towards Oman and the fertile crescent where they settle in due course (Lawrence 1935). If this theory of the 'transitional nature' of nomadic populations is correct, then in the Middle East the factors in favour of the sedentarization of nomads have strengthened, whilst those in favour of the generation of new nomadic groups have become weaker. The dynamic equilibrium that used to govern the size of the nomadic population has been disturbed.

The question to be addressed is whether or not this balance has been irrevocably lost, i.e. whether or not nomadic society is inevitably approaching its end or, alternatively, whether it is possible that the recent and present degree of sedentarization is merely part of the establishment of a new equilibrium. In order to balance with new economic factors and changing ecological conditions, new animal, and consequently human, populations might be much smaller than those prevailing in the past. Could the decline in numbers of nomads be a healthy response in the establishment of a new self-sustaining nomadic or semi-nomadic order in the Arab world? In order to cast light upon this, some facets of the spontaneous processes which reduce the numbers of nomads are examined.

Processes of sedentarization

In some cases of sedentarization, the family remains in the area of the grazing grounds but ceases seasonal or occasional movements of dwelling site, or at least no longer moves site in order to accommodate the needs of their animals. In others, the nomadic family moves away from the area of pastures which were grazed when animals were the major source of livelihood; this is a facet of rural–urban migration (Cole 1975). In practice, these might be considered consecutive facets of rural–urban movement; after a period of sedentary existence in the grazing grounds, families move to the growing Middle-Eastern towns.

In many instances, the overt cause for abandonment of nomadic pursuits, with consequent sedentarization either in the area of the grazing grounds or in movement to a town, might seem to be drought. It is often, for example, in periods following drought that farms are established by sedentarizing nomads, and after periods of low rainfall that extra numbers of nomads accumulate on the edges of towns. The deduction has been that it is easier and quicker to establish a farm or to obtain employment in urban areas than to build up a viable herd of livestock to its pre-drought size. This is true, but nomads, having lost their herds and flocks, have in the past successfully nurtured replacement stocks of animals. The impact of rainfall shortage and pasture deterioration upon nomadic society has recently been dramatically evidenced in the African Sahel and northern savannas (Markham 1977). The effect of the more prolonged drought (often called the 'seven-year drought') of the late 1950s and early 1960s in Saudi Arabia upon animal numbers and nomadic populations is well known if not accurately documented. These droughts did indeed spawn 'bedouin farms' – dispersed irrigated gardens in or near the pastures – and increased rural–urban migration of nomads.

It is asserted here, though, that the shortages of rainfall in the Arab world over the past two decades have affected nomadic societies disproportionately. It would appear that the nomadic pastoralists have become more susceptible to rainfall shortages than in the past. Resilience that used to be displayed by nomadic societies in the face of drought has been lost (Johnson 1973). Moreover, nomads have been selling their stocks, reducing their transhumance orbits, taking up farming, moving to towns and exhibiting other aspects of sedentarization without being threatened by drought. It is therefore necessary to look beyond drought in the search for underlying causes of pastoralism's present-day decline.

What all these sedentarizing pastoral societies have in common is the fact that they supply labour to the manpower-deficit areas of the Middle East. The pastoral societies, initially because of the surplus of male labour that their traditional economy allowed (i.e. time remaining after the flocks had been attended to), readily became source populations for migrant labourers who spent periods of work in the economically more developed areas of the Middle East. Nomads had some peculiar advantages as migrant labourers particularly in the Arabian peninsula. Despite the remote locations in which they lived, they often gained experience of wage-paid labouring during oil exploration phases. Possibly more important than the formal work experience was their introduction to the wage packet. Perhaps in contrast to the popular image of nomads, many groups, for generations rather than just a few years, have maintained a strong desire for money, if not a greed. This has been nurtured recently by receipt of wages and salaries, and has contributed to a high propensity amongst nomads to migrate on a temporary basis in order to obtain work.

So nomads (as well as village dwellers) migrate in large numbers from the western provinces of Sudan, Kordofan and Darfur, and from Chad (in particular Wadai and Bagirmi) to the Gezira and other irrigated agricultural schemes in eastern Sudan. Nomads from Jordan feature large in labour migrations on an international scale, travelling to Saudi Arabia in particular, where they work in a variety of largely unskilled operations. Within the Arabian peninsula, nomads form an integral part of the complex of migrant labour movements. Thus menfolk from the Al Murra, for example, can be found temporarily resident in many towns in Saudi Arabia, where they work for both the government and the private sector on a periodic basis. Duru' men from interior Oman travel to Saudi Arabia and all the countries of the Gulf as migrant labourers and in order to enlist in the armies of the capital-rich states (Birks and Sinclair 1977).

This has occurred to the degree that in many nomadic societies it is the exception rather than the rule for an adult male not to have spent at least one period away from the tribal homeland as a migrant worker.

The consequences of nomads' supplying labour

It is this role of nomadic populations as labour suppliers within the Middle East that is the key to understanding the decline of pastoralism in the area. The adaptation of nomadic society to facilitate migrant labour movements causes a series of set and predictable reactions amongst pastoralists. At first, migrant labouring is viewed as a valuable extra form of income to augment family budgets. It is seen in much the same light as, for example, hunting, trading, or haulage and transportation (whether originally by camel or latterly by vehicle). In its early stages, working away from home is one of the several ways in which the 'surplus' time available to adult males of pastoralist families is absorbed in an economically productive manner.

As time passes, however, the wage packet and the consumer goods associated with it come to feature larger in the budgets and lives of the nomadic family. The desire for a wage packet becomes the overriding motive behind the economic pattern of life for the household head, and perhaps other male members of the family. Progressively more of the efforts of the men are directed towards earning a cash wage; a diminishing amount of effort and concern is directed towards the mainstay of the domestic economy – the livestock. In consequence, partly out of neglect, and partly as a result of a deliberate policy of selling, numbers of stock begin to diminish.

This process becomes of critical importance as the household head, on becoming a migrant labourer after several temporary moves away from his home area in search of wage-employment, changes a fundamental perspective: until this point, he considers the working trips away from his tribal area as temporary absences, seeing his primary role to be his home tent and herd. Eventually, however, the migrant labourer considers these periods spent with the herds as mere interruptions of a new primary role – that of wage earner in the more economically developed areas to which he migrates (Birks 1978).

It is after this attitude has developed that the most significant decline in the pastoral system occurs. The substantial withdrawal of male inputs renders the traditional system unworkable. The compensating increase in the role of women does not make up for the loss of men's contribution to the local economy. As a consequence, the pastoral system collapses, numbers of stock diminish because of large-scale sales (this phase might well be accompanied by inflation and rising meat prices which reinforces the desire to sell) and more extreme neglect of the animals. Economically viable herds or flocks are reduced to a few animals which supply only a proportion of the immediate needs of the family, and which provide for the demands that might be made upon the household to give hospitality to visitors. This reduction in the numbers of animals is facilitated by one of the most profound changes undergone by the nomads – a change of diet away from local dairy and livestock products to canned food. Thus the reorientation is completed: the family is no longer dependent upon the livestock, previously the focus of their economic lives, but is now almost entirely reliant upon the cash income derived from migrant labouring. It is during this period of reorientation from a pastoral economy to one based upon exports of labour that sedentarization occurs.

Moves of household become less frequent and more sporadic as the numbers of stock decline and their requirements become both less pressing and less readily attended to. This abandonment of transhumance movements might be an almost entirely negative reaction, or could have some positive attributes, depending on whether or not the settling process is associated with a turn to agriculture. One of the reasons often given by migrant labourers for their departure is the desire to earn enough to invest in a small irrigated garden. In these instances the new settled location of the house is determined by a suitable garden site. If the settling is not associated with the establishment of a well-garden, then it is more a willy-nilly process.

Once the number of animals is reduced, several factors militate against continued mobility. The household becomes cluttered with consumer goods and so more difficult to move. At the same time the increasingly prolonged absences of the menfolk make the transfer of site more difficult to effect, in terms both of physical removal and of making the decision to shift.

Often a vestige of household mobility remains even after the number of animals has fallen to single figures. These 'relict moves' are over a small area and may be associated with abstract factors such as a 'tradition of movement' or 'bedu mentality', as well as more concrete influences such as the need to move to new sites with better firewood supplies and cleaner environs. They do not invalidate the assertion that, with the cessation of the nomadic transference cycle and sedentarization of households, the pastoral economy has collapsed. Its ending is a consequence of the stresses put upon it by the exports of human capital from the nomadic society.

Often, after a period of sedentary existence in the pastures, when the pattern of migrant labour movements of the menfolk continues, the family moves to an urban centre. This might be a direct move, or a series of stepped movements with periods of residence in small oasis villages near the grazing grounds. Alternatively, some members of the family might move some time before the whole household transfers to an urban centre. With the removal of the family to a town the links with pastoralism are all but completely broken. The only livestock the household is likely to keep after this stage are a few goats left to rout around the streets.

Some general implications

The non-sedentary populations of the Middle East, as they settle, are exchanging a pattern of exploitation of marginal regions for a marginal position within a wider economy. They are shunning dependence upon the vagaries of climate in the arid lands over which they migrated with their animals for a rather more subtle, but nevertheless pervasive, dependence upon the modern economic system that is expanding in the Middle East. The irony is that, in gross terms, most have not reduced the uncertainties of their existence. As suppliers of migrant labour, the nomads are the most easily dispensable element of the labour supply. They are therefore particularly vulnerable to downturns in the unstable Middle-Eastern economies. Quite apart from their being at risk from a fall in the rates of development, they are also prone to political factors when they migrate internationally – instances of repatriation of both Arab and non-Arab migrant labourers from the labour-importing countries of the Middle East are becoming increasingly common.

In short, the nomads have been drawn into the expanding modern economy of the Middle East, but have remained on the margin. They are situated geographically, economically and socially on the periphery of the central or core areas of development, namely the developing urban areas of the capital-rich states. This is true of most former nomads, with the exception of a few who are fortunate enough to comprise part of the core national populations of some of the capital-rich states. These sedentarized groups have, by virtue of their social position, moved into the very heart of the government sectors of these rapidly growing economies.

Non-sedentary peoples are not, of course, the only societies from which the urban core areas have drawn labour. Nomads, though, tend to have been some of the most profoundly affected for two reasons: first, the pastoral groups tend to have exported proportionally more labour than most of the remote agricultural settlements; secondly, the economies of sedentary agriculture-based societies have proved to be rather more durable in the event of their becoming source populations of migrant labourers. This is because farm infrastructure is able to lie dormant with only minimal deterioration if left idle. There are exceptions to this – terraces collapse in the Yemens and south-west Saudi Arabia, and the infrastructure of water supplies decays in Oman and Iran – but, generally speaking, in the event of the return of labour it is possible to organize a recrudescence of agricultural society (Birks and Letts 1977).

Not so with a nomadic pastoral economy. Once a whole tribe's herds and flocks are lost it becomes difficult to build them up again. An individual family can do so, given a healthy nomadic environment in which to re-establish itself, for animals can be borrowed to enable numbers to build up, the family being cushioned by the community in the meantime. It is much more difficult for a complete society to re-erect a pastoral economy.

A rather surprising aspect of the inability to turn back to pastoral activities more than a short period after their abandonment is the rapid rate at which essential expertise and husbandry appear to be lost. Once the number of animals becomes small, children are simply not taught the skills they need to manage a large herd or flock. Men do not wish to deal with animals after a number of years as migrant labourers – when they return home finally they do so to retire, not to take up an economically active traditional role; they therefore take little trouble to instruct younger family members in matters pertaining to livestock. Rather, they converse on the attributes of various makes of motor vehicle. The sum effect of this is that, over a surprisingly brief period of time, pastoralist groups in the Middle East are losing their expertise in dealing with animals – having left nomadic life, they cannot easily return to pastoralism.

A factor of considerable significance in assessing the relationships between nomadic groups peripheral to the modern economy and the developing labour-deficit core areas to which the menfolk migrate is the unprecedented development in the capital-rich countries since 1973, where acute tensions have been set up between the relatively developed urban cores and the traditional periphery. This widening disparity in development has caused especially strong demands for labour in the peripheral societies – there has been an explosion in the demand for manpower that has been felt throughout virtually all societies in the Middle East, even the most remote. So it is that export of human capital from nomadic societies has expanded recently. Individual societies which have exported labour for some time have found their menfolk departing (and also whole families leaving for the towns) at a new rate, which has set in train the more profound effects upon their local economies, with deep-seated structural effects.

Moreover, societies which previously experienced little out-migration because of remoteness and difficulty of access have been drawn more firmly into the economic spheres of the core urban areas of the capital-rich states. For instance the Beja in the Sudan, who, prior to 1975, supplied small numbers of labourers for the docks in Port Sudan, now send large numbers of men to work in Jeddah.

In short, the frontier of modern economic forces has now encompassed the whole of the Middle East, its progress having been accelerated recently by the new style and rate of growth facilitated by the price increases of oil in 1973.

The future

Given the contracting numbers of nomads, and the growing demand for labour in the expanding modern sector, can animal husbandry on a large scale survive in the Middle East? Syria appears to be the only country in the Arab world where serious and possibly successful efforts are being made to strengthen and adapt a nomadic mode of life to modern conditions. Livestock co-operatives are being developed to facilitate improvements in stock (Morassem 1972), and more recently efforts directed at fattening and selling livestock and improving the nomadic cycle look like resulting in some success. No other rural development scheme is directed towards ensuring survival of pastoral nomads.

What, then, are the prospects of livestock rearing as carried out by settling nomads? It is not axiomatic that when a household sedentarizes the interest in livestock is lost completely. Small stock, particularly sheep, often remain important. Many former camel-herders change to keeping sheep at the time of sedentarization. This is often in conjunction with development of a garden, fodder being grown for the sheep (and goats, if they are kept) in order to augment the limited browse available in the area around the household. Such a change accords well with a widening role of women, who take over much of the running of the farm, and who deal capably with the increasing numbers of small stock. This is often portrayed as a viable future form of livestock production in the Middle East, which might be the case, but some very profound reservations must be expressed about how such spontaneously developed adaptations of pastoral life are evolving. First, the more limited grazing patterns resulting from such farm-based flocks and herds leave wide areas of range untouched, even when, on occasion, animals are taken to the more remote areas by vehicle to browse. Whilst, in the short term, this allows these ranges to recover from past over-grazing (if this has occurred), in the longer term it means the neglect of a major asset of the Middle East. Moreover, in reality, this system of producing small stock with only very limited or non-existent transhumance movements appears itself to cause critical over-grazing problems particularly because the animals tend to be run in what were the best ranges. The very heart of the pastures of the Middle East are therefore put at greater, not lesser, risk.

A second factor that casts doubts upon the long-term viability of this sedentary fodder-based animal production is the extra demand for water. Examples of water tables in the Middle East being drawn down by medium-lift pumps are numerous, widespread and well known. The gardens run by former nomads in some of the pasture areas are amongst the worst offenders in this respect.

These factors, which suggest that this system is not viable in the long term and approaches a dangerous use of resources, are all compounded by the critical fact that many of these new ventures are not economically self-sustaining, but rely upon continued investment of remittances sent back by the migrant labourers.

In short, the modern evolution of formerly nomadic pastoral society does not represent change leading to a viable society in the face of widening modern horizons; rather it represents the deepening dependence of the peripheral areas upon the urban cores of the Middle East. Not only do these marginal regions supply labour to the more developed capital-rich areas of the Middle East, but their local economies have become entirely dependent upon continued migrant labour movement and the remittances these generate. The problem is that, should a downturn occur in the core areas – a not unlikely prospect, given the nature and resource base of economic development in the

Middle East – then the newly evolved peripheral economies will collapse also.

Yet it is towards these vulnerable, dependent societies and economies that virtually all non-sedentary peoples in the Middle East are evolving today. Once the wider field of strong modern economic forces is felt by a traditional nomadic society, it breeds migrant labourers rather than animals. What results is a dangerously close dependence upon continued boom in the core areas of development with no recourse to a local economy should the vacillatory demand for labour fail.

References

BARTH, F. (1962) 'Nomadism in the mountain and plateau areas of South West Asia', *The Problems of the Arid Zone*, Paris, UNESCO.

BIRKS, J. S. (1976) 'Overland pilgrimage from West Africa to Mecca', unpublished PhD thesis, University of Liverpool.

—— (1978) 'Development or decline of nomads: the Bani Qitab', in Serjeant, R. J. and Bidwell, R. D. (eds) *Arabian Studies*, vol. 4, 7–19, London, Hurst.

BIRKS, J. S. and LETTS, S. E. (1977) 'Diqal and Muqaydah: dying oases in Arabia', *Tijdschift voor Economische en Sociale Geografie*, 68, 143–9.

BIRKS, J. S. and SINCLAIR, C. A. (1977) *Movements of Migrant Labour from Part of the Sultanate of Oman*, International Migration Project Working Paper, Department of Economics, University of Durham.

COLE, D. P. (1975) *Nomads of the Nomads: the Al Murrah Bedouin of the Empty Quarter*, Chicago, Aldine.

DYSON-HUDSON, N. (1972) 'The study of nomads', *Journal of Asian and African Studies*, 7, 2–29.

ENGLISH, P. W. (1973) 'Geographical perspectives on the Middle East: the passing of the ecological trilogy', in Mikesell, M. (ed.) *Geographers Abroad*, University of Chicago, Department of Geography, Research Paper No. 152.

FYFE, A. (1977) 'Subtle attempts to settle wanderers', *The Times, Focus on the United Arab Emirates*, Thursday 21 June, xix.

GEORGE, A. R. (1972) 'Processes of nomadic sedentarization in the Middle East', unpublished MA thesis, University of Durham.

HAALAND, G. (1970) 'Economic determinants in ethnic processes', in Barth, F. (ed.) *Ethnic Groups and Boundaries*, London, Allen & Unwin, 58–73.

JOHNSON, D. L. (1969) *The Nature of Nomadism: A Comparative Study of Pastoral Migrations in South Western Asia and Northern Africa*, University of Chicago, Department of Geography, Research Paper No. 118.

—— (1973) *The Response of Pastoral Nomads to Drought in the Absence of Outside Intervention*, Special Sahelian Office, New York, UN.

LAWRENCE, T. E. (1935) *The Seven Pillars of Wisdom*, republished 1977, Harmondsworth, Penguin.

MARKHAM, P. (1977) *Nomads of the Sahel*, Minority Rights Group.

MORASSEM, M. (1972) 'Livestock co-operatives in the Syrian Arab Republic', *World Animal Review*, 1, 37–41.

SWIFT, J. (1978) 'Marginal peoples at the modern frontier in Asia and the Arctic', *Development and Change*, 9, 3–20.

TURSUNBAYEV, A. and POTAPOV, A. (1959) 'Some aspects of the socio-economic and cultural development of nomads in the USSR', *International Social Science Journal*, 11, 559–71.

UNITED ARAB EMIRATES (1976) *Population Census, 1975*, Abu Dhabi.

WILKINSON, J. L. (1977) *Water and Tribal Settlement in South East Arabia*, Oxford Research Studies in Geography.

CHAPTER SIX

The oil industry in the Middle East

KEITH McLACHLAN

The Middle East can be defined in a number of quite different ways, though, for the international oil industry, the area is perceived to be that occupied by the oil-producing states surrounding the Arabian–Persian Gulf. Such a definition would omit those countries of North Africa, particularly Algeria and Libya, which are relatively large producers of crude oil and natural gas, which generally act in concert with the Middle East oil states in oil affairs through the Organization of Petroleum Exporting Countries and through the Organization of Arab Petroleum Exporting Countries, and which have close political ties with the majority of the Middle-Eastern states. The omission of the North African oil exporters from the Middle East does have the great advantage that it enables the Gulf area to stand out clearly as a unique geographical concentration of hydrocarbon resources (Fisher 1963:232) and for this reason may be justified. The main concern of this study will be with the oil industries of Saudi Arabia, Iran, Iraq, Kuwait, the United Arab Emirates, Qatar, Oman and Bahrain. Other areas of the Middle East are bound to the main oil producers physically by the pattern of pipelines, through which a proportion of crude oil and natural gas is exported, and by less tangible but equally important economic links made possible by the special financial strengths of the oil states. Pipelines carrying oil exported from the Gulf cross Syria, Lebanon, Egypt, Turkey and Israel, giving each an immediate interest in the oil industry of the Middle East which will bring them into the scope of this study from time to time.

Evolution

The Middle East has long been a region of contrasts (Beaumont *et al.* 1976:3). Traditional and modern aspects of society and economy provide sharp variations of pace and style and, in the words of Freya Stark (1945:12),

> like two periods of history meeting, Viking and Wall Street, united perhaps by a tradition, but with all the circumstances of time and place to separate. In that meeting was embodied the fundamental problem of all Arab government – the immense gamut of civilization which it has to cover.

So it is with the oil industry in the Gulf area, where geological and commercial chronologies are to be seen against such different scales. Even in the short history of the world oil industry, the Middle East and particularly the Arab Middle East is a comparative latecomer. While the petroleum industry in the United States of America may be dated from 1859 (British Petroleum 1977:27), commercial production in the Middle East began as late as 1913 from the Iranian south, where oil had been discovered in 1908. Set against a long and varied history of civilization in the area, the Middle East oil industry is very recent indeed. The first commercial production of oil in the Arab world began in Egypt in 1909, though on a very small scale, and it was not until Bahrain became an oil producer in 1934 that the petroleum resources of the Arab states took on international importance. Development proceeded slowly and by the outbreak of the Second World War Iran provided the largest proportion of output of crude oil in the Middle East (Ferrier 1977a), supplemented with relatively insignificant production from Bahrain, Iraq and Saudi Arabia.

Inevitably the preoccupation with war and its aftermath retarded expansion of the embryonic petroleum industry. New capacity was installed in Kuwait, which came on stream in 1946, and in Qatar, where commercial lifting began in 1949. Exploitation of reserves in Turkey for local consumption was started in 1949. An important economic and political threshold was crossed in 1951, when a crisis in relations between the Iranian government and the Anglo-Iranian Oil Company led to the virtual closure of production, refining and export facilities in Iran for a damaging three-year period until a compromise agreement was reached between the two sides in October 1954 (Ferrier 1977b). On the one hand, the crisis was resolved with an acceptance of the nationalization of Anglo-Iranian and thus the seeds were set for a new and changing balance between host countries and international oil companies; and on the other hand, the closure of the Iranian export terminals gave impetus for the major oil companies to develop more rapidly their existing discoveries and to search with greater resources in areas as yet unsurveyed. The levels of production and exports in Kuwait, Saudi Arabia and Iraq were much enhanced as a consequence of the difficulties affecting the Iranian oil industry. Elsewhere new discoveries of oil were made and commercial production began in the Kuwait/Saudi Arabia Neutral Zone in 1954, in Libya in 1961 and in Abu Dhabi in 1962. Continuing exploration during the 1960s had its successes in Oman, Syria and Dubai, all of which had begun production by 1969, but the 1970s were less encouraging since only insignificant finds were made in Sharjah and Ras al Khaimah during this decade.

It will be apparent that Iran was the sole state to have a major oil industry in the years before 1951. Other regional producers have gained importance in the very recent past and emphasize the obvious but often forgotten characteristics of the Middle East oil industry, including its late emergence in, sudden effects on, and rapid growth within deeply traditional societies.

Location

The concentrated chronology of the development of the oil industry in the Middle East has a parallel in geographical terms. The incidence of reserves and production capacity is highly concentrated in a confined geographical area and, not without its political connotations, the greatest abundance of oil resources is to be found within the territory of a small number of states clustered around the Arabian–Persian Gulf. The Gulf geosyncline and its adjacent formations have been aptly described as uniquely perfect for oil accumulation (Longrigg and Jankowski 1970:253), with two distinct areas aligned along its axis together providing one of the world's greatest reservoirs of crude oil and natural gas. To the north of the main axis of the geosyncline the larger oilfields occur in a zone of folding and fissuring, where reservoirs are trapped in limestone formations, the structures often long and narrow lying more or less parallel to the axis in a dislocated string running from approximately Bushire in the east to Ain Zalah in the west. Most fields have been located in the Asmari limestones, though the Cretaceous limestone is also known to be oil bearing (Harrison 1968). South of the main axis there has been less vigorous folding and here large oilfields are disposed in a zone from Abu Dhabi to Basra, the oil and gas occurring as reservoirs in sandstone formations as in Kuwait or in limestones as in the majority of fields in Saudi Arabia.

Geological conditions are everywhere complex in the Middle East (Fig. 6.1), and the resulting distribution of oilfields means that by far the largest proportion of oil reserves of the area and at the same time one of the world's main sources of crude oil are located beneath the territory of eight countries at the most. Ironically but understandably, there is no correlation between wealth in oil reserves and population size nor a link between income from oil revenues and the financial requirements of the regional states for development funds. With the notable exception of Iran, the oil-rich nations are those with relatively small populations often of the most simple bedouin origins.

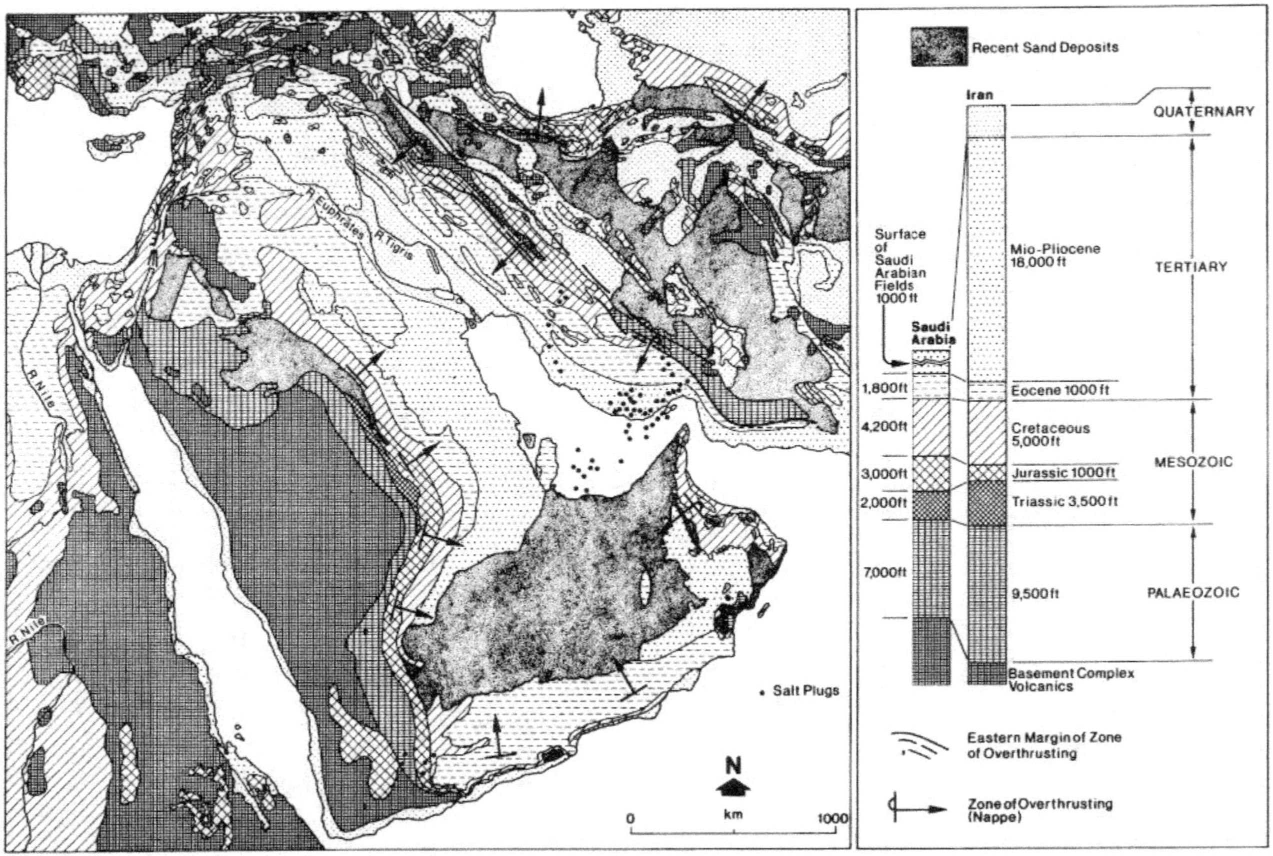

Fig. 6.1 Geology of the Middle East.

Effects

The effects of the oil industry in the Middle East are to be seen in two quite distinct aspects, one highly specific in geographical impact, the second general to the respective national territories of the oil-exporting states. Specific or direct effects arise from the operations of the national and international oil companies in the oilfield regions, along the lines of oil and natural gas transit and at the export terminals. The activities of the oil companies in the course of exploration, development, refining and maintenance have often made them one of the main investors and consumers within the oil economies, though the nature of their operations tended to create geographically distinct enclaves with limited linkages into the greater national economies (Mahdavy 1970). The oil companies, whether national or international, have become important employers of labour and frequently have the largest industrial workforce in their countries of interest. Additionally, oil operations have demanded the consumption of goods and services from the domestic economies, including foodstuffs, transport and specialist contracting facilities. Despite the rise of the national oil companies and the vigorous attempts by the oil states to assert their sovereignty in recent years, the oil industry still tends to treat its employees in a paternalistic way because it is demanded by the employees themselves. In consequence, the oilfield and related areas gain socially from the presence of the oil industry, with superior provision of health, educational and housing facilities compared to the country at large.

The operations of the oil companies in the producing countries were most strongly felt during early exploration and development when local expenditures exceeded oil revenues in importance. As exports of oil augmented and as the unit price of oil and its products rose, so the effects of the direct activities of the oil sector diminished relative to the indirect impact deriving from expenditures of oil revenues by central governments to cover the foreign exchange components of ordinary budgets, development programmes and defence allocations. General budgets of governments of the oil-exporting states provide for national economies as a whole, and geographical variations resulting from expenditures of oil revenues tend to arise between states and are at their most obvious when expressed as difference in standards of living between adjacent oil and non-oil states. Inherent bias in financial allocations within each oil-exporting state together with ineffective regional planning procedures have also tended to widen disparities in economic and social conditions between areas and between urban and rural populations, in a way that has imposed another and in many aspects more pernicious geographical pattern on the Middle East as a result of the development of the oil industry (Penrose and Penrose 1978:486).

Other resources

The rise in the importance of the oil industry in the Middle East and its outstanding international significance as an oil-exporting region and recipient of oil revenues has tended to obscure underlying realities concerning the quantity and quality of other natural resources. As early as 1963 Fisher warned

> natural resources are in general very small as compared with those of other regions, and the sparse natural wealth that does exist is only too often inefficiently or incompetently utilized. A mere 5 per cent of the total area is regarded as cultivable, while mineral resources, with the exception of oil, are scanty. Thus the economic potentialities of the Middle East are not large and it is regrettable that . . . there should have appeared . . . inflated estimates of possible future development.

Yet at that time the scale of operations of the oil industry in the area and oil income accruing to the oil-exporting governments were of comparatively modest proportions. Indeed, until the end of the 1960s few countries of the region were so wealthy from oil income alone as to have neglected other sources of revenue and employment. Even so, the proportion of national income contributed by the oil sector tended to rise steadily over the period for all oil-exporting countries.

The geography of resources strongly influences the relative decline in traditional economic activities and the degree of reliance on or independence from the oil sector. The smaller the domestic endowment with land, water and mineral resources, the greater and sooner the appearance of the dependence on oil production and exports. Despite conscious attempts to develop economic assets other than oil, Kuwait, Qatar and Abu Dhabi were all impeded in their efforts towards economic diversification by the absolute lack of natural resources and the problems arising from smallness of scale whether of surface area or of market size. For other oil-exporting states of the region the flow of oil revenues provided funds for welfare programmes and also for intensive development of existing areas of production and services. Even in Saudi Arabia, where agricultural land was limited in extent and water resources scarce, it appeared during the 1960s that a combination of improved irrigated agriculture, commercialization of livestock herding, development of import substitution industries and expansion of the pilgrim traffic to Mecca could offer a means of employment of the Saudi population and an appreciable source of national income for the future. At that time, the rapid increases in investment in industry and agriculture in Iran and Iraq seemed to presage an era in which relatively abundant natural resources would be brought into use to augment employment in and output from enterprises independent from the oil sector (Issawi 1971:61; Sayigh 1978:76). Dubai and Bahrain, using skills in trade and services rather than domestic resources, also appeared well placed to convert their oil income into viable alternative assets, while Oman, riven by internal divisions, was distracted from development of its economy.

The economic philosophies elaborated in the oil states of the Middle East during the 1960s had, without exception, the common objective of diversifying economies away from reliance on exports of crude oil and natural gas. Hydrocarbon resources were, after all, being depleted rapidly, the domestic oil industry provided only minor opportunities for employment of nationals, and petroleum exports were seen as hostages to the fortunes of international trade and the vagaries of Middle-Eastern politics. Many oil states were eager to diminish the role of oil exports and the activities of the international oil companies in order to minimize what were deemed to be neo-colonial pressures and infringements of national sovereignty. As a result of political attitudes and economic aspirations, few development plans formulated by governments in the area did not aim, as a priority, to achieve significant diversification away from oil.

The 1970s

Events affecting the petroleum industry during the early 1970s rendered the programmes for economic diversification largely ineffective. As a result of perceptible imbalance between supply and demand for crude oil on the international market, exacerbated by growing inability on the part of the western hemisphere countries to provide outside markets with augmented supplies at short notice, unit prices of crude oil improved gradually, beginning with the Tehran and Tripoli agreements between OPEC and the international oil companies in 1971. Disruption of oil supplies at the time of the Arab–Israeli war in 1973 precipitated a fourfold rise in oil prices by early 1974.

The incomes of the oil-exporting states rose steadily from 1971, and extremely rapidly in the months of crisis within the oil industry after October 1973, with a consequence that petroleum became the dominant contributor to the national incomes of all the Middle East oil-exporting states. For the desert oil economies such as Kuwait the insignificant gains made in creation of alternative sources of domestic income were dwarfed by the growth of oil income. States with modest non-oil resources, of which Saudi Arabia in the Middle East and Libya in North Africa are instances, experienced so large a shift in the relative contribution of oil to the economy that their hopes for non-oil developments as sources of alternative future income were all but extinguished, and in most respects their economic structures and degree of oil-dependence differ but little from that exemplified by Kuwait. It may be argued that Iraq, too, despite its apparent wealth in natural resources, has become a country in which the economy is increasingly oil-dependent (Penrose and Penrose 1978:542), possibly to the satisfaction of its current rulers (McLachlan and Ghorban 1978). In Iran the contribution of oil to national income as expressed in Gross Domestic Product rose above 40 per cent during the height of the oil boom and seriously undermined the positions of other productive sectors of the economy (Mabro 1976:6–7).

The rise in oil income in the Middle East that culminated in the crisis of 1973–4 served to make the area increasingly dependent on its one major internationally demanded resource – crude oil. Indeed, considerations of oil price and supply became the preoccupations of most governments of the region, and from 1973 the oil industry dominated economic and political life of the Middle East to an unprecedented extent. Other productive sectors lost relative importance and it is likely, too, that there was an absolute decline in levels of immediate output from and in the future prospects for non-oil sectors of the petroleum-exporting states (Mabro 1977; McLachlan 1978).

Reserves and production

The Middle East has been known for many years to contain a large share of the world's reserves of crude oil and natural gas. Exploration activity indicated that in 1938 oil reserves amounted to 5000 million barrels, some 15 per cent of the estimated world total; but the acceleration in exploration in the years after the Second World War soon proved that the Middle East was among the main oil-bearing regions of the world and by 1951 oil reserves there were estimated at 51,300 million barrels, approximately half of the world total (British Petroleum 1962). There has been only slight change in this situation over the years and by December 1977 the Middle East held 56 per cent of published proved world reserves of crude oil and about 28 per cent of reserves of natural gas. Although the commanding position of Middle-Eastern reserves of oil and gas has been known for almost 30 years, internal and international appreciation of this wealth has changed markedly over the period as a function of changes in control over the oil resource and the yield of income to the host countries. Development of new production capacity and export facilities has substantially increased the part played by Middle East oil in world trade.

Before the Second World War the Middle East had less than 6 per cent of installed oil-production capacity and by 1938 the area accounted for only 321,000 barrels/day of world output. In the period after 1945, despite the massive increase in reserves, exploitation went ahead relatively slowly so that by 1951 the area contributed 15.6 per cent of world production and as late as 1961 accounted for less than a quarter of global output. By 1978, 34.1 per cent of the world's oil production came from the Middle East, and though the full capacity of the region was not being exploited at this level it is clear that there remained a more favourable balance between the rate of output and the availability of reserves than that prevailing in most other oil-producing areas. The ratio between production in 1977 and the published proved reserves in the same year was estimated at 1:45 for the Middle East in comparison with a world average of 1:29 (Table 6.1). Important within this situation has been the very rapid and sustained rate of discovery of new oil reserves in the Middle East. In the decade 1968–77 some 24 billion tons of reserves were located in the Gulf area, against 15 billion tons in the decade 1958–67.

Table 6.1 Production to reserves ratios by main oil-producing areas in 1977

Middle East	1:45
North America	1:10
Latin America	1:24
Western Europe	1:53
Africa	1:26
USSR	1:19
Eastern Europe	1:20
China	1:33
Other areas	1:19
World average	1:29

Source: Calculated from British Petroleum, *BP Statistical Review of the World Oil Industry 1977*, London 1978

Within the Middle East region there were variations in the distribution of oil reserves between countries and the pattern of comparative strength in oil reserves has changed significantly over time. In the first place, the last decade or so has seen an increase in the number of countries in which oil has been discovered: notably Dubai, Oman and Sharjah. Meanwhile, continuing exploration and reappraisal of resources in the established oil states has brought even more radical shifts in the pattern of ownership of oil reserves. Saudi Arabia has emerged as the principal area of reserves of crude oil with more than 40 per cent of the regional total in 1978 against only 30 per cent a decade earlier. The rise of Saudi Arabia was mainly at the expense of Kuwait and the Neutral Zone (Table 6.2). New reserves discovered in Iraq during 1977 and 1978 would appear to have substantially improved its standing, and estimates by the Iraq National Oil Company put Iraqi reserves somewhat higher than those of both Iran and Kuwait.

More than a third of the world's oil production takes place in the Middle East, of which all but 1.1 per cent is located in those states adjacent to the Gulf.

Table 6.2 Estimated reserves of Middle East oil states by % of regional total

	end-1967	*end-1977*
Saudi Arabia	30.6	41.2
Kuwait	28.1	18.3
Iran	18.1	17.1
Iraq	9.8	9.2
Abu Dhabi	5.1	7.9
Neutral Zone	5.3	1.7
Oman	0.2	1.6
Qatar	1.6	1.6
Syria	0.6	0.6
Dubai	—	0.4
Sharjah	—	0.2
Turkey	0.4	0.1
Bahrain	0.1	0.1
Israel	0.1*	0.0*
Total	100.0	100.0

Sources: *Oil and Gas Journal*; *Oil in the Middle East*; Longrigg and Jankowski 1970

* Less than 0.01 %

Growth in output of crude oil in the Middle East in the period since 1913 is illustrated in Figure 6.2, which indicates the extreme concentration of output in a small number of countries, with Saudi Arabia (8,270,000 barrels/day in 1978), Iran (5,235,000 barrels/day), Iraq (2,600,000 barrels/day), Kuwait (1,865,000 barrels/day) and Abu Dhabi (1,450,000 barrels/day) together accounting for 91 per cent of all production. Saudi Arabia alone produced 39 per cent of all crude oil in the Middle East in 1978, followed by Iran with 25 per cent.

Changes in the contributions to output became particularly marked from the mid-1960s, when Saudi Arabia and Iran drew clearly ahead of Kuwait, and a major discontinuity is apparent from 1973 as production throughout the area faltered or fell, with the sole exception of Saudi Arabia. Whereas economic events outside the Middle East, and not least the world depression in trade, reduced severely the level of demand for oil on the international markets with increasing effect from 1974, changes in attitudes of the Middle-Eastern oil producers were important factors in reducing the rate of output. As a result of fears that known reserves would not sustain augmenting volumes of production, legal constraints were put on oil production in Kuwait even before the 1973 oil crisis. A conscious decision was made in Kuwait to maximize the length of time over which its reserves would last rather than optimize immediate income. In Iran, too, the government came round to the view that a ceiling should be placed on production at approximately the

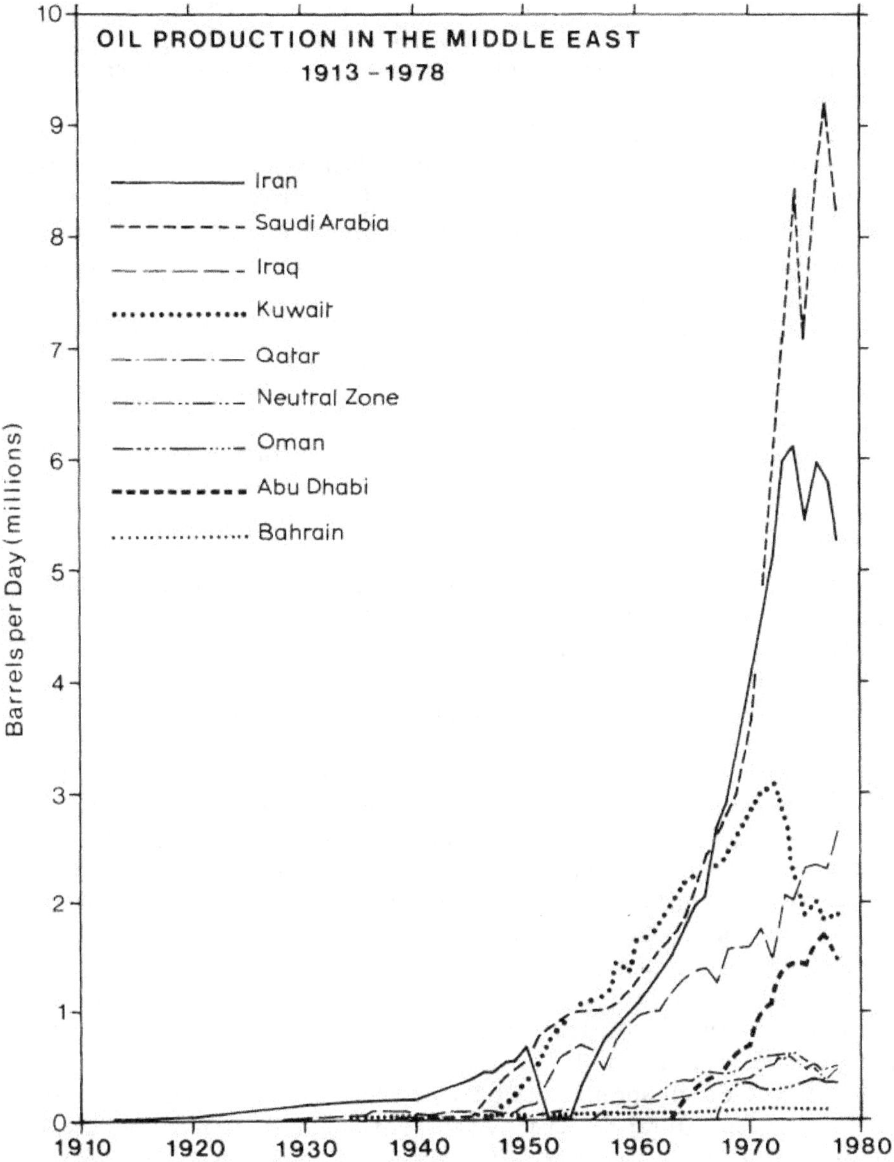

Fig. 6.2 Oil production in the Middle East, 1913–78.

6,500,000 barrels/day level as a means of conserving reserves and permitting introduction of sophisticated oil-recovery methods. Following the Iranian revolution of January 1979, the new regime instituted controls on production, which was not allowed to exceed some 4,000,000 barrels/day. Elsewhere in the Middle East, a combination of improved prices after 1973 together with declining demand coincided with a growth of insecurity over the longevity of oil reserves. Production to reserves ratios had deteriorated perceptibly throughout the area in the decade 1964 to 1973 except in Saudi Arabia, Kuwait and Iraq, and governments were sufficiently concerned for the future of their countries to restrain further rises in production. An expression of changed market conditions and producer attitude is the extent of under-utilization of capacity in the region: in Kuwait only slightly more than half of installed production capacity was exploited in 1978, in the United Arab Emirates 80 per cent and in Qatar 70 per cent. For the Middle East area as a whole, Saudi Arabia and Iraq were the only two states in the late 1970s that were planning substantial investments in new production capacity.

Exports

The Middle East is clearly a major area of crude-oil production, but its importance on a global scale arises from the large proportion of oil that is available for export. In 1978 almost 65 per cent of crude oil and 16 per cent of oil products entering international trade were derived from sources in the Middle East, which may be said to dominate the market in the world's main traded commodity. Changes are taking place in the abilities of the producing states to provide a disposable portion of oil output to the export market. Mention has already been made of the self-imposed ceilings on production instituted by a number of countries, of which Kuwait is the outstanding example, and the deteriorating reserves position in other states such as Bahrain. Yet another constraint on exports is coming into play as the domestic economies of the regional oil exporters begin to develop and internal consumption of oil and its products rises. In the case of Iran, where the process of economic modernization and development was most in evidence during the late 1960s and early 1970s, domestic consumption grew by as much as 15 to 20 per cent each year, and by 1977 one barrel of oil in every ten produced was absorbed by the domestic market while a parallel situation existed in respect of consumption of natural gas. The joint effect of stabilization of oil production to conserve oil reserves, together with rapid growth of domestic use of oil and its products, have tended to reduce the Iranian role as an oil exporter and producer with profound implications both for the structure of the domestic economy and for the pattern of oil supply to the international market. Rates of internal economic development elsewhere in the Middle East oil-exporting region have been less than those experienced in Iran, though all states, in their various ways, are seeking the means to elaborate their economies away from the oil base and the export of crude oil especially must be expected to suffer a perceptible decline.

Exports of oil from the Middle East almost doubled from 10,445,000 barrels/day in 1968 to 18,910,000 barrels/day in 1978. In the former year western Europe purchased approximately half of all exports, followed by Japan with a quarter, and these two customers remained in 1978 in first and second places respectively, though both substantially increased the absolute volumes of their liftings from the area (Table 6.3). The most significant change in the pattern of crude-oil demand in the decade was the emergence of the United States as a large-scale buyer of crude from the Middle East, its share of purchase rising eightfold to 2,300,000 barrels/day.

Entry of the USA as an importer of Middle East crude oil signalled an important change in the world oil industry, in which it became apparent that the Americans were not only unable to provide other consuming areas with supplies in the normal commercial course or at times of international oil crisis but were also unable to meet regional demand from internal sources. Such a situation gave added strength to OPEC in general and to the Middle-Eastern producers in particular, and was an important contributing factor to the emergence of the oil crisis that culminated in late 1973.

New developments

Expansion of oil production in the Middle East has been underpinned by heavy investments in new and improved production, transport and export facilities. The distribution of installations closely follows the pattern of commercially developed oilfields, their linkages through pipeline systems and the export terminals from which exports of crude oil take place (Fig. 6.3). The

Table 6.3 Main purchasers of Middle East oil 1969–78

	% of total		*% increase 1967–78 by volume*
	1969	*1978*	
Western Europe	50.6	44.0	49.5
Japan	26.4	20.3	32.1
USA	1.9	11.7	932.7
South-East Asia	3.1	8.2	139.4
Latin America	1.3	7.6	926.2

Source: Calculated from *BP Statistical Review of the World Oil Industry 1969 and 1978*

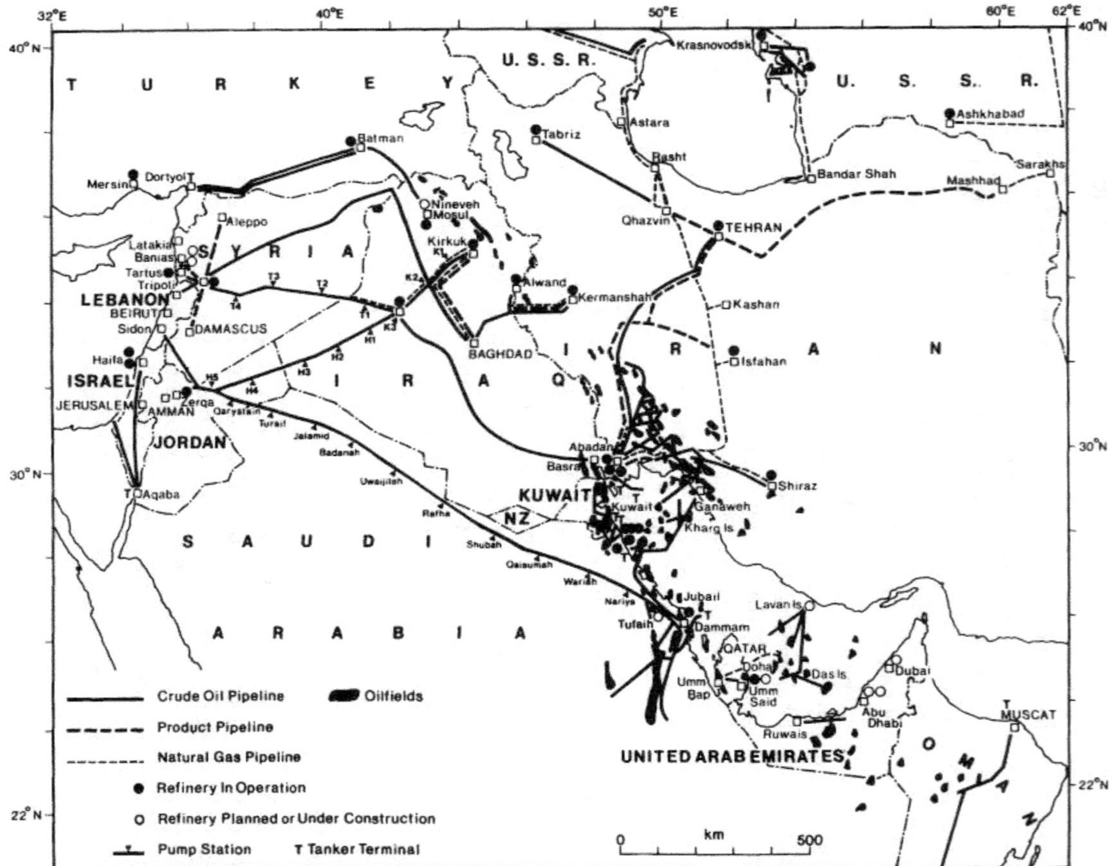

Fig. 6.3 Oilfields and oil installations in the Middle East.

shallow waters of the Gulf enabled relatively early development of offshore oilfields, beginning with the Safaniya field in Saudi Arabia in 1951. Settlement of offshore boundaries between national states with shorelines on the Gulf enabled most of the reservoirs to be exploited, with the notable exception of the head of the Gulf where agreement between Iran, Iraq and Kuwait on division of the undersea oil wealth is yet to be achieved. A small number of offshore oilfields are jointly owned, as in the cases of al Bunduq, jointly held by Abu Dhabi and Qatar, and Abu al Bakush, shared between Abu Dhabi and Iran.

A persistent trend towards increased domestic consumption of oil products has led to the growth of refining activity throughout the Middle East region, including both oil-exporting and non-oil states. Concurrently, the oil-exporting states have sought to increase the proportion of crude-oil production which is processed through domestic refineries before export as a means of raising the element of value added. Yet there has been a continuing decline in the relative importance of refinery capacity of the region as compared to the rest of the world, and in the decade 1969–78 the share of the Middle East in refinery capacity fell from 4.7 to 4.4 per cent of the world total. By 1978 only 16.5 per cent of Middle East production was refined within the area, when Iran was the main refiner at slightly less than one million barrels each day on average, the greater part of which throughput was used domestically. Iran also had the most comprehensive of regional refinery and oil-products distribution systems in the Middle East.

One of the most appreciable changes affecting the oil industry in the Middle East in recent years has been the re-evaluation of natural gas production. Natural gas produced in association with crude oil has traditionally been flared off as an unusable waste product with small volumes utilized within the oilfield and adjacent industrial and housing complexes for electricity generation, oilfield recharge or domestic purposes. Estimates vary, but it is unlikely that more than 10 per cent of total production of associated gas is utilized in the Middle East, largely because of the high costs of transporting this commodity to the main consuming areas of the industrialized countries. Current and future shortages of energy have brought about reconsideration of the use of natural gas for reservoir recharge, as a raw material for the petrochemical industry and as an export product. By the late 1970s, small-scale projects had begun in most oil-producing states with a view to reducing the waste of gas, though only Iran had made measurable progress towards developing its natural gas resources through the Iranian Gas Trunkline (IGAT) export facility to the USSR, commissioned in 1970. Associated gas may be used in the future to increase throughput of gas to the USSR by some threefold via a parallel IGAT project (for which the USSR will dispatch comparable quantities of natural gas from its northern fields to countries in eastern and western Europe), and to feed a programme of improved recovery of crude oil from existing fields. Gasfield development in Iran has begun at Sarakhs in the north for internal use, and in the south for exports of liquefied natural gas. Whether other oil states of the Middle East follow the Iranian model of gas utilization will depend on levels of prices for alternative energy sources and the amount of investment they are willing to undertake in this expensive new branch of the regional oil industry. The Iranian programme itself is less certain since the revolution.

Ownership

Change and development within the oil industry has had inevitable impact on its structure of ownership. Before 1951 it could be said with some truth that the oil resources of the Middle East were the financial, technical and political domain of powers situated geographically at a great distance from the area. Concession agreements prevailed between the oil-producing states and the international companies with their headquarters in the USA and western Europe, and many of the concessions gave away considerable sovereignty to the companies (Odell 1970:31). The Anglo-Iranian crisis in 1951, precipitated by nationalization of the Iranian oil industry but reflecting political and economic conflicts of long standing, and the earlier revision of the Aramco–Saudi Arabia concession in 1949, paved the way for a slow elaboration of arrangements between the international oil companies and the oil-producing states, which increasingly favoured the latter. Concessions, with their colonial and imperialistic overtones, were not extended to newcomer companies and in their place various joint-venture agreements, including national oil companies as equal partners with the foreign interests involved, were adopted in a number of states, beginning with Iran in 1957. Further improvements of conditions for the producers arose through use of service contract arrangements with internal oil companies under which foreign expertise and capital were mobilized on a purely contractual basis either alongside or on behalf of the national oil company. Iran utilized service contracts from 1966 and Iraq from 1968, in both cases to improve the terms of operation for the national companies and to diversify their oil sectors away from dependence on single concessionaires.

The most radical change in the structure of the oil industry came in the early 1970s as individual states negotiated a phased take-over of all oil operations within their national territories. Saudi Arabia and Kuwait acquired a 25 per cent share of the major concessionaire companies in 1973 and 60 per cent in 1974, Kuwait taking up total ownership of Kuwait Oil Company from 1975. A similar pattern of participation in the operations of the oil companies was followed by Abú Dhabi, Oman and Qatar. In Iraq, the Iraq Petroleum

Company was nationalized in 1972, Mosul Petroleum Company in 1973 and Basrah Petroleum Company in 1975, while in Iran the National Iranian Oil Company, which was already the owner of all installations, took on responsibility for oil operations from 1973. But whereas the Middle East oil-exporting states have been successful in assuming control of their national oil industries, with the possible exception of pre-revolutionary Iran, they have not yet been able to integrate their oil industries into the marketing and processing systems in the consumer areas, and many, indeed, still rely heavily on the international oil companies for assistance in oilfield, transit and export operations within their national boundaries.

Conclusion

The geography of the Middle East oil industry is characterized by a high degree of concentration of reserves and production: the number of countries producing and exporting oil in the area has increased in the last two decades, but strength in depth of reserves and production capacity has tended to become increasingly the prerogative of fewer states. This situation is accentuated by attitudes towards conservation of resources in a number of formerly leading oil-exporting countries, such as Kuwait, and by rapid growth in domestic consumption of hydrocarbons in Iran.

Important changes were set in motion in the early 1970s in the world prices for and division of profits from Middle-Eastern oil exports, which resulted in sharp rises in the oil revenues of the oil-exporting states. In effect, the oil industry became the overwhelmingly dominant economic sector in the oil-based countries, in many cases to the detriment of developments in the non-oil areas of activity which in a relatively near future must provide alternative sources of personal and national income, employment and foreign exchange earnings. Extreme variations in income and welfare have arisen between those states with oil and those without, with consequent political and migration stresses dealt with elsewhere in this volume. Within the oil states, imbalance between depressed rural and more prosperous urban areas have similarly been exacerbated and have provoked a rapid and major shift in patterns of employment and settlement.

The structure of the oil industry has been altered gradually over the last 25 years from one in which host countries and concessionaire companies existed in uneasy and unequal relationships in a horizontally integrated operation, to one where national oil companies are ascendant in their own states but the international oil companies still control a large measure of downstream activities.

Change and development in the Middle-Eastern oil industry has brought about a reappraisal of the region's importance, especially in the years since 1970. The region is unchallenged in possessing disposable oil and natural gas resources at the present time, and is the main if not the only reservoir of incremental energy supply over the period to the turn of the century. But the history of the oil industry in this area is one of apparent continuity broken from time to time by profound discontinuities and, as the pre-eminence of the Middle East as a supplier to the international market in oil has grown, so the impact of erratic deviations in oil supply or price has had greater effects upon other economies. Economic difficulties deriving from the oil boom and its aftermath of business recession have tended to exacerbate endemic political instability in the area: considerable oil wealth has done little to solve the burning political conflicts that divide nations in the Middle East, and it must be expected that the short and important life-span of oil production in the region will be interrupted in the future just as in the past.

References

BEAUMONT, P., BLAKE, G. H. and WAGSTAFF, J. M. (1976) *The Middle East: A Geographical Study*, London, John Wiley & Sons.

BRITISH PETROLEUM (1962) *Statistical Review of the World Oil Industry 1961*, London.

—— (1977) *Our Industry Petroleum*, London.

FERRIER, R. W. (1977a) 'A brief history of British Petroleum', in *Our Industry Petroleum*, London.

—— (1977b) 'The development of the Iranian oil Industry', in Amirsadeghi, H. (ed.) *Twentieth Century Iran*, London, Heinemann, 105–9.

FISHER, W. B. (1963) *The Middle East*, 5th edn, London, Methuen.

HARRISON, J. V. (1968) 'Geology', in Fisher, W. B. (ed.) *The Cambridge History of Iran*, vol. 1, London, CUP.

ISSAWI, C. (1971) *Iran Faces the Seventies*, New York, Praeger.

LONGRIGG, S. H. and JANKOWSKI, J. (1970) *The Middle East – A Social Geography*, London, Duckworth.

MABRO, R. E. (1976) *Aspects of Economic Development in Iran*, London, RIIA, Chatham House.

—— (1977) 'Development defects in OPEC's fast growth strategy', *Middle East Annual Review*, 1977, 31–40.

McLACHLAN, K. S. (1978) Iran, *Middle East Annual Review 1978*, 203–14.

McLACHLAN, K. S. and GHORBAN, N. (1978) *Economic Development in the Middle East Oil-Exporting States*, Economist Intelligence Unit.

MAHDAVY, H. (1970) 'The patterns and problems of economic development in rentier states – the case of Iran', in Cook, M. A. (ed.) *Studies in the Economic History of the Middle East*, Oxford, 428–67.

ODELL, P. R. (1970) *Oil and World Power*, London, Chivers Penguin.

PENROSE, E. and PENROSE, E. F. (1978) *Iraq*, London, Benn.

SAYIGH, Y. A. (1978) *The Economies of the Arab World*, London, Croom Helm.

STARK, F. (1945) *East and West*, London, Murray.

CHAPTER SEVEN

Offshore politics and resources in the Middle East

GERALD H. BLAKE

Interpenetrated by five seas, and fronting the Atlantic Ocean in the west and the Indian Ocean in the east, the Middle East and North Africa have long been subject to maritime influences, but never more so than today. For the first time, the seas are being assimilated into the political and economic systems of coastal states, and maritime zones and boundaries are becoming an integral part of the map. This chapter summarizes the nature of offshore sovereignty, and considers the complex problems associated with offshore boundary delimitation. Viewed from the perspective of 1979 such themes may seem rather academic, but within a few years numerous offshore disputes could well have added a new and dangerous dimension to the 'troubled triangle' where at least fifteen disputes over maritime boundaries, claims to islands, and fisheries remained unresolved in 1979 (Fig. 7.1).

Access to the sea

None of the 22 states of the region is landlocked, but Jordan and Iraq are among the world's 'geographically disadvantaged' states in having limited access to the sea. Although coastlines vary considerably in length (Table 7.1), most other states enjoy favourable access, and coastal waters represent economic and sometimes political assets of considerable significance.

Access to the sea has given rise to at least two major disputes, between Iraq and Iran, and between Israel and Egypt. Iraq has very restricted access to the Gulf, and disputes with Iran over their boundary along the navigable river Shatt al Arab have arisen over several decades. In 1913 the Turkish administration was given effective control of the river estuary, with the boundary following low-water mark along the Iranian bank. In 1937 the boundary opposite Abadan was shifted by agreement to the *thalweg* of the river. Iran now wants a similar mid-stream boundary throughout, while Iraq

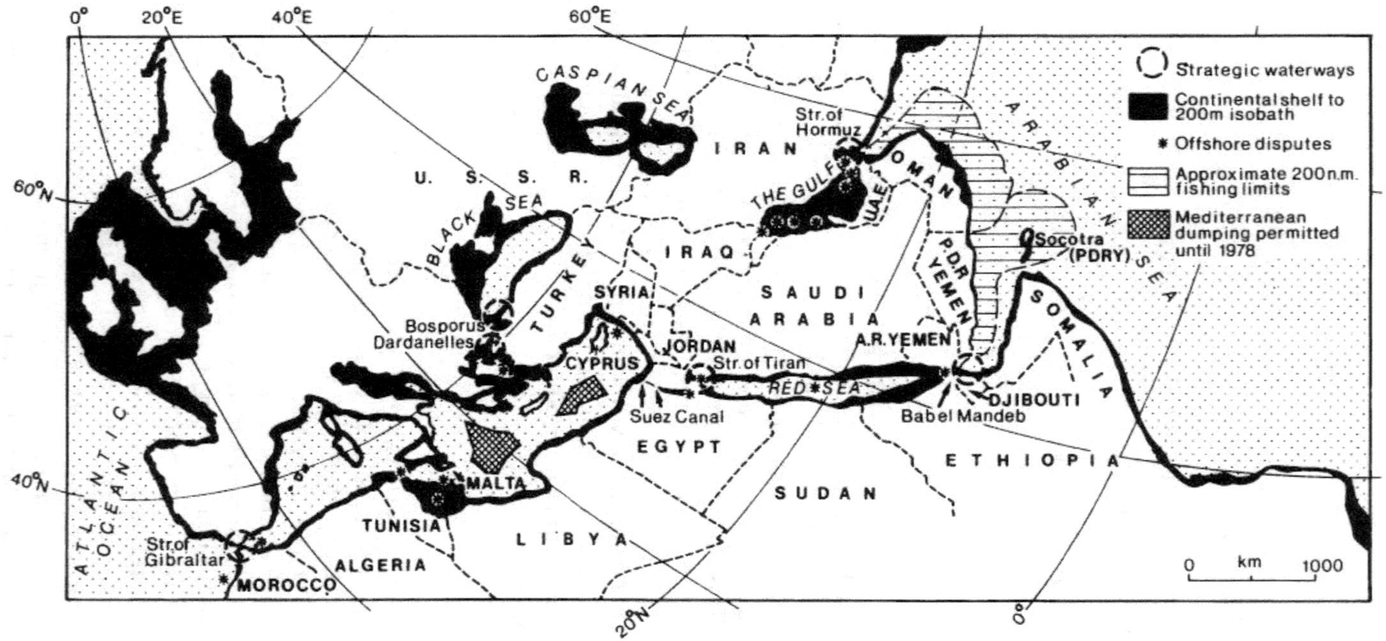

Fig. 7.1 Some offshore political features in the troubled triangle of the Mediterranean and Middle East.

Table 7.1 Maritime statistics of the Middle East and North Africa

	Number of sq km of land per km of coastline [a]	*Length of coastline (km)* [b]	*Territorial water claim* [c] *(nautical miles)**	*Exclusive fishing zone* [d] *(nautical miles)*
Bahrain	4.7	126	3	as TW
Cyprus	17.2	537	12	as TW but 20 in north-east
Qatar	27.4	378	3	as TW
Lebanon	54.0	194	undeclared	6
Kuwait	83.0	213	12	as TW
Israel	90.0	230	6	as TW
United Arab Emirates	123.0	700	3	as TW
(Sharjah)	—	—	(12)	(as TW)
Tunisia	159.0	1,028	12	as TW
Oman	161.0	1,861	12	200
Turkey	219.0	3,558	12	Black Sea 12 Aegean 6
Morocco	277.0	1,657	12	70
PDR Yemen	320.0	1,211	12	200
Egypt	414.0	2,421	12	as TW
Yemen AR	431.0	452	12	as TW
Iran	897.0	1,833	12	50
Saudi Arabia	920.0	2,437	12	as TW
Libya	1,044.0	1,685	12	as TW
Syria	1,218.0	152	12	as TW
Algeria	2,157.0	1,104	12	as TW
Sudan	3,490.0	717	12	as TW
Jordan	3,533.2	27	3	as TW
Iraq	22,842.0	19	12	as TW

Source: [a] Land areas from *Middle East Yearbook* (1978), IC Magazine, London;
[b] Office of the Geographer, US Department of State (1969), *Sovereignty of the Sea*, Geographic Bulletin No. 3, Washington DC, 19–22;
[c] *The Statesman's Yearbook 1978–79* (1978), Macmillan, xxiv–xxvi;
[d] Office of the Geographer, US Department of State (1974) 'Limits in the Seas No. 36', *National Claims to Maritime Jurisdiction*, pp. 1–140.
See also, Open University (1978), *Oceanography: Law of the Sea*, Milton Keynes, Tables 1 and 2.
* 1 nautical mile = 1.516 statute miles = 1853 m.

maintains the right to effective control of her only access to the sea. Israel's dispute with Egypt was different. From 1948 until 1956 the Arab states denied Israeli shipping access to the port of Eilat via the Gulf of Aqaba, on the grounds that the Gulf constitutes a closed sea or *mare clausum* in which the principle of innocent passage does not apply. In 1955 the Egyptians blockaded the Strait of Tiran, but in 1956 the Sinai campaign opened the Strait to Israeli ships. UN forces ensured free passage to Eilat until May 1967 when President Nasser ordered the UN to withdraw, thus precipitating a second Israeli invasion of Sinai in June 1967. Under Article V of the Egypt–Israel Peace Agreement of March 1979, the parties agreed to consider the Strait of Tiran and the Gulf of Aqaba as international waterways, open to all nations. Jordan and Saudi Arabia also have coastlines on the Gulf of Aqaba, but are unlikely to interfere with Israeli shipping. It is interesting to recall that in 1964 Saudi Arabia ceded Jordan 19 km of coast on the Gulf of Aqaba in return for large tracts of desert territory inland. Jordan's precious 27 km of coastline is meticulously planned in an effort to reconcile competing industrial, residential, recreational and fishing requirements.

The significance of offshore areas

Offshore resources

There are several reasons why the states of the region are becoming increasingly interested in their coastal waters and the oceans beyond. Most obvious is their considerable potential for oil and natural-gas production. Several offshore oil deposits were discovered in the shallow waters of the Gulf after 1945 and two-dozen fields were producing in the 1960s. Several more offshore fields began production of oil and gas in the 1970s and today offshore yields contribute significantly to the output of half a dozen Gulf states (Table 7.2). Outside the Gulf, only Tunisia, Egypt and Israel were offshore

Table 7.2 Offshore oil production in the Middle East (in barrels)

	Daily average for first 6 months of 1977			*Daily average for first 6 months of 1978*		
		Offshore			*Offshore*	
	Total	*Amount*	*%*	*Total*	*Amount*	*%*
Abu Dhabi	1,700,000	670,000	39	1,440,000	*	*
Divided zone (neutral zone)	334,865	166,298	49	375,000	*	*
Dubai	317,398	317,398	100	359,042	359,042	100
Egypt	394,612	297,498	75	456,312	399,955	88
Iran	5,617,906	507,955	9	5,688,229	654,498	11
Qatar	430,000	220,000	51	451,190	227,690	50
Saudi Arabia	9,180,382	*	*	7,574,257	2,265,121	30
Sharjah	*	*	*	25,442	25,442	100
Tunisia	86,960	45,537	52	99,255	45,234	45
Israel	663	Nil	0	6,670	6,000	90

Source: *The Oil and Gas Journal*, Worldwide Production, 26 December 1977, 103–47; 25 December 1978, 108–42.

* Not available

producers in 1978, the latter in a part of the Gulf of Suez claimed by Egypt. Offshore gas production is less important in the Middle East and North Africa than offshore oil, but it is nevertheless a very valuable resource for which there are plans to use more in future. Several countries are expected to begin offshore oil and gas production before long, including Libya, Yemen AR, PDR Yemen (off Socotra) and Kuwait, all of whom are currently conducting offshore exploration. Politically, the most important fact is that large areas already designated for concession have yet to be properly investigated and no state can afford to settle disputes too readily. A shift of a few hundred metres in the position of a boundary could result in massive gains or losses of revenue.

Knowledge of other minerals in Middle-Eastern and Mediterranean waters is sketchy. Exploration of the ocean floors beyond the continental shelves has scarcely begun; and even the continental shelves are incompletely explored and charted. Yet the 200 nautical mile Exclusive Economic Zone (EEZ), if and when it is introduced, would give states the exclusive right to exploit minerals up to 230 statute miles offshore. One region where valuable mineral resources are known to exist is in the Red Sea, whose non-oil mineral wealth was put in the vicinity of US $7000 million in 1975 (Arab Economist 1975). Heavy metal deposits were first located in the deepest parts of the central areas of the Red Sea in 1963 at depths of 2000 m. Substantial quantities of highly concentrated metals were found, including copper, zinc, lead, silver, gold and iron. Most lie on the Sudanese side of the median line, but Saudi Arabia claims an equal right to exploit them since they lie beyond Sudan's continental shelf jurisdiction, and 50 miles beyond Sudanese territorial waters (Fielden 1978). It is likely that more oil and gas deposits will also be found in the shallow waters of the Red Sea, but political disputes such as that between the Yemens over the Kamaran Islands will have to be resolved to expedite exploitation. Apart from the celebrated Red Sea deposits, it is difficult to find reports of minerals elsewhere in Middle East waters, though manganese nodules are known to exist in the Mediterranean between Cyprus and Turkey (Albers *et al.* 1973).

Sea fishing has long been an important activity in certain parts of the Middle East and Mediterranean though the scale of the fishing industry varies greatly from country to country (Table 7.3). Statistics do not always distinguish between freshwater and marine catches, but apart from Egypt, Iran and Israel, figures shown in Table 7.3 are almost entirely marine catches. In addition, Kuwait, Bahrain and Qatar all have commercial fisheries specializing in processing Gulf shrimps and prawns for export.

Anxiety over the protection of fisheries has been the most important reason for the extension of territorial water claims in recent years. Fish stocks are not inexhaustible, and there has been a big increase in the catches of most states, and several states are expanding their commercial fishing activities. Iran,

Table 7.3 Fish catches of Middle-Eastern countries, 1965–75 (in thousand tonnes)

	1965	*1975*
Turkey	135.8	259.4
Morocco	206.4	210.5
PDR Yemen	52.0	127.7
Egypt	94.0	106.6 (¾ freshwater)
Oman	*	100.0
United Arab Emirates	*	68.0
Tunisia	22.8	42.7
Algeria	18.2	37.7
Saudi Arabia	*	30.0
Israel	19.5	24.2 (¾ freshwater)
Iran	*	20.0 (⅓ freshwater)

Sources: FAO (1968) *Yearbook of Fishery Statistics* 1967, vol. 25, Rome, United Nations, 6–15; UN (1977) *Statistical Yearbook* 1976, New York, 146–7.
* Not available

Morocco, Oman and PDR Yemen have already claimed fishing jurisdiction far beyond their territorial waters (Table 7.1). So far, disputes over fishing grounds have been few in the Mediterranean and Middle East, but they could easily become more common. The dispute between Spain and Morocco illustrates what can happen. Faced with declining catches up to 1972, Morocco declared an extension of her fishing jurisdiction from 12 to 70 nautical miles in 1973. Foreign vessels would require permission to fish within this limit. Spain refused to accept these measures since they repudiated a 1969 fishing agreement and several incidents occurred involving naval units. In 1974, however, Morocco agreed to allow 200 Spanish vessels to fish within 70 miles of the coast as part of a joint Moroccan–Spanish company, but Moroccan seizures of Spanish vessels continued. The situation is further complicated by the existence of the small Spanish enclaves of Ceuta and Melilla on the north coast of Morocco, and exacerbated by disagreements over the seabed boundary between Morocco and the Canary Islands (Buzan 1978).

Environmental management

A second reason for growing political interest in offshore areas is international concern over problems of environmental management. The Mediterranean has become one of the world's most seriously threatened seas. In 1978 the Barcelona Convention on control of pollution came into force, signed by 15 countries. Albania, Algeria and Syria were not signatories. Since 1975 some important steps have since been taken towards tackling pollution through monitoring and research, and the establishment of a regional oil-combating centre in Malta. Oil pollution is the most conspicuous type of pollution, and tar balls in the water and tar on sandy beaches have become familiar to Mediterranean holidaymakers. But oil is not the only pollutant. A human population of 100 million around the Mediterranean basin, supplemented by annual influxes of a similar number of tourists, discharges 90 per cent of its sewage into the sea untreated. Industrial waste and agro-chemicals washed down by rivers add to the problem. Evaporation exceeds the gains from rivers and precipitation in the Mediterranean, the only significant water exchange being through the Straits of Gibraltar (Ambio 1977; Mirabelli 1977).

Although marine pollution is less serious in the Red Sea and Gulf, the coastal states of both seas have already taken steps to protect their offshore environments. The worst threat is to the Gulf, with its numerous oil terminals and growing range of petrochemical complexes. The oil companies established a regional organization to deal with oil spills in 1972, and in 1976 the Gulf states agreed on a draft convention for the protection of the marine environment, which may have come just in time. In mid-1975, for example, pollution of the beaches of Kuwait was so bad that swimming was officially banned for some time.

Marine environmental management is still in its infancy in the Mediterranean and Middle East, but so far there has been a surprising degree of co-operation in a region where political disagreements seem endemic. The exercise of effective controls will ultimately mean more boundary lines and zones; a good example is that prohibitions on dumping in the Mediterranean left only two areas where dumpers could escape heavy penalties (Fig. 7.1).

Strategic and military interests

It is impossible to do justice to the strategic and military importance of the region's seas in a few paragraphs. The Mediterranean, the Gulf and the Red Sea are the foci of superpower rivalries and local conflicts in which the influence of naval power is of paramount importance. Regional politics are characterized by the dangerous interaction of global and local interests as classically shown by the Arab-Muslim conflict with Israel. The two dozen or so states of the region are highly susceptible to external influence by virtue of their differing ideologies and geopolitical locations. There is, therefore, ongoing competition for access to bases, anchorages, and to markets for the sale of weapons. Arms acquired in recent years have included many ships and naval weapons, with Israel, Egypt, Turkey and Iran building significant naval forces. Under the Shah Iran was developing one of the most powerful navies in the world, including a large hovercraft fleet. At the same time, local merchant and fishing fleets have grown substantially in the last decade.

The six strategic waterways of the region are key features in maritime strategy (Fig.7.1 and Table 7.4), and international political events are often indirectly or directly concerned with their control. Their importance may be summarized as follows:

(a) The Bosporus and Dardanelles provide the only outlet for the Soviet Black Sea fleet which operates in the Mediterranean and the Indian Ocean. The passage of warships is limited under the Montreux Convention of 1936, notably aircraft carriers. With Turkey in NATO, and theoretically the gatekeeper of the straits, the Soviet Union would like to see the Convention revised. Meanwhile the potential for serious disagreement is present, as when the Soviet aircraft-carrying cruiser *Kiev* passed through in 1976. In the event of hostilities, the Russians would no doubt seize the straits.
(b) The Strait of Tiran controls the entrance to the Gulf of Aqaba, and access to Eilat (Israel) and Aqaba (Jordan). Already a cause of two wars, the strait is a potential flashpoint, defused at present by the Egypt–Israel accord.
(c) Bab-el-Mandeb controls passage through the Red Sea to the Suez Canal. Future Arab attempts to prevent access to Eilat could occur here; Israel's formidable long-range gunboats and F-15 fighters are said to be

Table 7.4 International straits in the Middle East and North Africa

Strait	*Sovereignty and territorial sea claims (nautical miles)*	*Least width (nautical miles)*	*Depth in feet*	*Length (nautical miles)*
Gilbraltar	Spain(12) Morocco(12)	8.0	70–1,800	33
Dardanelles	Turkey(6)	0.5	150–300	36
Bosporus	Turkey(6)	0.5	240	17
Tiran	Israel*(12) Saudi Arabia(12)	4.0	240–600	7
Bab-el-Mandeb	Djibouti(12) Yemen AR(12) PDR Yemen(12) Ethiopia(12)	9.0	40–600	50
Hormuz	Iran(12) Oman(12)	21.0	180–300	17

Source: Smith, R. W. (1974) 'An analysis of the strategic attributes of International Straits', *Maritime Studies and Management*, 2, 88–101.

* Israel until east Sinai is returned to Egypt in 1982

intended to counteract this threat. Much of the geopolitics of the Red Sea and Horn of Africa are associated with this vital waterway. By the end of the 1980s, the Suez Canal will be wide enough to admit supertankers and a greater proportion of Europe's oil will pass via Bab-el-Mandeb to Suez or to Egypt's SUMED pipeline.

(d) The Strait of Hormuz is perhaps the most important of all, since through it passes a high proportion of oil destined for western Europe, Japan and the USA. It would be difficult for any power or extremist group to interrupt shipping in the Strait of Hormuz, but the fear is widely felt.

Tangible evidence of the importance attached to the maritime Middle East and Mediterranean is the number of naval bases and facilities available to the United States, Soviet Union, Britain and France (IISS 1971). While the pattern of access has changed over the years, there has been no diminution of interest, as exemplified by overtures to Malta after Britain's withdrawal in 1979. It must also be noted, however, that the proposed increase in maritime jurisdiction by local states will necessitate effective control by naval forces up to 230 statute miles offshore. At both levels, therefore, strategic and military interests will continue and possibly intensify in future.

National jurisdiction offshore

Coastal states may exercise jurisdiction over the following zones: internal or inshore waters; territorial waters; contiguous zones for the exercise of customs and sanitary controls; exclusive fishing zones; and the continental shelf. A sixth type of offshore jurisdiction, the Exclusive Economic Zone (EEZ), is likely to emerge from the Third UN Conference on the law of the sea. Offshore jurisdiction differs markedly from onshore jurisdiction in a number of respects. The sovereignty exercised by the coastal state is absolute only in internal waters. The rights of the coastal state and other maritime users are governed by conventions which may be variously interpreted. Similarly, conventions for delimiting maritime boundaries are open to differences in interpretation.

Only a minority of states ever ratified the conventions drawn up at Geneva in 1958 at the first UN Conference on the Law of the Sea (UNCLOS I). Nor can maritime boundaries be demarcated, and even today they rarely feature on maps and charts. Over the next few years, however, national offshore rights and responsibilities will become more clearly defined and efforts will be made to fix boundaries more precisely. The following paragraphs outline the characteristics of offshore jurisdictions in the modern Middle East. Internal waters are not discussed, as having little international significance.

Territorial waters

While the concept of the territorial sea goes back to the seventeenth century in Europe, its application in the Middle East and North Africa has been largely confined to the last few decades. A few states claimed territorial waters some time ago, including Turkey (1914), Palestine (1924), Iran (1934), Cyprus (1935), Jordan (1943), Egypt (1951) and Tunisia (1951). Most, however, made claims in the late 1950s and 1960s after UNCLOS I recognized the right of coastal states to territorial waters in which sovereignty is exercised over the waters themselves, the seabed beneath, and the airspace above. Ships of all states, however, enjoy the 'right of innocent passage' through the territorial sea. The right of innocent passage has been the subject of much debate in international law, and practices differ considerably between states in spite of certain rules laid down in 1958. Thus Algeria requires foreign warships to obtain prior permission to pass through her territorial waters and PDR Yemen requires prior permission for all foreign ships.

The need for territorial waters has become increasingly pressing in modern times; to control smuggling and immigration, to control pollution and dumping, to enforce health regulations, to regulate shipping and for national security. Above all, offshore resources of all kinds need to be protected as improved technology makes their exploitation increasingly attractive. Territorial waters may be claimed to discharge these functions, but there is no international consensus as to a standard width (Table 7.1). The Third UN Conference on the Law of the Sea (UNCLOS III – still in session in 1979) might agree to a standard 12 nautical miles (or 13.8 statute miles). Half a dozen Middle-Eastern states would appear to benefit from this, but in reality the politico-economic gains would be limited.

Delimitation of territorial waters is complicated by the adoption of straight baselines by a number of states. These are permitted where coastlines are deeply indented, or if there is a fringe of islands along the coast. Baselines are normally at the low-water mark along the coast, though there are rules for using closing-lines across the mouths of bays and other inland waters. Thus the straight baseline can greatly extend the outer limit of territorial waters, but conventions agreed in 1958 to discourage its abuse are frequently disregarded or misinterpreted. Although several proposed straight baselines in the Middle East have yet to be charted, some have been described and mapped in the US Department of State's authoritative series on maritime boundaries (Office of the Geographer 1970, 1975). By liberal interpretation of such terms as 'bay' and 'island', Saudi Arabia and Oman appear to have claimed large areas of inland waters, which the 1958 Convention does not strictly allow. In the case of Oman's Musandam peninsula the results are striking; the principal navigation channel of the Strait of Hormuz is enclosed within Oman's internal waters. The 1958 Geneva Convention stipulates that where straight baselines take in what had previously been high seàs or territorial waters, innocent passage must be allowed through such internal waters. Unfortunately, Oman is not a party to this particular convention.

The international debate concerning the introduction of a standard 12-mile limit has focused a great deal on possible threats to freedom of navigation as a result of the extension of territorial waters into international straits. It is therefore worth noting that all waterways in the Middle East are already territorial seas (Table 7.4).

Contiguous zones

Beyond the territorial waters, a state may exercise the control necessary to prevent infringement of customs, fiscal, immigration or sanitary regulations. According to the 1958 Convention, such 'contiguous zones' should not extend beyond 12 miles from the baseline from which the territorial sea is measured. In fact, several Middle East and North African states claim contiguous zones for customs and sanitary purposes beyond 12 miles, a common claim being 18 miles (e.g. Egypt, Saudi Arabia, the Yemens). Lebanon has no territorial-water claim, and, in the absence of such, claims to 20 km for customs and criminal jurisdiction and to 6 nautical miles for security are of interest (Office of the Geographer 1974).

Exclusive fishing zones

States have exclusive rights to fish in their own territorial waters. Beyond territorial waters they may claim complete control over the exploitation of fish stocks in an exclusive fishing zone. Only seven such claims appear to apply in the region (Table 7.1), though more can be expected if the 200-mile EEZ fails to materialize. The Gulf states in particular will be tempted to claim fishing zones, since catches of shrimps and prawns have greatly declined as a result of over-fishing. Coastal states rarely insist on the right to exclusive fishing, but on the right to control fishing and thus preserve stocks. Concession agreements with second parties have been granted by PDR Yemen and Oman within specified areas in their 200-mile exclusive fishing zones. PDR Yemen had agreements with Japan and the Soviet Union in the 1960s and with Iraq in 1977. Oman had similar agreements with Japan in 1976 and 1977 and with South Korea in 1978.

The continental shelf

The continental shelf refers to the seabed beyond the outer limit of the territorial sea, where the coastal state has exclusive right to the mineral resources and sedentary species of marine life such as oysters and sponges. Rights to the continental shelf do not extend to the water or airspace above, and the laying of cables and pipelines by other states is permitted. Continental shelves vary greatly in their physical properties, and there is confusion as to where the exclusive rights of the coastal state end. The Convention of 1958 specified that rights would extend:

(a) to the seabed and subsoil of the submarine area adjacent to the coast but outside the area of the territorial sea, to a depth of 200 metres or, beyond that limit to where the depth of the superadjacent water admits of the exploitation of the natural resources of the said area;
(b) to the seabed and subsoil of similar submarine areas adjacent to the coasts of islands (Lay *et al.* 1973).

Since exploitation in depths above 200 m is now quite feasible, there is scope for national claims to seabed which the UN regards as 'the common heritage of mankind', though this is uncommon in the Middle East or North Africa.

Continental shelves in the region vary considerably in width, representing natural inequalities with regard to offshore resources (Fig. 7.1). The whole of the Gulf is continental shelf averaging 20–60 m in depth, and effectively partitioned between coastal states to a degree only rivalled in the North Sea. Continental shelf boundaries are based on the application of median lines, though the principle of equity may be applied by agreement of states concerned. In practice many problems arise and several continental shelf boundaries in the Gulf have yet to be agreed. Some are a matter of minor adjustment, as between Iraq and Kuwait, Iran and Kuwait, and Iran and the UAE, but Ras al Khaimah and Oman dispute a 10-mile coastal strip. Libya and Malta, Malta and Tunisia, and Tunisia and Libya have also failed to agree on a median line between them. In all these disputes substantial oilfields are known or said to be involved.

The dispute between Greece and Turkey is potentially the most dangerous in the Mediterranean. In 1973 and 1974 Turkey made *de facto* unilateral delimitations of a continental shelf boundary in the Aegean by granting concessions to the Turkish Petroleum Company (Fig. 7.2). The dispute brought the countries close to war in 1976 and remains unresolved. Oil and gas deposits were located in Greek waters in 1973 raising hopes of further finds in the Aegean, but resources are only part of the problem. Turkey is also sensitive about access to the Turkish Straits and to the ports of Izmir and Istanbul. Law of the sea conventions unfortunately provide only vague guidelines for continental shelf delimitations which cannot cope with the geographical and political complexities of the Aegean. Greece owns 3054 islands, many close to the Turkish mainland, while Turkey possesses only two. Territorial waters extend by mutual consent to only 6 miles in the Aegean, and

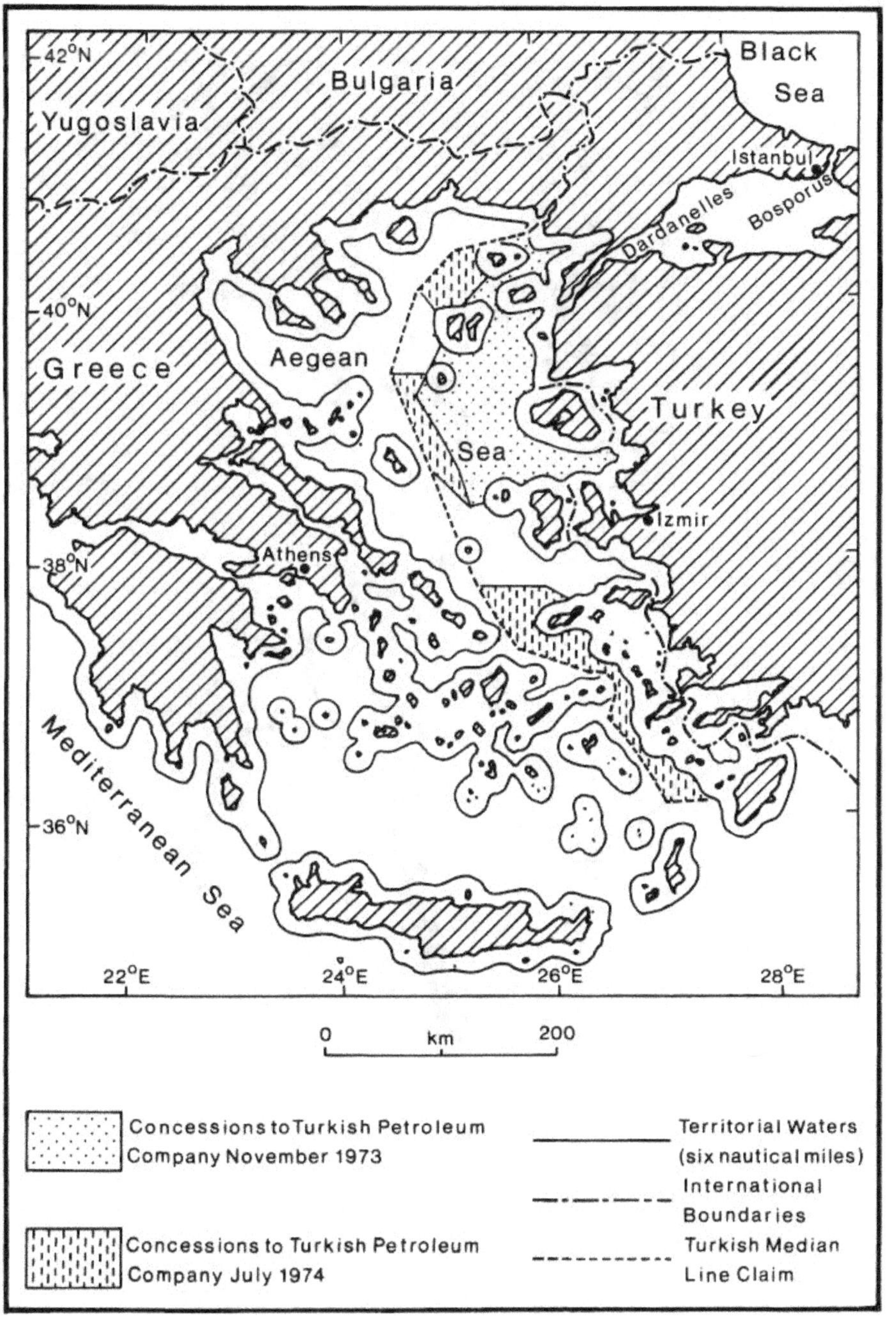

Fig. 7.2 The Greek–Turkish dispute in the Aegean Sea.

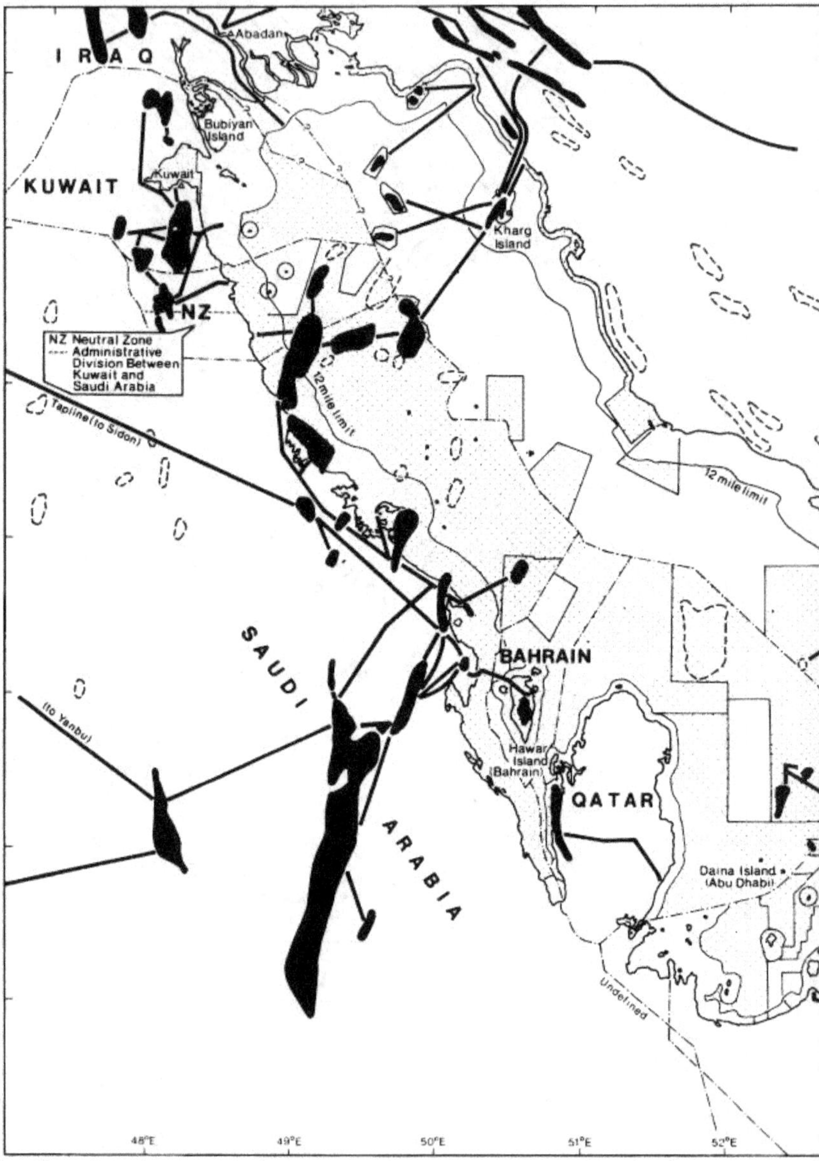
IRAQ
Abadan
Bubiyan Island
Kuwait
KUWAIT
Kharg Island
NZ
NZ Neutral Zone
---- Administrative Division Between Kuwait and Saudi Arabia
12 mile limit
Tapline (to Sidon)
12 mile limit
SAUDI
ARABIA
BAHRAIN
(to Yanbu)
Hawar Island (Bahrain)
QATAR
Daina Island (Abu Dhabi)
Undefined
48°E
49°E
50°E
51°E
52°E

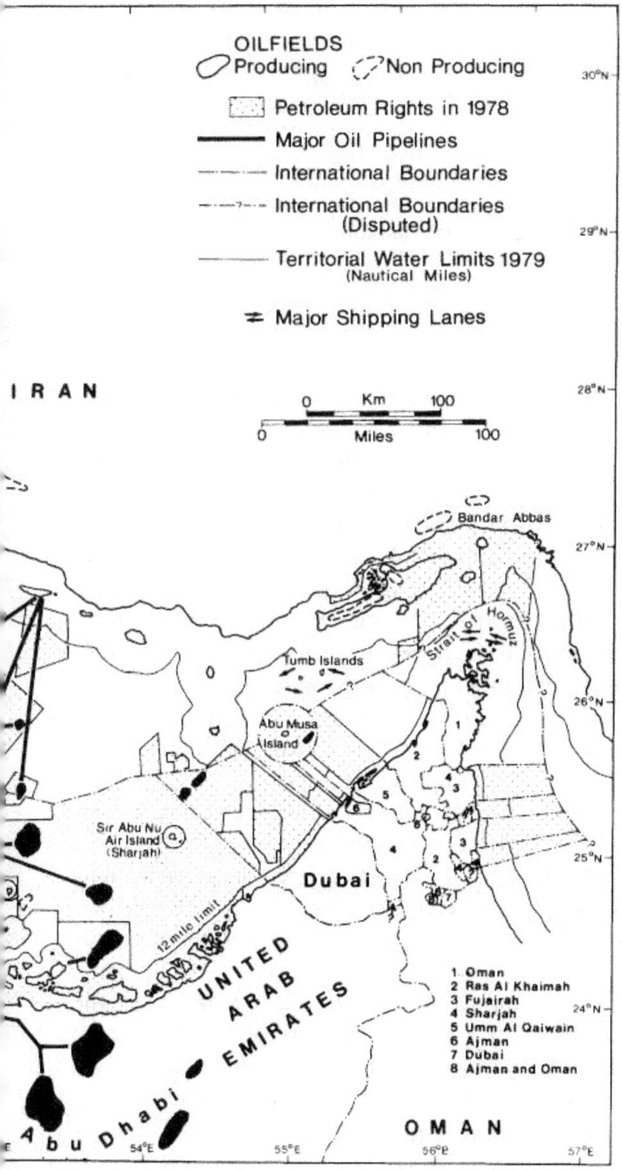

Fig. 7.3 Maritime boundaries of the Gulf, 1978.

any increase by Greece would turn the Aegean into a Greek lake. As seen above, however, the Greek islands give Greece the right to continental shelf resources adjacent to the islands. Turkey did not sign the 1958 Convention and regards the islands as 'protuberances' on the natural extension of the Turkish mainland (Rozakis 1975). The Turks wish to establish a median line regardless of islands, whose location is suggested by the 1973–4 concessions (Fig. 7.2). It is not easy to see a solution to this dispute. Joint exploitation of oil could be achieved by a Greek–Turkish company, and revenues divided, but the issue is really one of national sovereignty associated with rivalries going back many centuries.

The Exclusive Economic Zone

The concept of the EEZ has been discussed at various sessions of UNCLOS III since 1974 and seems likely to be adopted in 1980 in a historic international treaty (Financial Times 1979). If it is not, there will undoubtedly be numerous unilateral declarations of 200-mile zones akin to those of Oman and PDR Yemen. Many claims have been reported in recent years, but there is no standard definition of what is being claimed. Oman declared 'exclusive fishing' to 200 miles on 16 June 1977, whereas PDR Yemen claimed an 'integrated economic zone' of 200 miles on 18 December 1977 (Keesing 1978). The UNCLOS definition is much clearer concerning functions of the EEZ, national jurisdiction extending to all living and non-living resources and to all activities relating to these resources. Pollution control is clearly implicit in this. Freedom of navigation and traditional rights to cable and pipeline laying are unaffected. It is generally accepted, but not certain, that the 200-mile EEZ would be accompanied by a standard 12-mile territorial water limit, which would be *included* in the 200 miles. The EEZ would thus be measured from the baseline of the territorial sea. The technical and practical problems associated with the delimitation and charting of the necessary boundaries are colossal (Kapoor 1977). In addition, many general questions will have to be resolved, but they cannot be discussed here.

Delimitation of 200-mile EEZs in the Mediterranean and Red Sea might precipitate considerable political acrimony in the region and possibly low-level conflict. For example, boundary drawing could be bedevilled by the claims of the UK Sovereign Bases in Cyprus, which have their own territorial seas, the possibility of Palestinian autonomy in Gaza, and the claims of the Turkish Cypriot state. The Gulf is already divided between coastal states into one of the most clearly defined offshore political mosaics in the world and yet it has several unresolved problems associated with the role of islands in boundary delimitation, rival claims to islands and the use or otherwise of the straight baseline (Fig. 7.3). Such difficulties will be widely encountered when

EEZs come to be drawn. Eventually, unfamiliar-looking maritime maps will appear in our atlases, a hint of which is given by the 200-mile zones of south Arabia (Fig. 7.1). Whether or not the new regime of the oceans will be benign or malignant remains to be seen. Either way, it deserves our attention.

References

ALBERS, J. P. *et al.* (1973) *Summary of Petroleum and Selected Mineral Statistics for 120 Countries, Including Offshore Areas*, Geological Survey Paper, 817, Washington, US Govt Printing Office, 13–121.

AMBIO (1977) Special issue on the Mediterranean, 6 (6).

ARAB ECONOMIST (1975) 'Who owns the Red Sea?', 7 (74): 16.

BUZAN, B. (1978) *A Sea of Troubles? Sources of Dispute in the New Ocean Regime*, Int. Institute for Strategic Studies, Adelphi, Paper 143, 26–8.

FIELDEN, D. G. (1978) *The Political Geography of the Red Sea Region*, University of Durham, Department of Geography, Paper No. 13, 2–8.

FINANCIAL TIMES (1979) 'Sea-bed mining accord close', 4 May, 3.

IISS (1971) 'The Soviet Union in the Mediterranean', 30–4, and 'Power in the Persian Gulf', 39–45, *Strategic Survey 1971*, London, International Institute for Strategic Studies.

KAPOOR, D. C. (1977) 'The delimitation of Exclusive Economic Zones', *Maritime Policy and Management*, 4, 255–63.

KEESING CONTEMPORARY ARCHIVES (1978) 'Unilateral declarations of economic or fishing zones', 31 March, 28907–8.

LAY, H. S., CHURCHILL, R. and NORDQUIST, M. (eds) (1973) *New Directions in the Law of the Sea*, vol. 1, *Documents*, London, Oceana Pubs, 101.

MIRABELLI, T. (1977) 'Mediterranean and Gulf pollution: Domesday warning?', *The Middle East*, 31, 14–20.

OFFICE OF THE GEOGRAPHER (1970) 'Limits in the Seas No. 20', *Saudi Arabia*, Washington, US Department of State.

—— (1974) 'Limits in the Seas No. 36', *National Claims to Maritime Jurisdictions*, 2nd revision, Washington, US Department of State, 68.

—— (1975) 'Limits in the Seas No. 61', *Oman (Hypothetical)*, Washington, US Department of State.

ROZAKIS, C. L. (1975) *The Greek–Turkish Dispute over the Aegean Continental Shelf*, Rhode Island, Law of the Sea Institute, Paper No. 27, 1–17.

CHAPTER EIGHT

Population growth in the Middle East since 1945 with special reference to the Arab countries of West Asia

ALLAN G. HILL

Around the time of the Second World War, official reports on the human resources of the Arab countries were characterized by an air of general pessimism concerning the potential for future economic growth and the expansion of population and settlement. The poor health and nutritional status of the Arab populations was seen as a constraint on their economic growth, and the smallness of the populations of countries like Iraq, Libya and Syria was viewed as a problem of central importance in the introduction of modern agricultural and industrial methods. Both the discovery of oil and the Arab–Israeli conflict changed these views entirely and made the region a focus of Great-Power concern. In addition to the new oil income, especially since 1973, massive amounts of civil and military aid poured into the region from both the west and the communist countries.

Although health conditions have been improved and mortality rates have fallen rapidly since 1945, fertility has remained high with few signs of decline in most of the Arab countries. It appears that improvements in mortality rates have been achieved quickly and easily by a combination of national, local and household-level efforts since everyone's interests are served by a reduction in mortality and morbidity. Fertility changes, on the other hand, involve a basic revision of traditional attitudes and behaviour, in particular towards women and children, and separate interest groups assess the situation differently. Attempts to speed up this process of change have encountered opposition or indifference from several quarters in different parts of the Arab world. Whilst knowledge of the way in which attitudes and behaviour related to fertility and childbearing are determined is still imperfect, it seems plain from experience in countries like Kuwait that increases in wealth alone are not sufficient to alter radically these attitudes.

Despite the high rates of natural increase which have resulted from falling mortality rates and small (sometimes positive) changes in fertility levels, there appears to be restricted private or official concern with rapid population growth and its consequences in the Arab countries. First, there is still a feeling that the natural-resource base, either agricultural land or mineral wealth, is sufficiently large to accommodate further population growth. Second, internal and international migrations, including important refugee movements, have relieved some of the population pressures which would have otherwise built up in countries like Egypt and Jordan or parts of rural Syria and Iraq and diverted attention from natural increase as a factor of population growth. Finally, high rates of economic growth sustained by the oil moneys and foreign aid have created a buoyant demand for labour and a growth in real wages which have discouraged the development of negative attitudes towards high fertility and rapid population growth. Until recently, problems which had an important population component, such as overcrowded housing conditions and inadequate public services in urban areas, were not analysed as demographic problems at all with the result that, for example, in national economic plans and public life generally demographic issues were not given serious consideration.

Nevertheless, some important changes in fertility have taken place both at the country level and for sub-populations in the Middle East. The situation on the ground has become much more complicated since Kirk (1967) was able to write about Muslim natality and fertility in terms of their uniformity and high and unchanging levels throughout the Muslim world. Some dimensions and causes of these changes are explored below.

Recent trends in mortality

The post-1945 improvement in mortality in all Middle-Eastern countries is well documented but some recent estimates for selected countries are presented in Table 8.1. There are two interesting questions which arise in connection with mortality declines. First, did mortality fall as a result of capital investment in public utilities (water supply, sewage and waste disposal); other public health measures such as vaccination and provision of basic health care facilities; or the broader process of development which has raised incomes and the standard of living in the region to high levels? Second, did mortality improvements consist primarily of preventing early childhood deaths rather than improving the health conditions of the rest of the population? The first question is difficult to answer since the historical evidence on mortality-declines for the region is rather thin. One view expressed by WHO (n.d.) is that modern medicine at least removed the threat of the occasional catastrophic plague and severely limited the effect of the periodic epidemics in the region. More general improvements in morbidity and mortality are probably attributable as much to improved education,

Table 8.1 Mortality levels and trends in selected countries of the Middle East

	Year	*Crude death rate (per thousand)*	*Life expectancy at birth*	*Infant mortality rate (per thousand)*
Egypt	1966	16	50	127
	1970	15	51	116
	1976	12	54	100
Iraq	1965	18	49	137
	1973	11	53	92
	1975	10	54	86
Jordan (East Bank)	1930	19	38	222
	1961	19	46	154
	1972	10	59	73
	1976	9	60	89
Jordan (West Bank)	1961	19	46	167
	1974	16	53	132
Kuwait (Kuwaitis only)	1965	8	65	56
	1970	6	69	43
	1975	6	70	43
Lebanon	1970	9	64	65
	1976	8	66	41
Palestine (Arabs)	1924–6	25	38	160
	1940–2	19	50	130
	1976	5	70	37
Syria	1960	21	38	215
	1970	12	58	115
	1976	9	63	86

Sources: Abu Jaber (1979) and UNECWA (1978)

rising standards of living and improved levels of nutrition as to investments in clean water supplies, sewage systems and the like (Preston 1978).

Some pieces of information from the region provided support for this generalization. The first point of note is that the age pattern of mortality in the region has changed very markedly as a result of more rapid falls in infant and childhood mortality relative to the mortality levels of adults. In terms of the Princeton model life tables, mortality in the Middle East was characterized by relatively heavy childhood mortality, so much so that the age pattern of mortality was 'more South than South'. In some instances infant mortality has fallen so fast that the pattern is at least comparable to the South model pattern and, in some instances (e.g. for Palestinians in Israel) closest to the international average, to the West pattern with its more moderate levels of childhood mortality relative to those of adults.

Without doubt, the principal factor causing the excessive childhood mortality of the past was gastro-enteritis and diarrhoea, interspersed with epidemics of the common infectious diseases of childhood such as measles. Whilst the infectious diseases are no longer responsible for large numbers of childhood deaths, intestinal infections are still important. The transition from a 'South-plus' pattern of mortality to a West pattern is related to a changing set of causes of death. In 18 towns of Palestine in 1930, 24 per cent of all deaths were due to enteritis and diarrhoea; 19 per cent in Palestine as a whole. By 1976 just over 4 per cent of all deaths amongst the non-Jewish population of Israel were due to these causes. In other Arab countries even with quite low infant mortality levels as in Kuwait, the importance of enteritis and diarrhoea remains. On the East Bank of Jordan in 1974, 34 per cent of all infant deaths were due to these causes and the comparable figure for Kuwaitis in Kuwait was 28 per cent for 1975. Both populations still had a South type of mortality pattern in the 1970s as a result.

Why has the incidence of these intestinal infections been so quickly reduced? First, it seems that mothers are better informed about the causes of infection than ever before. Despite the widespread substitution of bottle for breast feeding, general levels of hygiene have never been better. But it is apparently not simply income and rising standards of living which affect childhood mortality, for in Kuwait, for example, the Kuwaiti population has similar levels of childhood mortality to the immigrants. It seems that female education is more important in preventing childhood deaths than high standards of living. The addition of health-care facilities can make a further contribution to preventing childhood deaths. A crude measure of the importance of medical facilities, such as re-hydration and inoculation centres, in improving childhood mortality can be obtained by comparing infant mortality rates in the UNRWA Palestinian refugee camps (where such facilities are available) with rates for the surrounding population (Table 8.2). Here we see the effect a basic health-care system focused on the first year of life can have on childhood mortality.

Fertility levels and trends

A summary of the best available recent fertility estimates for selected countries of the region is shown in Table 8.3. Two main points emerge: first, there is considerable diversity in fertility levels, contrary to the situation described some years ago. Second, it seems that fertility is changing in some parts of the region but often in unexpected ways. It appears that in a number of Arab countries fertility has risen recently prior to beginning any sort of decline (Hill 1977a). By far the best illustration of this trend is Algeria, but Syria and Jordan seem to have shared the same experience although to a lesser degree.

Table 8.2 Infant mortality rates in Palestinian refugee communities compared with host country rates

	Infant mortality rates (per thousand)
Jordan (West Bank)	
Nationals (1974)	132
Refugees (1970–2)	69
Lebanon	
Nationals (1970)	65
Refugees (1970–2)	45
Syria	
Nationals (1970)	115
Refugees (1970–2)	50

Sources: UNRWA (1974) and Abu Jaber (1979).

Naturally, the fertility of some sub-populations (the educated residents in capital cities) is declining quite rapidly, but over all there appears to be little sign of significant national changes in fertility in the Middle East with the exceptions of Lebanon, Syria, Tunisia and Turkey.

Why are these fertility changes relatively slight in the region as a whole, given both the very rapid rises in standards of living and other related socio-economic changes? First, as mentioned above, some groups (the older generation, possibly employers and some members of minority groups) have an interest in sustaining fertility at high levels. Wives, especially the uneducated, probably feel pressurized by society into having large families both because this is the traditional social norm and because it provides some protection against early divorce. Secondly, it may make good economic sense to have a large family, especially in rural areas where family labour is still more important than wage labour. In addition to these labour-force considerations, a large family can also provide insurance against destitution in the face of mishaps, illnesses or retirement in countries where social security provisions are just beginning.

These first two sets of factors primarily affect attitudes and values to fertility and childbearing. There is a third set of more tangible factors directly connected to fertility which in practice fix the level of fertility in all human populations. These 'proximate determinants of fertility', as referred to by Bongaarts (1978), mediate between the biological process of reproduction and attitudes and values determined by the factors referred to above. Evaluation of the effect of these intermediate fertility variables in selected instances gives

Table 8.3 Fertility levels and trends in selected Middle-Eastern countries

	Year	*Crude birth rate (per thousand)*	*Total fertility rate*
Algeria	1915	35	*
	1930	42	*
	1955	47	8.3
	1965	48	9.0
	1970	48	8.8
Egypt	1947	44	*
	1966	41	6.1
	1970	35	5.1
Gaza	1968	43	7.0
	1976	53	7.2
Iraq	1965	49	*
	1973	44	7.1
	1975	42	6.9
Israel (non-Jews)	1955	46	7.3
	1965	51	8.4
	1976	43	6.9
Jordan (East Bank)	1961	49	7.3
	1972	49	7.6
	1976	48	7.8
Jordan (West Bank)	1961	53	7.5
	1968	44	7.6
	1976	47	7.3
Kuwait (Kuwaitis)	1957	43	7.0
	1965	51	7.5
	1970	50	7.3
	1975	51	7.1
Lebanon	1955	41	5.8
	1960	40	5.8
	1970	34	5.5
	1975	33	4.6
Syria	1960	43	6.6
	1970	48	8.1
	1976	41	7.5
Tunisia	1960	46	6.8
	1973	36	6.1
Yemen Arab Republic	1975	51	6.9

Sources: Abu Jaber (1979); Allman (1978); Bachi (1976); UNECWA (1978) and UN *Demographic Yearbooks*.

* Not available

us some new insights into the causes of high and sustained fertility in the Middle East. The process by which the social and psychological factors which, through attitudes and behaviour, fix the values of the intermediate fertility variables is still an area in the early stages of exploration.

Fertility determinants: the intermediate fertility variables

There have been a number of attempts to classify and analyse the socio-economic and biological factors affecting fertility, beginning with the pioneering work by Davis and Blake (1956). The general framework can be summarized in a simple diagram:

Indirect determinants		*Direct determinants*		
Socio-economic, cultural and environmental variables	→	Intermediate fertility variables	→	Fertility

Attempts to quantify some of the connections between the sets of factors shown above have been legion, and some of the models connecting the intermediate fertility variables and fertility have been highly complex mathematical formulations. A new and simpler system developed by Bongaarts (1978) provides a method for quantitatively assessing the strength of the most important determinants of fertility and for parcelling out the separate influences of each of the intermediate fertility variables. The intermediate fertility variables themselves can be listed under three broad headings:

I Exposure factors
  1. Proportions married
  2. Spousal separation

II Deliberate marital fertility control factors
  3. Contraception
  4. Induced abortion

III Natural marital fertility factors
  5. Lactational infecundability
  6. Frequency of intercourse
  7. Sterility
  8. Spontaneous intra-uterine mortality
  9. Duration of the fertile period

Indices of the intermediate fertility variables for the Middle East

The effect of each of these factors on fertility can be quantitatively assessed by a combination of direct survey methods and inference from models. It can be shown that only the first five factors cause significant variations in fertility levels within sizeable population groups. Thus, Bongaarts developed four indices – spousal separation (C_s) was added later by Hill and Shorter (1979) – to measure the independent effect on fertility of marriage (C_m), contraception (C_c), induced abortion (C_a) and lactational infecundability (C_i). The indices range from 0 to 1 and are interpreted thus: 1 minus the index yields the proportional reduction in fertility which can be attributed to that variable alone.

The extent of contraceptive use, the duration of breastfeeding or of post-partum amenorrhoea and the induced abortion rate are not readily obtainable for all the countries of the Middle East since the collection of this information does require the conduct of a specialized survey; the index of the proportion married can be readily obtained from census and similar statistical surveys. None the less, we obtain an encouragingly complete picture of the relative importance of three of the most important intermediate fertility variables from Table 8.4.

Table 8.4 Values of the indices of three intermediate fertility variables in selected Middle-Eastern populations

Populations	*Marriage* C_m	*Contraception* C_c	*Lactational infecundability* C_i
Egypt (1977)			
(Menoufia Governorate)	0.75	0.82	0.62
Iraq (1973–4)	0.69	0.71	*
Jordan (1976)			
(East Bank)	0.73	0.79	*
Lebanon (1976)			
(South)	0.51	0.63	0.78
Syria (1973)	0.73	0.76	0.73
Turkey (1968)	0.78	0.69	0.73
(1973)	0.74	0.62	*
Yemen Arab Republic (1977)			
(Rural)	0.89	0.85	0.66

Sources: Hill and Shorter (1979); Myntti (1978); Nosseir (1979); Özbay and Shorter (1977); Özbay (1978); Zurayk (1979).

* Not available

Marriage as a factor affecting fertility

As the Bongaarts framework clearly shows, the total fertility rate (TFR) in a population is simply the total fecundity rate (TF – assumed to lie between 13.5

and 17 in almost all human populations), times the product of the indices of the intermediate fertility variables:

$$TFR = TF \times C_m \times C_s \times C_c \times C_i \times C_a$$

When extra-marital fertility is negligible and the majority of married couples are not deliberately controlling their fertility, C_m assumes major significance in the determination of fertility levels. In general, the probable range of C_m is wider than that of C_i and hence its effect on fertility becomes more pronounced. This is especially true in the Middle East where the duration of breastfeeding in the past appears to have been remarkably uniform in a wide variety of locations.

The direct effect of the pattern of marriage, including the mean age of women at marriage, the proportion of women who ever-marry and the rate at which couples 'destined' to marry enter marriage, can be readily demonstrated by multiplying one marital fertility schedule by two contrasting nuptiality schedules. Using values from Syria in 1976 as an illustration, multiplication of the marital age specific fertility schedule for all women by the proportions of women with a secondary level of education currently married produces a 39 per cent reduction in the total fertility rate. In reality, fertility differences between the two groups of women used in this illustration are wider than those shown since educated women are also more likely to be using contraception in addition to marrying later and less often.

A more mathematical expression of the significance of the marriage pattern in determining the level and age pattern of fertility is contained in the fertility models developed at Princeton (Coale 1971; Coale and Trussell 1974). These models are based on a series of observed fertility schedules and three of the four principal parameters describe the pattern of marriage:

a_0 – the age at which women begin to marry;
k – the rate at which couples due to marry enter marriage;
C – a scale factor which describes the fraction of women who ever-marry.

The fourth variable describes the extent of the departures from natural fertility by age when contraception or other forms of fertility regulation are a consideration.

Lestaeghe (1971) was able to use the Princeton fertility models to show how much alteration there had to be in existing patterns of marriage in the Middle East to obtain certain reductions in the level of fertility. Using the parameters a_0 and k, he showed that:

1. a rise in the value of a_0 alone from 13.5 (its approximate current value in the Middle East) to 16 would produce a 15 per cent fall in fertility, *ceteris paribus*;

2. a rise in the value of k alone from 0.4 to 0.6 (i.e. a slowing in the rate at which couples enter marriage – equivalent to spreading the process of marriage over a wider age span) would also reduce fertility by 15 per cent;
3. both changes occurring simultaneously would bring about a 30 per cent drop in fertility levels.

Age at marriage and proportions ever-marrying in the Middle East

Using data from censuses and national surveys, it is possible to derive estimates of the singulate mean age of marriage and of the proportions of women ever-marrying for a number of countries in the Middle East, including Iran and Turkey (Table 8.5). Here we see the extent of the variation in different parts of the Middle East and, for countries where trend data are available, the direction and speed of trends in recent years. These national data conceal a substantial amount of sub-national variation in the three selected measures shown in Table 8.5; statistics for regions or classes would substantially expand the range of the variations. Even with regard to the national data, there appears to be a substantial amount of variation – certainly enough to refute Kirk's assertions about the uniformity of Muslim marriage patterns.

Trends are difficult to describe because of the degree of variation from one country to another, but over all it appears that women in the Middle East are marrying later than ever before and that a growing but still small fraction never marry. There are some exceptions to these generalizations: in Iran and Algeria the age at first marriage has fallen rather than risen, and in some cases the proportions of single women in the age range 40 to 44 have altered very slightly. None the less, the broad trends remain even though the variation in the age at first marriage was as wide as five years (from 18 in North Yemen to over 23 in Lebanon) in the mid-1970s.

Determinants of marriage in the Middle East

In view of the importance of the pattern of marriage in affecting fertility, it seems worthwhile reviewing some of the factors which have brought about a significant change in nuptiality in the Middle East in recent decades. Whilst for modelling purposes it is useful to separate out the three parameters referred to above (a_0, k and C), in practice we can regard the pattern of marriage as a single phenomenon since there is a high degree of association between the three components; that is, in populations where marriage occurs at an early age, a large fraction of all women marry at some time during their lifetime and they also marry within a narrow age span. What determines whether a population adopts this marriage pattern, or an alternative? In particular, what factors have contributed to a modification of the traditional

Table 8.5 Age at first marriage and proportions never married for selected countries in the Middle East (females only)

Country or region	*Year*	*Singulate mean age at first marriage*	*Percentage never married at age: 15–19*	*40–4*
Algeria (Muslims)	1948	19.8	*	2.6
	1966	18.3	53.5	1.4
Egypt	1960	19.7	66.9	1.8
	1966	20.1	69.8	2.1
Iran	1956	19.1	*	1.2
	1966	18.5	53.9	1.0
Iraq	1957	20.1	*	4.6
	1965	20.6	67.9	4.4
	1974	22.4	77.9	4.2
Palestine (Muslims)	1931	19.5	65.7	2.4
Israel (Muslims)	1961	20.8	78.3	2.3
Israel (non-Jews)	1975	21.6	83.1	7.7
Jordan	1961	20.3	72.0	2.8
Jordan (East Bank)	1972	19.3	69.6	1.8
Jordan (East Bank)	1976	21.5	80.5	2.0
Kuwait (Kuwaitis)	1957	17.4	30.0	3.0
	1965	18.9	65.3	2.1
	1970	19.8	66.1	2.2
	1975	20.9	71.4	2.1
Lebanon	1970	23.2	86.8	7.6
Libya	1964	18.8	25.7	0.7
Morocco	1960	17.5	45.0	1.7
Saudi Arabia (Saudis)	1974	18.9	58.4	1.5
	1977	19.9	68.1	1.6
Syria	1960	20.4	58.9	3.0
	1970	20.7	72.3	3.2
	1976	20.8	75.9	6.9
Tunisia	1956	19.5	58.1	1.4
	1966	21.0	81.0	1.8
	1975	23.3	93.9	1.7
Turkey	1965	19.3	72.3	1.6
	1975	20.4	78.1	2.1
Yemen Arab Republic	1975	18.0	49.7	1.3

Sources: Data for the North African countries are taken from Dixon (1971 and 1976). The remainder were calculated from censuses or nationally representative surveys.
* Not available

pattern of marriage in the Middle East, which for women was almost universal and for most took place at a relatively early age?

A useful framework for the analysis of the determinants of the marriage pattern is provided by Dixon (1971). She separates these factors into three groups: factors affecting the availability of mates, the feasibility of marriage and the desirability of marriage.

Availability of mates

Two factors affect the sex ratios in an adult population: mortality differentials by sex due to such factors as sex-selective infanticide or war, and migration. A third factor, an imbalance between the sexes due to a preference of males for very much younger brides, has been observed in some Muslim populations in Russia but this factor is of diminishing importance in the contemporary Middle East.

Whilst sex differentials in infant and childhood mortality due to differences in the attention paid to young girls and boys in the Middle East are well documented (Harfouche 1965), the extent of these differentials has been narrowing gradually. In any case, the sex differentials in childhood mortality in the Middle East appear to have been much less severe than in countries such as India, but any recent change has been in the direction of reducing this factor as a spur to early marriage.

Despite the relatively frequent wars which are a feature of the Middle East, the Arab countries have not experienced military and civilian losses on the scale recorded, say, in Turkey during the two world wars. In the June War of 1967, for example, Jordan's losses totalled about 6000 men, 3 per cent of the total population at that time. Although sizeable and concentrated in the young marriageable ages, these losses would hardly have a very strong direct effect on nuptiality although the indirect effects of such wars can be significant. There seem to be reasonable grounds for accepting that the 1973 war and its accompanying recession had a powerful effect on marriage patterns in Egypt. The economic recession and the continuing mobilization of the armed forces after the end of the war forced many couples to delay their marriages.

Migration is the factor of greatest significance in determining the availability of mates in the Middle East. In recent years, the continuing disturbance of the Palestinian population, the massive migration of male workers from Egypt, North Yemen and the northern Arab countries to Saudi Arabia and the Gulf states, and civil wars and political disturbances in a number of countries have all contributed to the separation of potential spouses.

A few examples serve to illustrate the magnitude of this separation effect due principally to migration. In the Yemen Arab Republic, a country virtually isolated from the outside world until the 1970s, there were approximately 635,000 emigrants abroad at the time of the first census in February 1975. Of this total, 364,000 were short-term migrants, most of whom consisted of young men, many of them unmarried. This movement of males, equivalent to a migration of 7 per cent of the *de jure* population, created very unbalanced sex ratios in a number of rural and urban districts in North Yemen. In twenty-eight of the 171 census districts (containing 18 per cent of North Yemen's population), the proportion of men abroad was equal to 12 per cent of the districts' population. In Ta'iz Province, four districts had sex ratios (males/females) of below 70 per 100; in the country as a whole, the sex ratio of men in the age range 15–59 was 77 per 100 (Steffen *et al.* 1978). A survey of some rural areas of North Yemen revealed that the husbands of 30 per cent of the married women were absent at the time of interview, so that both the marriage pattern and the exposure to the risk of conception were being affected by emigration (Myntti 1978).

In Kuwait, 59 per cent of its 523,000 immigrants in 1975 were males of whom 34 per cent were unmarried. For this male population the implied singulate mean age at marriage was almost 27 years which compares with a mean age at marriage for males in the countries from which they were drawn of under 25 years. In other words, assuming that the immigrant males were directly comparable in social status to the populations from which they were drawn, the effect of migration was to raise the mean age at marriage by over two years and presumably a roughly similar increase was being experienced by their brides-to-be.

The Tunisian experience is important since it demonstrates how a 'marriage squeeze' can develop because of the high degree of age–sex selection implied by emigration. Half the Tunisian emigrants in 1972 were aged 20–9 and three-fifths were single. As a result, sex ratios in some age groups fell below 60 and Tunisian girls experienced great difficulty in finding suitable mates throughout the 1966–75 period (Duza and Baldwin 1977).

A second dimension of the feasibility of marriage concerns the process of mate selection. As indicated by Goode (1970), Prothro and Diab (1974) and others, a system in which parental control is important in the marriages of sons and daughters tends to produce an early age of first marriage and a high proportion ever-marrying. This strong control is well documented, but gradually couples have been given more say in the selection of their mates. In a survey of the desert areas of Jordan in 1976 and 1977, very few mothers had been given a say in their choice of mate whereas many more of the daughters had been given this freedom (Abu Jaber *et al.* 1978). Thus, the decline of parental control affects the marriage pattern in two ways. First, women presented with the choice between early marriage or later marriage along with the opportunity to engage in worthwhile activities before marriage will generally choose the latter. Almost by definition, societies in which this choice is available are those in which power and status for women are allocated on the basis of factors other than the traditional ones such as high achieved fertility. Second, the search process for a suitable mate further postpones the date of marriage. This effect is difficult to measure since rising education has the effect of increasing the age at marriage and of providing couples with greater freedom of choice.

Feasibility of marriage

Over the last thirty years, the traditional pattern of marriage has been altered by the transfer of resources and responsibility for decisions concerning marriage from parents to children. In general terms, this trend can be seen as a modification of the 'inter-generational contract', described by Ryder (1976) and Caldwell (1976), from the traditional form in which 'resources' are transferred upwards from children to parents to the modern form in which the net movement of 'resources' is in the opposite direction.

Marriage has always been expensive in the Middle East and it has become no less expensive in real terms since inflation in bride prices and the costs of the ceremonies themselves have kept pace with, or even overtaken, the general increase in wealth. Still the most expensive item in the establishment of a separate household for the new couple is the cost of the dwelling and its furnishings. In urban areas, rents and house prices have been driven upwards at a fast rate owing to migration from rural to urban areas and to surges in income with which the construction industry has been unable to keep pace. As a result, very crowded conditions have emerged in many of the larger cities so that young couples cannot be sure of beginning their married life in a separate dwelling despite the almost universal desire for residential independence.

Desirability of marriage

This factor is the hardest to measure since, as Dixon (1971) points out, the proportion of women never marrying is probably the best index of the desire for marriage despite the element of circularity implied. In traditional society, where the roles of women are strongly circumscribed and restrict them to tasks centred on the home, including childbearing and rearing, women have almost no social function or role outside the household which is independent of their husbands or families. For a woman, access to resources, to power and to participation in a community was through marriage and the production of children. There are certain exceptions to this generalization since, in some of the poorer agricultural parts of the Middle East, women's labour on the farm is a vital contribution to the production process but payment for the work is not involved. In most cases, however, parents feel the need to marry off their daughters as soon as possible since there is virtually no social role for spinsters. An additional factor making for early marriage of girls is the great importance attached to virginity up to the time of marriage.

The expansion of work opportunities for women coupled with the shift of people from rural to urban areas has begun to create new sources of income and power for women, sources previously available only to a minority of well-educated upper-class women. As a result, the proportion of women never marrying is rising steadily and even those who do marry are marrying at older ages than before.

Additional factors affecting fertility

Contraception and abortion

Steadily rising proportions of couples are using some form of deliberate control of marital fertility (Table 8.6). The proportions of married women reported to be currently using a contraceptive in the various fertility surveys in the region are probably under-stated both because there is some resistance to declaring the fact to an outsider and also because there is considerable use of the traditional methods, principally withdrawal. Abortion is less widely approved although some surveys have shown that significant proportions of women have actually resorted to this method to prevent a further birth. In the Jordan survey of 1972, for example, 57 per cent of the women questioned approved of the use of abortion in certain circumstances whilst 6 per cent had themselves experienced an induced abortion. Most of these women were poorly educated, lower-class women, which gives some indication of the magnitude of the pressures felt by poor and quite traditional sections of society as a result of the large size of many families.

Table 8.6 Contraceptive use in selected Middle-Eastern countries

	Year	*Current users as % of married women 15–44*
Egypt	1975	21
Iran	1977	23
Iraq	1973–4	21
Jordan	1972	21
	1976	24
Lebanon (South)	1976	35
Syria	1973	22
Tunisia	1977	over 17
Turkey	1963	22
	1968	32
	1973	38
Yemen Arab Republic	1977	15

Sources: Iraq (1976); Jordan (1979); Myntti (1978); Nortman and Hofstatter (1978); Syria (1979); Watson (1977); and Zurayk (1979).

Despite the rising contraceptive use rates, fertility is still high for two principal reasons. First, the majority of contraceptive users are women more than halfway through their reproductive span who want to stop having children. Young women, and especially newly married couples, rarely use contraception to space the first births. From a demographic standpoint, this pattern of contraceptive use has relatively little impact since the couple will have almost completed their childbearing lifespan before beginning to contracept seriously. A second reason for the limited effect of contraception so far is that the fertility reductions achieved by current use rates are offset by gains in fecundity and fertility due to a curtailment of breastfeeding. Usually, it is the younger women with jobs and other commitments (as well as being more susceptible to changing fashions) who are weaning earlier or not breastfeeding at all. The same group is less likely to be using a reliable contraceptive so that the net effect can be a measurable increase in fertility for women in the age range 20–34. The marital fertility schedules for the Kuwaiti population of Kuwait show some evidence of trends which would fit this pattern of reduced periods of lactational amenorrhoea for younger women and increased use of contraceptives by older women (Hill 1977a). Eventually, contraceptive use will obliterate this increase in fecundability, but for the present, curtailment of lactation is preventing fertility falling as fast as it would have if contraception had been the only factor to change in recent years.

Breastfeeding and lactational amenorrhoea

The connection between the duration (and intensity) of breastfeeding (L) and the duration of the period of post-partum amenorrhoea (i) is now well established, although the precise physiological links are as yet incompletely understood. The relationship can be neatly summarized by an equation arrived at by Corsini (1977) as a result of the examination of twenty-eight sets of survey data:

$$i = 1.5 + 0.56\ L$$

From this equation we can readily demonstrate that, for example, a 10 per cent reduction in the duration of breastfeeding will produce a 3 per cent rise in the total fertility rate, assuming that all other factors remain constant. This effect is sizeable but of course it only applies in populations where contraception, especially in the post-partum period, is not significant. Survey data from around the region show that a 10 per cent reduction in the mean duration of lactation is well within actual experience and sometimes very much larger changes have been recorded. However, the effect of this factor on fertility is probably short-lived if we assume steadily rising contraceptive use rates and stabilization of the duration of lactation at some lower level.

Migration

In this chapter, the stress has been on accounting for some of the very high rates of natural increase observed in several Middle-Eastern populations. In practice, the fastest rates of overall growth are seen where migration is a significant factor. Some of the small Gulf states have experienced overall population growth rates in excess of 10 per cent per annum for considerable periods; Kuwait's population grew at a rate of 10.4 per cent between 1957 and 1961, for example. Chapter 9 deals with the effect of urbanization on city population growth rates; here, attention will be focused on two important international movements of people – the emigrants and refugees from Palestine and the in-migration of workers and dependants to the Arab oil-exporting states.

Palestinian emigration

As a result of the establishment of Israel in 1948, approximately 900,000 Palestinian Arabs were displaced or lost their source of livelihood in Palestine (Abu Lughod 1971: 161). Nearly 300,000 people are estimated to have moved within Palestine; the remainder left Palestine altogether. Many became dependent on the United Nations Relief and Works Agency (UNRWA) for Palestine Refugees which concentrated its efforts in Jordan, Syria, Lebanon and Gaza in an attempt to alleviate the immediate suffering. Those with resources migrated to the major urban centres of the West Bank and the countries surrounding Palestine, beginning new lives in a variety of locations. The demographic impact was greatest in Jordan since over half the refugees moved into Jordanian territory prior to 1967, but the growth in the economies of the Arab oil-exporting states started to attract immigrants from the northern Arab countries in large numbers from the early 1950s. In 1957 there were already over 13,000 Jordanians and Palestinians (mostly the latter) in Kuwait, a number which has risen steadily owing to both immigration and a high rate of natural increase to total 204,000 in the 1975 census. In the countries surrounding Palestine the natural increase rate of the Palestinian population has been high, in some cases higher than that of the host population, in part because of the reduction in infant mortality which was especially marked in the population served by the UNRWA health-care system. In June 1978 1.76 million Palestinians were registered with UNRWA in its areas of operation; of this total, 614,000 were still living in camps, many by choice. The camp shelters, although very basic, are also cheap and access to the UNRWA health, education and welfare services in the camps is assured. The camps themselves are now surrounded by a number of light industries and many are situated in areas where agricultural work is available. The social solidarity enjoyed in the camps is also another factor which acts as a deterrent to re-location outside the camp.

The regional total of Palestinians is difficult to estimate for two main reasons: the reluctance of some countries (e.g. Jordan and Saudi Arabia) to estimate and publish the size of their Palestinian populations, and because intermarriage and prolonged residence outside Palestine can make the definition of the nationality of offspring difficult at times. For the forthcoming census of Palestinians in the Middle East, to be conducted by the Population Division of the Economic Commission for Western Asia, the PLO has adopted a standard definition which is based on residence and descent through the father's line. Kossaifi (1976) and others have made some estimates of the Palestinian population in the Middle East in advance of the census, principally by making certain assumptions about the nationality composition of the East Bank of Jordan and the proportion of the Palestinian and Jordanian population in Kuwait who are in fact Palestinians. Their figures have been considerably revised in Table 8.7. The most problematic figure is that for the East Bank of Jordan, but in Lebanon, too, the numbers registered with UNRWA are probably a substantial underestimate of the actual total. The global figure, based on forward projection from 1948 using natural increase rates estimated by both direct and indirect methods, is also subject to error since the fertility of the Palestinian population in Saudi Arabia and the Gulf states excluding Kuwait is virtually unknown. There are, however, strong indications that Palestinian fertility in the rest of the region is very high and subject to little change so far. The evidence from the 1972 Jordan fertility survey indicates higher fertility in camps compared to the average levels on the East Bank as a whole. It seems that in 1978 only about 37 per cent of the total Palestinian population were living in what was Palestine and the Gaza Strip. This is not the place to review the implications of this major movement of people and their conditions in exile; here we can simply indicate the numerical importance of the migration.

Worker migration

As a result of the steady increase in wealth of the Arab oil-exporting countries in the post-1945 period, economic opportunities in all sectors of the rapidly growing economies attracted migrant labour initially from the northern Arab countries and to a lesser extent from the Indian sub-continent, Europe and North America. Initially the migrants were males in the prime ages, but later the migrant streams included more and more dependants (Hill 1972). The

Table 8.7 An estimate of the Palestinian population in the Middle East and abroad in mid-1978

	Thousands	%
All non-Jews in Israel	588.6	12.8
West Bank of Jordan (Arabs)	681.5	14.8
Gaza (UNRWA registered = 350.1)	445.5	9.6
East Bank of Jordan (UNRWA registered = 682.6)	1,200.0	26.0
Kuwait	231.8	5.0
Lebanon (UNRWA registered)	211.9	4.6
Syria (UNRWA registered)	198.4	4.3
Saudi Arabia	50.9	1.1
Egypt	45.0	1.0
Rest of the world, including allowance for under-estimation in countries listed above	962.4	20.8
Estimated world total derived by forward projection from 1948	4,616.0	100.0

Sources: Calculated from census statistics, UNRWA records and other official sources using known rates of natural increase.

Notes: For all countries except Palestine (Israel and Occupied Territories) the numbers shown are subject to unknown margins of error because of definition problems.

Palestinian population in particular was the first to assume the characteristics of a settled population in countries like Kuwait, but other nationalities have shown similar tendencies as a result of both the natural increase of the immigrant population and changes in the character of the immigrant streams. Certain nationalities (e.g. Iranis, Omanis, Yemenis) retain their predominantly male character, but for many other immigrant groups (e.g. Lebanese, Syrians, Pakistanis, Indians and Egyptians) natural increase exceeds immigration as a factor affecting the community's overall growth in the country of immigration.

The nationality mix of the immigrant populations in Arabia can be only partly understood in terms of economic factors since political (the Palestinian refugees) and cultural (preference for Arabic-speaking Muslims) considerations are significant (Hill 1977b). In 1975, about three-quarters of the 1.6 million migrant workers in the Arab oil-producing states were of Arab origin; by far the most numerous were Egyptians and Yemenis working in Libya and Saudi Arabia respectively (Table 8.8). In recent years, Asian workers have been of growing importance especially in the United Arab Emirates and Saudi Arabia, partly because of the ability of Asian contractors to undercut competition from elsewhere through the use of cheaper labour. In addition, the labour-importing countries prefer contractors to import their own labour for a job since, at the end of the contract, they can be required to export their

Table 8.8 Arab migrant workers in the Middle East in 1975

Country or region of origin	*Country of work*								
	Bahrain	*Iraq*	*Jordan*	*Kuwait*	*Libyan Arab Jamahiriya*	*Qatar*	*Saudi Arabia*	*United Arab Emirates*	*Total*
Egypt	1,200	7,000	5,300	37,600	229,500	2,900	95,000	12,500	391,000
Iraq	100	—	—	18,000	—	—	2,000	500	20,600
Jordan (including Palestine)	600	5,000	—	47,700	14,200	6,000	175,000	14,500	263,000
Lebanon	100	3,000	7,500	7,200	5,700	500	20,000	4,500	48,500
Oman	1,400	—	—	3,700	—	1,500	17,500	14,000	38,100
People's Democratic Republic of Yemen	1,100	—	—	8,700	—	1,300	55,000	4,500	70,600
Somalia	—	—	—	200	—	—	5,000	1,000	6,200
Sudan	400	200	—	900	7,000	400	35,000	1,500	45,400
Syrian Arab Republic	100	—	20,000	16,500	13,000	800	15,000	4,500	69,900
Yemen Arab Republic	1,100	—	—	2,800	—	1,300	280,400	4,500	290,100
Maghreb	—	—	—	100	41,000	—	—	—	41,100
Total	6,100	15,200	32,800	143,400	310,400	14,700	699,900	62,000	1284,500

Sources: Birks and Sinclair (1978).
— indicates no migrants of this nationality recorded.

workforce. The labour force on this kind of contract usually consists of unaccompanied male workers – another point in favour of Asian contractors compared to their Arab counterparts, who more often have accompanying dependants who require housing, education, medical and other services, raising the indirect costs of Arab labour.

Future demographic prospects

The last three decades in the Middle East have lifted incomes and standards of living to extremely high levels in the oil-exporting countries. The non-oil countries have also benefited from this economic growth through direct payments and subsidies and the remittances of workers abroad. In 1978, for example, the Central Bank of Jordan estimated that remittances from Jordanian workers abroad alone amounted to $500 million. In addition to these earnings, the 'confrontation states' (Syria, Jordan and Egypt) have received massive aid and military subventions from Europe, the USA and, more recently, other Arab countries. In 1977, for example, $3.5 billion in aid was pledged to Egypt alone; $2 billion came from the Gulf Organization for the Development of Egypt, $1 billion from the USA and the rest from several western countries. In global terms of per capita external assistance, only Israel is ahead of Jordan and Egypt.

Despite these dramatic economic changes, there are very few indications that a transition to lower rates of natural population increase is under way in any major Arab country. In Lebanon, Tunisia and possibly Syria, fertility is falling, but in Egypt, although the evidence is conflicting, fertility still appears high and subject to annual fluctuations from which a definite downward trend is hard to discern. The outstanding demographic achievement has been the sharp reduction in mortality, especially infant mortality. Infant and childhood mortality levels in the Middle East are low compared to levels in other developing regions, some (like parts of Latin America) relatively developed. Further mortality declines are certain in the Arab countries in future, which means that, even if fertility does begin to decline, rates of natural increase will remain very high for some time to come.

Migratory movements are notoriously hard to predict, but continuing oil price rises ensure that the labour needs of the Arab oil exporters will grow rather than diminish in future. Many jobs in the civil service and education sectors of these countries can best be done by Arabs, but much of the contracting work may be taken over by Asian (especially Korean) contracting firms. Already Yemeni day-labourers are experiencing some difficulties in obtaining work in Saudi Arabia; it remains to be seen whether this trend away from the use of Arab labour in Arabia will continue in future. Whatever direction these changes in natural increase and migration flows take in future, it will become increasingly difficult to generalize about overall trends in the region since the developments over the last thirty years have created wide differentials in demographic characteristics and levels of living which are unlikely to disappear in the near future.

References

ABU JABER, K. *et al.* (1978) *Bedouins of Jordan: a People in Transition*, Amman, Royal Scientific Society.

—— (ed.) (1979) *Levels and Trends of Fertility and Mortality in Selected Countries of Arab West Asia*, Amman, University of Jordan.

ABU-LUGHOD, J. L. (1971) 'The demographic transformation of Palestine' in Abu-Lughod, I. (ed.) *The Transformation of Palestine*, Evanston, Northwestern University Press.

ALLMAN, J. (ed.) (1978) *Women's Status and Fertility of the Muslim World*, New York, Praeger Special Studies.

BACHI, R. (1976) *The Population of Israel*, Jerusalem, CICRED Monograph, Scientific Translations International.

BIRKS, J. S. and SINCLAIR, C. A. (1978) 'International migration in the Arab region: rapid growth, changing patterns and broad implications', paper presented at the UNECWA Seminar on Population and Development, Amman, November 1978.

BONGAARTS, J. (1978) 'A framework for analyzing the proximate determinants of fertility', *Population and Development Review*, 4, 105–32.

CALDWELL, J. C. (1976) 'Towards a restatement of demographic transition theory', *Population and Development Review*, 2, 321–66.

COALE, A. J. (1971) 'Age patterns of marriage', *Population Studies*, 25, 193–214.

COALE, A. J. and TRUSSELL, T. J. (1974) 'Model fertility schedules; variations in the age structure of childbearing in human populations', *Population Index*, 40, 185–258.

CORSINI, C. (1977) 'Is the fertility reducing effect of lactation really substantial?', paper presented at the IUSSP/INED Seminar on Natural Fertility, Paris, March 1977.

DAVIS, K. and BLAKE, J. (1956) 'Social structure and fertility: an analytic framework', *Economic Development and Cultural Change*, 4, 211–35.

DIXON, R. B. (1971) 'Explaining cross-cultural variations in age at marriage and proportions never marrying', *Population Studies*, 25, 215–33.

—— (1976) 'The roles of rural women: female seclusion, economic production and reproductive choice', in Ridker, R.G. (ed.) *Population and Development: the Search for Selective Interventions*, Baltimore, Johns Hopkins Press.

DUZA, M. and BALDWIN, C. S. (1977) *Nuptiality and Population Policy*, New York, Population Council.

GOODE, J. (1970) *World Revolution and Family Patterns*, New York, Free Press.

HARFOUCHE, J. K. (1965) *Feeding Practices and Weaning Patterns of Lebanese Infants*, Beirut, Khayat.

HILL, A. G. (1972) 'The Gulf States: petroleum and population growth', in

Clarke, J. I. and Fisher, W. B. (eds) *The Populations of the Middle East and North Africa*, London, University of London Press.

—— (1977a) 'The demography of the population of Kuwait', UNECWA *Population Bulletin*, Beirut, 13, 42–55.

—— (1977b) 'Les travailleurs étrangers dans les pays du Golfe', *Revue Tiers Monde*, 18, 115–30.

HILL, A. G. and SHORTER, F. C. (1979) 'Intermediate variables in fertility analysis: a practical guide', *Regional Paper*, Cairo, Population Council.

IRAQ CENTRAL STATISTICAL OFFICE (1976) *The Fertility of Iraqi Women*, 4 vols, Baghdad (Arabic).

JORDAN DEPARTMENT OF STATISTICS (1979) *The Jordan National Fertility Survey of 1976*, Country Report 1.

KIRK, D. (1967) 'Factors affecting Moslem natality', *Proceedings of the World Population Conference*, Belgrade 1965, UN Department of Economic and Social Affairs, 2, 149–54.

KOSSAIFI, G. (1976) *Contribution à l'Etude Démographique de la Population Palestinienne*, Thèse de Doctorat de 3ème Cycle de Démographie, Université de Paris 1, Panthéon Sorbonne, Paris, Institut de Démographie.

LESTHAEGHE, R. (1971) 'Nuptiality and population growth', *Population Studies*, 25, 415–32.

MYNTTI, C. (1978) 'The effects of breastfeeding, temporary emigration and contraceptive use on the fertility of the Yemen Arab Republic', *Regional Paper*, Cairo, The Population Council.

NORTMAN, L. and HOFSTATTER, E. (1978) *Population and Family Planning Programs*, 9th edn, New York, The Population Council.

NOSSEIR, N. (1979) 'Measurement of intermediate variables affecting fertility in rural Egypt', paper presented at the 14th Annual Conference of the Institute of Statistical Studies and Research, March, Cairo University.

ÖZBAY, F. (1978) 'Türkiye de Doğurganlik Duzeyine ve Değişmelerine etki Eden Ara Değişkenler', in *Population Structure and Population Problems in Turkey*, Ankara, Hacettepe University Institute of Population Studies.

ÖZBAY, F. and SHORTER, F. C. (1977) 'Accounting for the trend of fertility in Turkey', paper presented at a UN Expert Group Meeting on the Demographic Transition and Socio-Economic Development, Istanbul (UNESA/P/AC8/CRP).

PRESTON, S. H. (1978) 'Mortality, morbidity and development', paper presented at the UNECWA Seminar on Population and Development, Amman, November.

PROTHRO, E. T. and DIAB, L. N. (1974) *Changing Family Patterns in the Arab East*, American University of Beirut.

RYDER, N. B. (1976) *Some Sociological Suggestions Concerning the the Reduction of Fertility in Developing Countries*, Honolulu, East-West Population Institute.

STEFFEN, H. *et al.* (1978) *Final Report on the Airphoto Interpretation Project of the Swiss Technical Co-operation Service, Berne carried out for the Central Planning Organization, San'a*, Zurich.

SYRIA CENTRAL BUREAU OF STATISTICS (1979) *Family Size and Family Health in Syria 1973*, Damascus.

UNECWA (1978) *Demographic and Related Socio-Economic Data Sheets for Countries of the Economic Commission for Western Asia*, Beirut.

UNRWA (1974) 'UNRWA services and the Palestine refugee population', paper presented at the UNECWA First Regional Population Conference, Beirut.

WATSON, W. B. (ed.) (1977) *Family Planning in the Developing World*, New York, Population Council.

WHO (n.d.) *Men and Medicine in the Middle East*, Geneva.

ZURAYK, H. (1979) 'A two stage analysis of the determinants of fertility in rural south Lebanon', *Population Studies*, 33, 489–504.

CHAPTER NINE

Contemporary urban growth in the Middle East

JOHN I. CLARKE*

Urban growth in less developed countries

Urban growth in the Middle East must be seen in the context of that in the world as a whole, which is even faster than the much more publicized population growth. While at the beginning of the nineteenth century only about 20 million people in the world lived in towns, say one in forty of the total world population, by 1970 the number had risen to 1500 million, two out of five (Goldstein 1973), with the highest proportions of town-dwellers in the most developed countries and the lowest proportions in the least developed countries. By the end of this century the number living in towns may reach 3000 million, roughly half the world population, and most of these will be living in less developed countries, which account for at least 85 per cent of world population growth and for an increasing proportion of its urban growth. Moreover, much of this mercurial urban growth in the world is localized in large cities; between 1950 and 1970 the number of cities with a million inhabitants or more rose from 75 to 162, nearly half of which were in the Third World. During the 1960s alone the number of cities with 100,000 inhabitants or more increased more than threefold from 249 to 837.

So the world is experiencing an 'urban explosion' of enormous magnitude, triggered off by the industrial and agricultural revolutions in nineteenth-century Europe and their subsequent diffusion around the world, assisted by the spread of Europeans overseas, the widespread improvements in communications and the ramifications of the world economy. The twentieth century has witnessed an acceleration of urbanization through diverse processes such as increasing human mobility, rapid population growth, the dispersal of industrial locations, the rise of the service sector and the waves of political

* This article was previously published in *The Changing Middle Eastern City*, R. W. Lawless and G. H. Blake (ed.), Croom Helm, 1980. Grateful acknowledgement is made to the publishers.

independence across the Third World. These varied processes have contributed differently to urbanization in time and space, so that cities and towns vary enormously in size and spacing, in form and function and in age and amenity, and consequently urban systems exhibit many contrasts. Even the terms 'town', 'city', 'urban' and 'rural' are difficult to define and delimit, and have particular connotations in different countries; and so does the process known as urbanization or urban growth, which may be variously defined as

(a) the growth in the number of people living in urban centres;
(b) the growth in the proportion of people living in urban centres;
(c) the growth in the number of urban centres;
(d) the socio-economic processes involving an increase in urban life;
(e) the physical extension of urban land use; or
(f) combinations of the above.

Obviously any one of the five processes may occur without corresponding change in the other four. Generally the term is used without precision, especially with reference to less developed countries, where some of the most rapid rates of urban growth are found among the least urbanized countries. So although contemporary urban growth is undoubtedly rapid, it is notoriously difficult to measure and compare internationally (Davis 1969 and 1972). Indeed, nowadays there are fewer efforts to attain precise statistical comparisons of urbanization and city size, since it is known that such attempts tend to be unsatisfactory.

Urbanization in less developed countries has been particularly rapid since mid-century as the gap in economic opportunities between urban and rural areas has widened and as government policies have favoured the localization of most modern economic activities in cities rather than in rural areas. So although during the period 1950–75 the total population of less developed countries increased at 2.4 per cent a year, their total urban population increased much faster, probably at about 6 per cent. However, 'less developed countries' is an all-embracing term which includes a wide variety of countries. In their fivefold classification, the World Bank categorizes some as low-income countries, middle-income countries, capital-surplus oil exporters or centrally planned economies, but none as industrialized countries. Table 9.1 reveals that these five categories have distinctive levels of urban population and rates of urban population growth, the most rapid rates being in the capital-surplus oil exporters and low-income countries, which are collectively the most feebly urbanized. Indeed, the five categories indicate almost an inverse relationship between the level of urban population and rate of urban population growth; a fact which will not surprise statisticians, but intensifies the problems of urban management in countries which can least afford to solve them.

Table 9.1 Urban populations of less and more developed countries

Countries	% of population in urban areas		% annual population growth		% annual urban population growth	
	1960	*1975*	*1960–70*	*1970–5*	*1960–70*	*1970–5*
1 Low-income	8	13	2.4	2.4	5.4	5.5
2 Middle-income	32	43	2.7	2.7	4.8	4.5
3 Industrialized	66	76	1.0	0.8	1.9	1.8
4 Capital-surplus oil exporters	23	31	4.0	4.2	6.6	6.3
5 Centrally planned economies	40	57	1.2	0.9	3.2	2.8

Source: The World Bank, *World Development Report 1978*, 100–3.

Urban growth in the Middle East

It is in these contexts that we observe contemporary urban growth in the Middle East, here defined as stretching from Morocco to Iran and from Turkey to Sudan. This region contains about 224 million people (1977 estimate) living on 8.75 million sq km – 5.5 per cent of the world's population living on 6.5 per cent of its land area. Its average density of population of 25.6 per sq km is therefore slightly less than the world average of 30 per sq km. Nowadays a facile physical explanation can be easily propounded for this relatively low population density, but we should recall that 'Mesopotamia at the beginning of the Christian era was undoubtedly one of the world's most densely populated countries' (Durand 1977), and that both North Africa and South-West Asia possibly had more people in ancient times than at the beginning of the modern period. Indeed, at the beginning of the Christian era they may well have contained 10–20 per cent of the total world population (Table 9.2), a situation which is unlikely to be retrieved in the foreseeable future, despite a recent upsurge in the average annual population growth rate

Table 9.2 Population estimates of North Africa and South-West Asia since the time of Christ (in millions)

	AD 0	*1000*	*1500*	*1750*	*1900*	*1975*
North Africa	10–15	5–10	6–12	10–15	53–5	80–2
South-West Asia	25–45	20–30	20–30	25–35	40–5	115–25
Combined total	35–60	25–40	26–42	35–50	93–100	195–207
World total	270–330	275–345	440–540	735–805	1,605–1,710	3,950–4,050

Source: Durand, J. D. (1977) 'Historical estimates of world population: an evaluation', *Population and Development Review*, 3,259.

to 2.6 per cent during the period 1950–75, well in excess of the world average. The long history of urban life in the Middle East must be seen within this setting of former numerical strength, which facilitated and even stimulated urban growth. Nowadays this region is not one of the major demographic concentrations in the world. Although its population is growing rapidly in response to high fertility and markedly declining mortality (see Chapter 8), the Middle East and North Africa have about as many people as South America and a few million more than the United States, but fewer than the Soviet Union and less than half as many as Europe.

On the other hand, the Middle East, with many capital-surplus oil exporters and middle-income countries and few low-income countries, is one of the most urbanized regions of the Third World, exceeded only by Middle and Tropical South America. It is also one of the most rapidly urbanizing regions. Precise figures are unobtainable, but Population Reference Bureau estimates for 1977 suggest that some 44 per cent of the population of South-West Asia and 39 per cent of North Africans live in towns (as defined by each country) and that these are higher percentages than those of other regions in Asia and Africa. A United Nations (1975) survey, however, suggested that in 1975 the urban population of North Africa and the Middle East as a whole accounted for 44 per cent of the total population (cf. 32 per cent in 1960), and that the average annual growth rate of the urban population during the period 1960–75 was 5 per cent.

Urban populations of the Middle East

For a variety of reasons the estimates of the urban populations of individual countries in the region (Table 9.3) are not very reliable, particularly for international comparisons. First, census data are distinctly uneven in quantity and quality; while countries like Algeria, Tunisia, Egypt and Turkey have had a long run of censuses, others like Lebanon, Oman and Qatar have never held a full census, Jordan has not held one since 1961 and Saudi Arabia never officially accepted the results of its only censuses. So Table 9.4, which lists censuses held in the Middle East since 1945, is more impressive than reality, for the censuses vary greatly in detail and accuracy. Another major difficulty for international comparability is the variation in definition of urban status, from country to country and from time to time. Table 9.5 reveals that even the criteria for definition vary, and it should be noted that Lebanon, Oman, Qatar, Saudi Arabia and Yemen AR have no official definitions at all.

Perhaps one of the most important reasons for low international comparability of levels and rates of urban population growth is the wide range of population sizes of states from 0.1 to 41.9 million inhabitants (1977 estimates). This results from the intense political fragmentation of the Middle East, one

Table 9.3 Urban populations of Middle-Eastern countries

	Total population (millions) 1976	*GNP per capita US $ 1976*	*Area (thousand sq km)*	*Percentage urban population 1960*	*1975*	*Percentage annual growth rate of urban population 1960–70*	*1970–5*
Turkey	41.2	990	781	30	43	5.2	4.2
Egypt	38.1	280	1,001	38	48	4.3	3.9
Iran	34.3	1,930	1,648	33	44	5.0	4.7
Morocco	17.2	540	447	30	38	4.2	5.1
Algeria	16.2	990	2,382	31	50	3.2	3.2
Sudan	15.9	290	2,506	9	13	6.2	5.5
Iraq	11.5	1,390	435	43	62	6.3	5.0
Saudi Arabia	8.6	4,480	2,150	12	21	6.6	6.3
Syria	7.7	780	185	37	46	4.8	4.2
Yemen AR	6.0	250	195	4	9	9.0	8.0
Tunisia	5.7	840	164	32	47	4.9	4.2
Israel	3.6	3,920	21	78	84	4.0	3.4
Lebanon	3.2	1,070	10	35	60	7.4	5.4
Jordan	2.8	610	98	43	56	5.1	4.9
Libya	2.5	6,310	1,760	23	31	5.8	5.0
Yemen PDR	1.7	280	330	20	29	5.5	5.4
Kuwait	1.1	15,480	18	69	89	13.0	8.2
Oman	0.8	2,680	213		5*		
UAE	0.7	13,990	84		84**		
Cyprus	0.6	1,480	9		42*		
Bahrain	0.3	2,140	1		80*		
Qatar	0.2	11,400	11		88*		

Sources: The World Bank, *World Development Report 1978*.
* United Nations Economic Commission for Western Asia, *Demographic and Related Socio-Economic Indicators for Countries of the ECWA Region*, 1975.
** Population Reference Bureau, *World Population Data Sheet, 1977*.

of the world's shatter belts. At the bottom end of the demographic scale are some small countries (e.g. Qatar, United Arab Emirates and Bahrain) which are little more than city states, while at the top of the scale the large countries (Turkey, Egypt, Iran) have complex urban systems and well-developed urban hierarchies including regional capitals. However, there are no clear relationships between the areas and populations of countries on the one hand and percentage urban populations on the other (Figs 9.1 and 9.2). It would appear that state boundaries are not very significant divides between urban systems. Some of the largest states (e.g. Saudi Arabia, Algeria, Sudan, Iran, Libya) contain several urban sub-systems, while some of the smallest states (e.g. the Gulf states) form part of multinational urban systems. Indeed, most of the largest cities in the Middle East, like Cairo, Tehran, Istanbul, Baghdad, Casablanca and Beirut, have an international significance far surpassing the boundaries of the countries in which they are situated. Cairo and Tehran each

Table 9.4 Population censuses of countries of the Middle East since 1945

Algeria	1948, 1954, 1966, 1977
Bahrain	1950, 1959, 1965, 1971
Cyprus	1946, 1960, 1973
Egypt	1947, 1960, 1966, 1976
Iran	1956, 1966, 1976
Iraq	1947, 1957, 1965, 1977
Israel	1948, 1960, 1972
Jordan	1952, 1961
Kuwait	1957, 1961, 1965, 1970, 1975
Lebanon	
Libya	1954, 1964, 1973
Morocco	1950, 1960, 1971
Oman	
Qatar	1970*
Saudi Arabia	1962–3,* 1974**
Sudan	1955–6, 1973
Syria	1960, 1970
Tunisia	1946, 1956, 1966, 1975
Turkey	1950, 1955, 1960, 1965, 1970, 1975
United Arab Emirates	1968, 1971,* 1975
Yemen AR	1946, 1955 (Aden colony), 1975
Yemen PDR	1973

* Enumeration considered largely incomplete.
** Only total population figures published.

has more inhabitants than over half of the countries in the region, including Israel, Lebanon, Jordan and Libya; Baghdad has as many as Lebanon, and Istanbul has as many as Libya.

In these circumstances statistical analyses of urban populations, however defined, are not particularly enlightening except in the broadest terms. Then it is apparent that there is a fairly marked north–south contrast between the more urbanized, longer-settled countries of the northern tier, the fertile crescent and the Atlas lands of the Maghreb, and the less urbanized, traditionally more nomadic countries of the Arabian peninsula, Libya and Sudan (Table 9.3). In nearly all the more urbanized countries at least 40 per cent of the population live in localities defined as urban by each country, while in Israel and most of the Gulf states over 80 per cent live in such localities. On the other hand, in the less urbanized countries less than 40 per cent live in towns, and in two countries, Yemen AR and Oman, probably less than 10 per cent.

We should remind ourselves, however, that the percentage living in urban localities is a rapidly changing statistic. The oil-rich countries of the southern zone, like Oman, Saudi Arabia and Libya, are undergoing very rapid urbanization and will not be permanently at the base of the league table of urban populations. Certainly high GNP per capita is not closely correlated with high urban population levels, but in time the correlation will be closer.

Table 9.5 Definitions of urban population in Middle-Eastern countries

Morocco	All municipalities, *centres autonomes* and other urban centres
Algeria	Centres of communes with 1000 or more population and at least 50 per cent or more of them in non-agricultural sectors. More than 75 per cent: urban; between 50 and 75 per cent: semi-urban
Tunisia	Population living in all communes
Libya	No specific definition. After the census was taken, populations of Tripoli and Benghazi *muhafadat* and the urban parts of the *mutassarifias* of Beida and Derna were treated as urban
Egypt	The governorates of Cairo, Alexandria, Suez, Ismailia and Port Said and the chief towns of provinces and districts
Sudan	Centres with 5000 or more population having some urban characteristics
Jordan	Urban includes the population resident in all localities of 10,000 or more population (excluding localities inhabited only by Palestinian refugees), all district capitals regardless of size, all localities of 5000 to 10,000 inhabitants in which two-thirds or more of the economically active males were reported in non-agricultural occupations and those suburbs of Jerusalem and Amman cities with at least two-thirds of males in non-agricultural pursuits
Israel	All settlements with more than 2000 inhabitants, except those where at least one-third of the heads of households, participating in the civilian labour force, earn their living from agriculture
Syria	Cities, district (*mohafaza*) centres and sub-district (*mantika*) centres
Cyprus	Six district towns and Nicosia suburbs
Turkey	Population of the localities within the municipality limits of administrative centres of provinces and districts
Iraq	Population living within the boundaries of municipality councils
Iran	All *shahrestan* centres, regardless of size, and all places of 5000 or more inhabitants
Kuwait	Kuwait city (Dasman Sharq/1, Sharq/2, Murgab, Salihia and Qibla) and Labourer's City
Bahrain	Towns of Manama, Muharraq (including Muharraq suburbs), Hedd, Jiddhafs, Sitra, Rifa'a and Awali
Yemen, People's Democratic Republic of	The entire former colony of Aden excluding the oil refinery and villages of Bureiqa and Fugum

Source: *UN Demographic Yearbooks*.

Urban growth rates in the Middle East

Some indication of recent changes in the levels of urban population may be gleaned from the estimates by the World Bank for 1960 and 1975 (Table 9.3 and Fig. 9.2). Unfortunately, urban growth rates are not available for some of the smaller states, but from those which are available it is clear that in almost

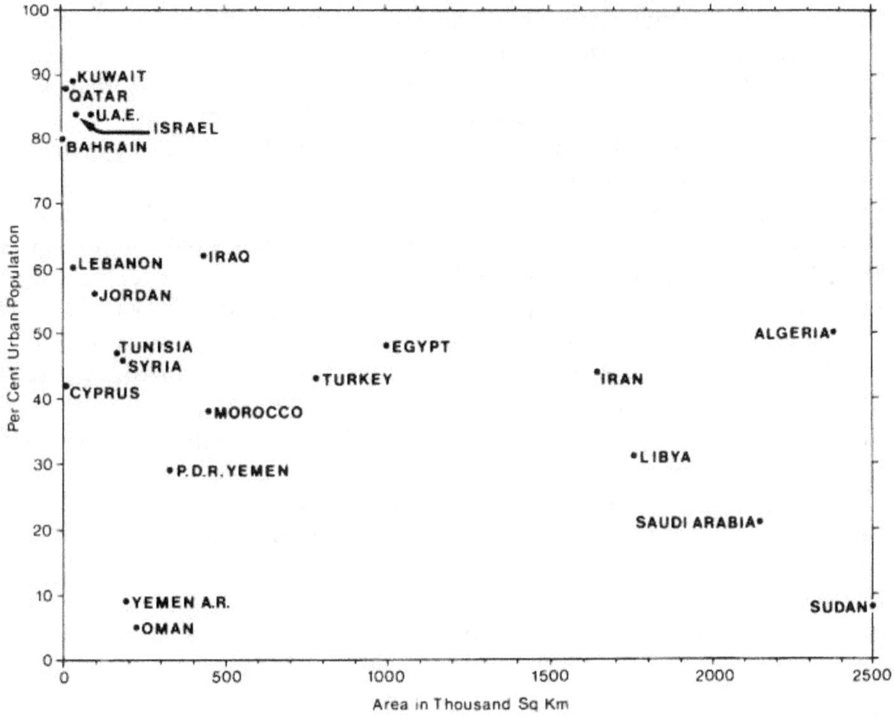

Fig. 9.1 Relationship between areas of Middle-Eastern countries and their percentage urban populations, 1975.

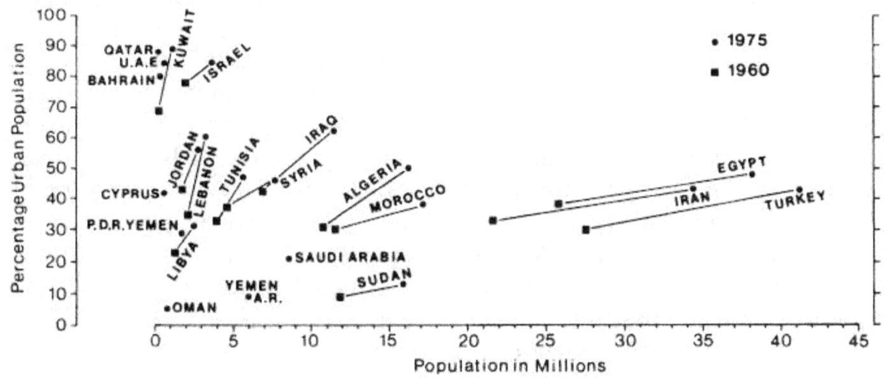

Fig. 9.2 Changes in urban populations in the Middle East, 1960–75.

every country of the region the annual rate of urban growth was quicker during the 1960s than during the first half of the 1970s and was rarely less than 4 per cent, the only major exception being Algeria. The average national rate for 1960–70 was 5.9 per cent and for 1970–5 5.1 per cent, which in more realistic terms implies that urban populations doubled every 12–14 years. In highly urbanized Kuwait and feebly urbanized Yemen AR the rates were even more spectacular, exceeding 8 per cent, but in Kuwait the total population was small and in Yemen AR the urban population was small, so absolute growth is less impressive than relative growth. Perhaps more impressive is the powerful growth (over 6 per cent annually) of Saudi Arabia's urban population from a previously low level, and of the urban populations of a number of countries which were already moderately urbanized: Algeria, Tunisia, Lebanon, Jordan and Iraq. Here political events have usually played an important role, sustaining the snowballing urbanization, though in Algeria this was in spite of more modest overall population growth.

In contrast, Egypt's urban growth has been more impressive in absolute terms than in relative terms, for with that in Turkey and Iran it constitutes the bulk of Middle East urban growth. Urban growth is also slower than average in Israel, partly a reflection of planning; but only little Cyprus has seen very little recent change in its numbers of town-dwellers.

Large city populations

One of the most noteworthy features of urban growth in the Middle East is its overwhelming concentration in large cities. In the mid-1970s there were more than 140 cities with 100,000 inhabitants or more, but over half of them were in Egypt, Iran or Turkey, the most populous countries. In a number of countries (Egypt, Iraq, Syria, Lebanon, Israel and Kuwait) these cities comprise at least one-third of the total populations. In contrast, middle-sized towns with 50,000–100,000 inhabitants are only numerous in the most populous countries, though the Middle East is not unusual in this respect.

Cities with at least a million inhabitants are becoming much more numerous. By the late 1970s there were at least a dozen: Cairo, Tehran, Istanbul, Alexandria, Baghdad, Casablanca, Ankara, Algiers, Beirut, Damascus, Greater Khartoum and Tunis. In addition, the agglomeration of Tel Aviv–Yafo contains more than a million, and it cannot be long before the list is joined by Riyadh and Amman. Moreover, by 1976 Cairo had over 5 million inhabitants, Tehran 4.5 million and Baghdad over 3 million, so all three were well into the ranks of major world cities. Figure 9.3, which is based upon census data, depicts the growth of most of the actual and near 'million-cities' of the region, and is perhaps more effective than a table of growth rates giving a spurious indication of accuracy. It may, however, give a false

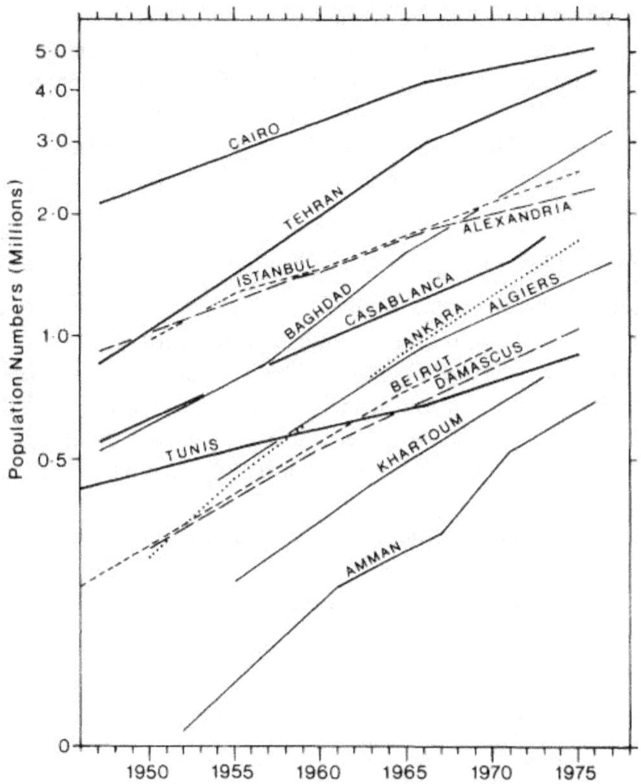

Fig. 9.3 Recent growth of some large cities of the Middle East.

impression of smooth growth when the reality, especially where in-migration is voluminous (e.g. Baghdad, Amman), is less even. Particularly striking growth may be observed for Tehran, Baghdad and Ankara, all interior capitals of highly centralized states, and therefore somewhat unusual among less developed countries (though Mexico City and Bogota are other examples). Tehran's population doubled from 1.5 million to 3 million between 1956 and 1966, and another 1.5 million were added by 1976. Even more remarkable was the growth of Ankara, which multiplied nearly six times in the 25 years 1950–75 to reach a total of 1.7 million, while the formerly very much larger city of Istanbul multiplied just under three times to attain 2.5 million. Baghdad also multiplied six times during the 30 years 1947–77. Lesser cities like Beirut, Khartoum, Riyadh and Amman – the last three again with inland locations – have also grown rapidly, faster than most of the major cities of North Africa (Cairo, Alexandria, Casablanca, Algiers and Tunis), some of which were affected by decolonization. Capital status is also important to the growth of large cities and all but three of the 'million-cities' are capitals, the exceptions being Istanbul, Alexandria and Casablanca, the first two curiously similar in size and rates of growth. Istanbul and Casablanca are still the largest cities in their respective countries, but the fact that there are two major cities in Turkey means that Istanbul contains a smaller percentage of the total population than most of the largest cities in countries of the Middle East.

City-size hierarchies

In most countries the largest city is incorporating an increasing proportion of the total population. Excluding the micro-states with less than one million inhabitants, the largest cities of most of the remaining countries comprise at least one-tenth of the total populations, but in Saudi Arabia and Yemen PDR they contain a fifth, in Iraq, Jordan and Libya a quarter, in Israel a third, and in Lebanon nearly a half (Table 9.6). Obviously there is a tendency for greater

Table 9.6 Urban primacy in Middle-Eastern countries

	Population of largest city (thousands)	*Percentage of total population*	*[Urban primacy] Two-city index*	*Four-city index*
Turkey	2,547 (1975)	6	1.50	0.91
Egypt	5,084 (1976)	14	2.19	1.33
Iran	4,496 (1976)	14	6.69	2.44
Morocco	1,753 (1973)	11	2.93	1.20
Algeria	1,523 (1977)	9	3.10	1.42
Sudan	800 (1973–4)	6	5.92	2.26
Iraq	3,206 (1977)	26	5.58 (1965)	2.41
Saudi Arabia	669 (1974)	18	1.19	0.59
Syria	1,042 (1975)	14	1.34	0.86
Yemen AR	139 (1975)	3	1.68	0.75
Tunisia	900 (1975)	16	3.46	2.12
Israel	1,181 (1975)	35	3.28	1.97
Lebanon	1,173 (1975)	46	5.97 (1970)	4.16
Jordan	691 (1976)	23	2.66	—
Libya	552 (1973)	24	1.96	—
Yemen PDR	285 (1973)	18	—	—
Kuwait	277 (1975)	28	—	—
Oman	—	—	—	—
UAE	141 (1975)	25	—	—
Cyprus	116 (1973)	18	1.45	0.84
Bahrain	111 (1975)	43	—	—
Qatar	145 (1975)	85	—	—

Note: Two-city index $= \frac{P_1}{P_2}$ and four-city index $= \frac{P_1}{P_2+P_3+P_4}$ where P_1 is the population of the first city, P_2 is the population of the second city, etc.

concentration in the largest cities of the less populous countries and more highly urbanized countries, but the correlation is not very clear. Apart from Istanbul, the two cities of Greater Khartoum and Sanaa stand out as containing only small proportions of the total populations of their respective countries, both of which are feebly urbanized.

On the other hand, analysis of two- and four-city indices (Table 9.6) shows that Greater Khartoum is very dominant in the urban hierarchy of Sudan, as Tehran, Baghdad and Beirut are in Iran, Iraq and Lebanon. All these cities are more than five times larger than their nearest rivals, and more than twice as large as the combined populations of the second, third and fourth cities. Tunis also exhibits marked primacy, and city-size distributions of these five countries (Fig. 9.4) clearly differ from those of most other countries in the Middle East, though Iran and Morocco have many more cities with at least 100,000 inhabitants than Iraq, Tunisia or Sudan.

The city-size hierarchies (Clarke and Murray 1973; Clarke 1978) of Egypt, Turkey and Algeria approximate to log-normal (i.e. $P_n = P_1/N$ where P_n = the population of the *n*th city, P_1 = the population of the first city and N = the number of cities in the array), a reflection perhaps of factors such as their large size, long history of urban life and economic diversity. Turkey, however, has a tendency towards a binary (or intermediate) city-size distribution with two main cities, a pattern manifested by Syria, Saudi Arabia and Yemen AR (Fig. 9.4), where no single city has achieved overwhelming preponderance.

These city-size distributions are not immutable; the growth of Ankara to offset the formerly strong primacy of Istanbul is indicative of the way that city-size distributions can change, as is the growing primacy of Tripoli in Libya. On the other hand, the evidence suggests that where the largest city is the capital the tendency for growing concentration is very strong, even when it is far from the sea. When it is also the major port (Algiers, Tripoli, Tunis, Beirut, Tel Aviv, Aden), primacy is often intensified. Casablanca is the only primate city which is a port but not a capital.

Components of city growth

Rapid city growth results from high natural increase, net in-migration and incorporation of surrounding villages. Natural increase averages 2.8 to 2.9 per cent per annum for the region as a whole, and is only substantially lower in Cyprus (1.5 per cent) and Israel (2.1 per cent) and substantially higher in Kuwait (4.2 per cent) and Syria (4.1 per cent). High rates are caused by sustained high fertility and declining mortality. In cities they are even higher because the difference between urban and rural mortality is greater than that

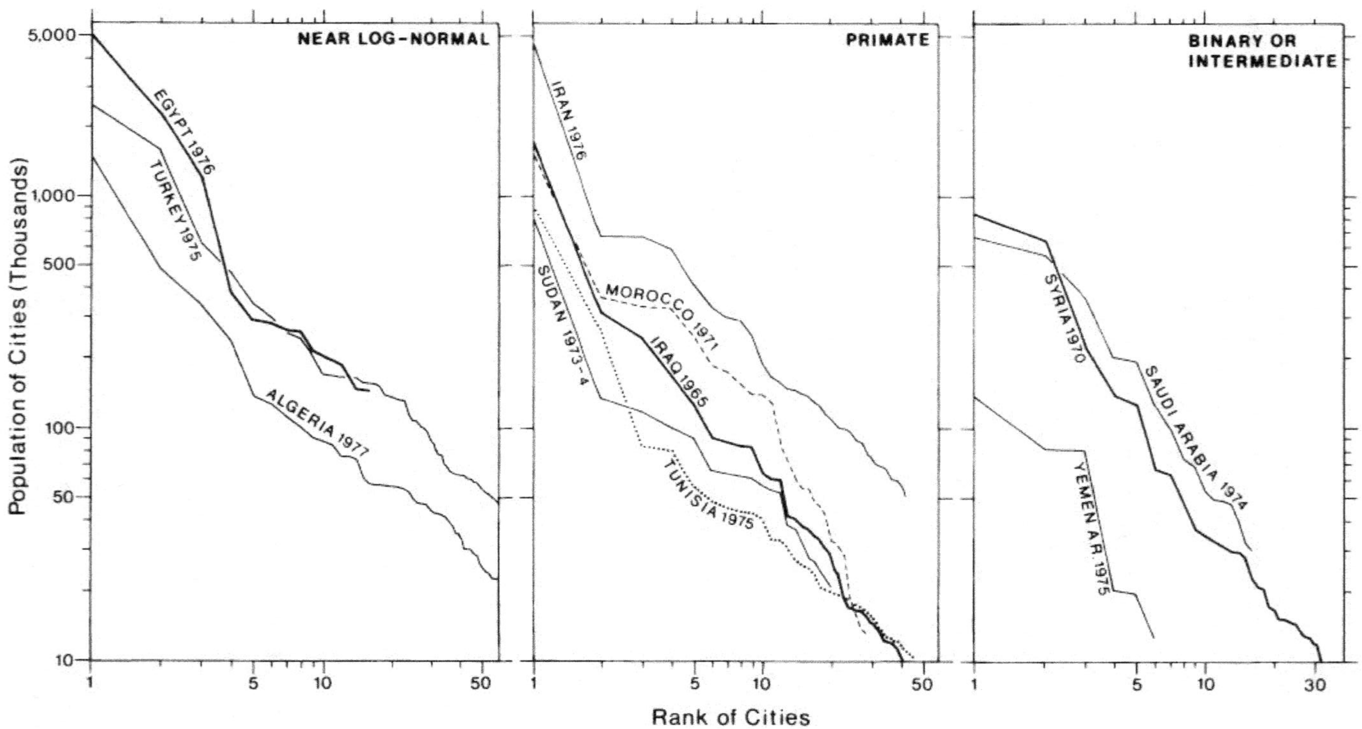

Fig. 9.4 City-size distributions of countries of the Middle East with at least 5 million inhabitants.

for fertility; frequently natural increase rates of cities exceed 4 per cent, which means a doubling time of less than 18 years.

In the period 1970–5 death rates of many countries were 14–16 per thousand, but higher (about 20) in Sudan, Saudi Arabia, Yemen AR, Yemen PDR, Oman and Qatar, and much lower (below 10) in Cyprus, Lebanon, Kuwait, Bahrain and Israel. For many reasons, mortality has declined more rapidly in cities than fertility. Although Middle-Eastern cities were formerly extremely unhealthy, and that is still the case in shanty towns and refugee camps, there have been great improvements in sanitation, hygiene, nutrition and in medical and pharmaceutical services, raising the life expectancies of city-dwellers. Naturally, the degree of mortality decline varies considerably in time and space and is less in Sanaa than Sousse, in Muscat than Meknes; but feebly populated oil-rich countries are able to effect more rapid declines than most other countries.

The situation is quite different as far as human fertility is concerned, as in this predominantly Islamic region the fertility levels of many countries are amongst the highest in the world, with birth rates of 45–50 per thousand (1970–5) and total fertility rates (TFRs) of about 7 (Cassen 1978) – i.e. a woman would have seven children if she gave birth at the prevailing age-specific fertility rates and survived to the end of her reproductive years. The major exceptions are Cyprus and Israel, with TFRs of 2.8 and 3.7 respectively, but some fertility decline has been experienced in Egypt (TFR of 5.2), Turkey (5.8) and Tunisia (5.9), especially in urban areas, and the partly Christian Lebanon has a lower level of fertility (TFR of 6.3) than more Islamic Arab countries. But even in countries with high total fertility rates there are distinct signs that fertility is beginning to fall among some city-dwellers, as in Tehran (Amani 1972), and that improved living standards, increased education and other aspects of socio-economic development are related to lower fertility (Audroing *et al.* 1975). Family planning is making limited progress among urban élites, and more widespread progress in Egypt, Turkey and Tunisia.

There still seems to be more immediate prospect of further mortality decline in cities than there is of fertility decline. Some of the oil states, for example, are not interested in family planning schemes, partly because they have experienced so much immigration of workers that they see their own nationals being outnumbered; yet they are able to establish the most modern of medical services. So high natural increase is likely to be the major and most persistent component of urban growth for some time to come.

Migration, on the other hand, is the cause of much spectacular growth of Middle-Eastern cities, and accounts for many of the differences between growth rates of large cities. It is a more spasmodic and less general component of urban growth than natural increase, partly because net-migration is the balance of in-migration and out-migration. Generally it leads to spurts of growth, as in the case of Amman, a city of migrants and refugees. Cities in oil-rich states have been particularly attractive to migrants, not merely the capitals but others like Benghazi, Jeddah, Dammam and Abadan. But perhaps most impressive is the huge volume of migrants into the major Middle-Eastern cities – Cairo, Tehran, Baghdad – where net in-migration frequently exceeds 2 per cent per annum (Cairo Demographic Centre 1973), despite their size.

Population concentration and redistribution

By 1985 37 per cent of the world's urban inhabitants may live in 'million-cities', and there may be fifteen cities with 10 million inhabitants, most of them in less developed countries. The Middle East is no exception to this growing concentration; Greater Cairo is expected to be one of these '10-million-cities', and there could well be '15-million-cities' by 1985. This mercurial metropolitan expansion poses urgent problems of many kinds, not least of which is the formulation of policies to cope with the spatial concentration of population.

Most modernization in the Middle East has been highly polarized and strongly linked with 'the world economy'. It has thus contributed substantially to urbanization, which has increased through initial advantage and cumulative causation. The often harsh conditions of the physical environment have made rural development expensive and difficult, and the huge revenues of the oil industry, the epitome of capitalistic multinational enterprise, are more easily spent in cities than in deserts or steppes. Industrialization, the panacea of the 1960s, was also urban orientated, so the contrast between cities and their rural surrounds was intensified. The cities attracted rural labour, but their links with the world economy meant that many of the traditional urban industries suffered from external competition, except where tourism enabled their survival, as in Tunisia.

There has been much discussion over whether these metropolitan cities are parasitic or generative, whether they drain their countries' lifeblood or whether they are centres of innovation and diffusion. In fact, it is unwise to generalize, as the situation varies from country to country, but there can be little doubt that many large cities have developed an enclave type of economy, especially in poor countries which suddenly acquire wealth. On the whole, the spread effects of large cities are greater in middle-income countries, where there is less spatial inequality than in low-income countries, and where there are stronger links between rural areas and major cities.

Despite much discussion over the economies and diseconomies of large cities in less developed countries, there is a growing consensus that the diseconomies are excessive and may culminate in 'ecodisasters' (Elizaga 1973). Many favour intermediate or middle-sized cities as potential growth centres

(Hansen 1971). Unfortunately, few population policies in less developed countries, and the Middle East is no exception, have paid much attention to spatial distribution or redistribution of population, but have focused on population growth (Bose 1977). The main policy approach to this question has been through urban and regional planning, which has mostly aimed at decentralization but has met with mixed fortunes. Israel is, of course, an unusual case, but a prime aspect of policy has been to limit the growth of the largest three cities and encourage the growth of new towns; but in Jerusalem this policy has succumbed to the desire for a Jewish majority. Algeria, on the other hand, applied growth-centre concepts to restrict concentration of development and to spread it to lagging regions, but success has been limited because much development has been based upon industrialization as the propulsive factor, thus causing further urban concentration even if in lesser cities. In Iran also, development plans have contributed to the evolution of regional capitals (Behnam and Amani 1974), but without slowing down appreciably the massive growth of Tehran. We shall have to see whether de-westernization in Iran will lead to changes in the urban system (see Chapter 19); so far decolonization of the Maghreb has had no significant effect.

Growth-centre policies have tended to reinforce economic dependency and existing urban spatial patterns. Time will tell whether political changes in the Middle East will provoke new approaches to regional and urban planning. These will not and should not be alike, because each country presents different historical and geographical conditions of urbanization; but it is very doubtful if population measures to arrest migration or control fertility will do much in the first instance except effect initial slowing down of metropolitan expansion. The real solution must be found in more effective spatial economic planning of each country as a whole. So far planners' dreams have done little to change the spatial patterns of urban growth.

References

AMANI, M. (1972) 'La population de l'Iran', *Population*, 27, 411–18.

AUDROING, J. F. *et al.* (1975) 'Recherche des correlations entre des variables démographiques, sociologiques et économiques dans les pays Arabes', *Population*, 30, 61–80.

BEHNAM, D. and AMANI, M. (1974) *La Population de l'Iran*, CICRED.

BOSE, A. (1977) 'Population policies and metropolitan growth', *International Union for the Scientific Study of Population, International Population Conference Mexico 1977*, 2, 225–40.

CAIRO DEMOGRAPHIC CENTRE (1973) *Urbanization and Migration in some Arab and African Countries*, Cairo, Research Monograph Series, No. 4.

CASSEN, R. H. (1978) 'Current trends in population change and their causes', *Population and Development Review*, 4, 331–53.

CLARKE, J. I. (1978) 'Primate cities of the world and the Middle East', in *The Population Framework: Data Collection, Demographic Analysis, Population and Development*, UN Economic Commission for Western Asia, 290–7.

CLARKE, J. I. and MURRAY, J. L. (1973) 'Population dynamics of large Middle Eastern cities', *International Union for the Scientific Study of Population, International Population Conference Liège 1973*, 1, 271–86.

DAVIS, K. (1969 and 1972) *World Urbanization 1950–1970*, 2 vols, Population Monograph Series, Berkeley, University of California.

DURAND, J. D. (1977) 'Historical estimates of world population: an evaluation', *Population and Development Review*, 3, 253–96.

ELIZAGA, J. C. (1973) 'The demography of metropolitan growth and planning', *International Union for the Scientific Study of Population, International Population Conference Mexico 1977, Proceedings*, 289–300.

GOLDSTEIN, S. (1973) 'An overview of world urbanization, 1950–2000', *International Union for the Scientific Study of Population, International Population Conference Liège 1973*, 1, 177–90.

HANSEN, N. M. (1971) *Intermediate-Sized Cities as Growth Centres*, New York, Praeger.

UN (1975) *Selected World Demographic Indicators by Countries, 1950–2000*, New York.

PART II

Case studies

CHAPTER TEN

The study of and a contribution to the geomorphology of the Arabian Gulf

TAIBA A. AL-ASFOUR

Geological and structural studies

Today the Arabian Gulf and its littoral are the scene of great economic developments and activities, this associated with the wealth of oil and natural gas found beneath the sea and its shores. The study of the geology and geomorphology of this region has in great part been encouraged by the quest for oil, but many aspects of such study are significant for other reasons, particularly those investigations concerned with the more recent geomorphological history of the Gulf. In this chapter a brief overview of these latter investigations is first attempted, followed by a summary statement of one particular relevant piece of research.

The Arabian Gulf is a tectonic basin whose general form was largely established in late Pliocene/Pleistocene times. It is now a shallow sea, aligned north-west to south-east and connected to the Arabian Sea by the Strait of Hormuz and the Gulf of Oman. The geology of the Gulf has for many years attracted the attention of scholars (Pilgrim 1908; Lees and Richardson 1940; Henson 1951; Mina *et al.* 1967; and Kassler 1973). In general, the Arabian Gulf is located between the north-west to south-east trend lines of the folded and thrust ranges of the Zagros mountains of Iran on the one hand and the Pre-Cambrian Arabian Shield on the other. The Zagros mountains emerged from a geosyncline squeezed between the relatively resistant plates of Iran and the Arabian foreland, these ranges representing a belt of folded and faulted Palaeozoic, Mesozoic and Cainozoic rocks. The area is still active as can be witnessed in the river terraces on the Makran coast (Lees and Falcon 1952; Falcon 1947). On the other hand, data from the western side of the Gulf indicate that the Arabian Shield is slightly tilted towards the north-east and is covered by Palaeozoic, Mesozoic and Cainozoic sediments (Powers *et al.* 1966). The low western coast of the Gulf is interrupted at its southern end by the Omani mountains (the Hajar ranges) which have resulted from the folding

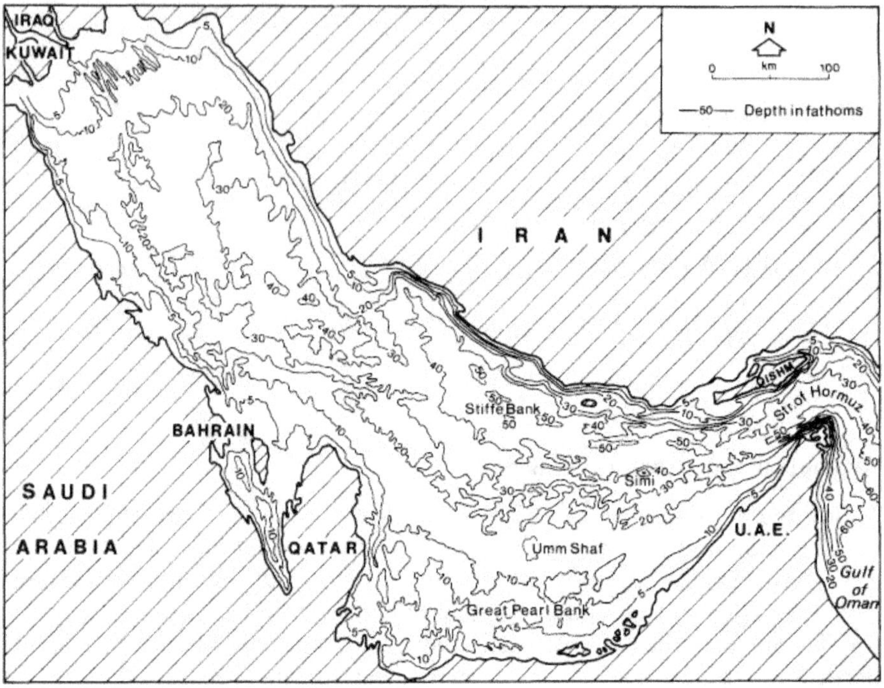

Fig. 10.1 Bathmetry of the Gulf, based on British Admiralty and US Hydrographic charts.

and crushing movements that started at the beginning of the Cretaceous period and were completed in the Mio-Pleistocene.

The Gulf is generally a shallow sea with average depth of 31 m and a maximum depth of 110 m near the Strait of Hormuz (Fig. 10.1), before the seafloor plunges to a depth of 900 m in the Gulf of Oman. The greatest depth alignment is offset to the north-east along the Iranian coast where the mountain ranges run parallel to the coast and in some cases form a cliffed coastline. The Arabian coast is low lying with shallow offshore stretches apart from the extreme south where the Oman mountains, which are in tectonic terms closely associated with the Zagros ranges, reach the sea around Ras Musandam. This south-eastern part of the Gulf lying between Qatar and Oman has received more attention and detailed studies than areas to the north (Houbolt 1957; Evans *et al.* 1964, 1969; Kendall and Skipwith 1969; Evans 1970; Shinn 1973; Falcon 1973, 1975; Vita Finzi 1973). The head of the Gulf, by contrast, is an area of deltaic sedimentation formed by the Euphrates, Tigris and Karun rivers.

Today, water temperature is high especially during the summer, the surface water temperature reaching 32 to 33 °C; the rate of evaporation is also very high. These conditions, together with the absence of large supplies of freshwater, except from the Tigris, Euphrates and Karun and some intermittent small rivers flowing from the Zagros (and these vary seasonally), produce a high degree of salinity (Emery 1956; Evans 1970). The replacement of evaporation loss occurs by a broad anticlockwise surface current which brings in new oceanic water, whilst water made dense by evaporation sinks to the bottom and leaves the Gulf by the deeps near the Musandam peninsula (Emery 1956; Purser and Seibold 1973). The tidal range in the Gulf is now between 2 and 3.5 m. Winds in the Gulf are bimodal in strength, south-easterly and north-westerly, the latter being predominant. These are very dry winds, blowing more frequently in summer than in winter and carrying terrigenous sediments to most parts of the Gulf, especially the Arabian coasts, and causing waves and surface currents.

The north-western part of the Arabian Gulf is an area of both structural instability and extensive deposition by the Tigris, Euphrates and Karun rivers. The geological history of this part of the Gulf has been interpreted by two contradictory points of view, those of the archaeologists and of the geologists. The archaeologists' thesis, based on the early work of De Morgan (1900), is, broadly, that the long continual deposition of the sediments brought down by the Tigris, Euphrates and Karun rivers has caused the north-western shoreline to advance towards the south and has, in general, left many near-littoral cities of older civilizations now remote from the sea (Beke 1835; Rawlinson 1857; Loftus 1857; Woolley 1938). On the other hand the work of, for example, Arthur Holmes (1944) and Sarnthein (1972), produced a contradictory point of view, that the northern shoreline of the Gulf has retreated northward, that the basin of the Gulf was dry land during the Pleistocene glacial epoch but after the melting of the ice sheet the water level rose and the sea advanced northward in the Gulf depression. In 1952, Lees and Falcon claimed that the picture was more complicated than the simple explanation presented by the archaeologists and that the head of the Gulf had never lain northward of its present limit since early Pliocene times. They put forward a geological point of view in which they demonstrated that the area of the head of the Gulf is structurally unstable and that the dominant vertical movement is one of coastal subsidence. They also maintained that the continued existence of the marshes and lakes at the head of the Gulf resulted from such tectonic movement, since otherwise these would have been eliminated by long continued siltation. Lees and Falcon (1952) recorded also that Bubiyan Island, Kuwait, indicated tectonic subsidence to the north and that this is shown by the island's drainage trend towards the north-west and the expanding area of land in the southern part of the island. More recently, however, Larsen (1975) has re-examined the Lees and Falcon results in the light of new data about the area and has concluded that the area at the head of the Gulf is structurally stable. Kassler (1973) seemed to support the argument for recent structural stability of the head of the Gulf, in particular calling attention to a marine platform at −16 to −20 fathoms (−29 to −37 m) which 'has been mapped' extending from offshore of Abu Dhabi north-westward to Kuwait. This is in disagreement with V. S. Colter's unpublished oil-company report (quoted by Kassler) that the amount of subsidence in the Kuwait part of the delta has been as much as 37 m during the last 5000 years. Kassler suggested that the age of this marine platform is about 9000 to 11,000 years BP. Larsen and Evans (1978) also dispute the Lees and Falcon argument for tectonic subsidence in the delta region and tend to agree with De Morgan's early hypothesis of an advancement of land towards the south over the last 4000–5000 years. They believe that the geological history of the delta has been controlled more by eustatic changes of sea level than by local or isostatic movements. Larsen and Evans noted that 150 to 200 km of progradation has almost certainly taken place since the attainment of a near-modern sea level about 5000 years ago. Brice (1978), referring to personal correspondence with Hausman and Falcon, has indicated that they believe that the topography of the delta is a result of a delicate balance of alluviation and subsidence.

Morphological studies

The landforms in and around the Gulf are relicts of topography largely developed during the late Pliocene/Pleistocene and Holocene under a variety of climates; landform developments have also been influenced by structure and eustatic movements of sea level. Humidity and precipitation in the region reached their climax during the pluvial periods which were contemporaneous with the last glacial period in Europe. As a result, an increase in runoff and stream flow led to increased alluvial erosion and large quantities of detritus were removed from the highlands by stream erosion (as the principal degradation process) and deposited as outwash fans or flow deposits, such examples being well documented in Oman, the United Arab Emirates (Stevens 1978) and in Iran (Vita Finzi 1978). In the drier periods or stages, weathering, pedimentation and related processes played a greater part in creating the landforms. Chapman (1971) has referred to the effect of climatic changes in the evolution of landforms in the Dammam dome and in the Shedgum area in the eastern province of Saudi Arabia; dendritic wadis in Arabia extend for hundreds of kilometres, representing channels of ancient streams. The Ad Dibdibba plain which extends north-east of Saudi Arabia, west and north-west of Kuwait, consists of sheet gravels which were carried from western Arabia through the Wadi-ar-Rimah–Wadi Al-Batin channel system during pluvial periods of the Pleistocene (Holm 1960; Powers *et al.* 1966). The rise or fall of sea level as a result of eustatic changes or tectonic movements has also left its mark on the coastal areas of the Gulf. Holm referred to marine transgressions along the Arabian coasts which occurred at the base of present east-facing cliffs and to wave-cut benches at an elevation of about 150 m above sea level. Sabkhas which stand at the same height have been found inland, which seems to indicate that they may represent early embayments of the Gulf at that height. Glennie (1970) has also suggested that the inland sabkha of Umm as-Samin in Oman may represent a relict arm of the sea at an elevation of less than 70 m above sea level. Kapel (1967) referred to C_{14} dating which gave an age of 39,800 years to shells collected from current-bedded formations of calcareous sandstone which occur as raised beaches in Qatar, suggesting a higher Pleistocene sea level at height of about 25 to 30 m. Voute (1957) also noted the occurrence of recent marine fauna on raised beaches of the Abu Dibbis depression in Iraq, requiring a higher sea level of about 25 m to connect Abu Dibbis with the sea (Larsen 1975). Powers *et al.* (1966) referred to a probable marine transgression of Pleistocene age and indicated that there are marine terraces which extend along the coast of Saudi Arabia from north of Jubail to Dawhat Salwah. These terraces are of sand and coquina shells and are found at a height of 1–2 m above present mean high tide. On the coast of Kuwait, Fuchs *et al.* (1968) pointed out that Pleistocene deposits are spread south of Kuwait Bay, along the spur of Ras Ashairij and Umm al Nemel Island. These deposits consist of fine-grained thin-bedded, partly cross-bedded, oolitic limestone of a thickness up to 10 m. Evidence of Quaternary tectonic movement in Kuwait has been recorded by Saleh (1975) from the coastal cliffs south of Al-Khiran where an ancient sabkha was elevated to about 8 m above sea level associated with a fault causing the uplift of the cliff and the exposure of the Tertiary basement.

From the Iranian coast of the Gulf, Harrison (1941), Falcon (1947) as well as Vita Finzi (1975) have referred to river terraces and raised beaches at various levels which are probably the result of uplift, and according to Falcon the Jaz Murian plain is an area of recent subsidence. The raised beaches which stand at heights of up to 120 m along the Makran coast indicate recent uplift with respect to present sea level. Falcon also referred to raised beaches at 80–90, 30 and 15 m above sea level. However, Butzer (1958) suggested that these were also affected by eustatic changes of sea level, and added an additional 60 m level; he further suggested that the whole sequence would then agree with the altimetric sequence from Sicilian to Monastirian. The same sequence occurs on Kharg Island, whilst shorelines of + 30 and + 15 m are found on Qishm Island and Bushire. The results of C_{14} dating of two shells of material collected from the coast of Makran zone, giving ages of 23,390 ± 400 BP and 25,610 ± 640 BP (these samples coming from the beach 8 m above highwater), are, however, suspect since at Jask the samples were probably contaminated by younger carbon (Vita Finzi 1975). According to Vita Finzi (1969), two major phases of alluvial deposition, tentatively correlated with the Würm glacial and the 'Little Ice Age', have been largely responsible for the infilling of basins and alluvium of western and northern Iran.

The work of Blandford (1872), which first suggested that the Persian coast of the Strait of Hormuz has been rising while that of Arabia was sinking, has been recently confirmed. More information has been added about the Quaternary history of the area, following notably the results of the scientific expeditions carried out on Musandam peninsula by the Royal Geographical Society (Cornelius *et al.* 1973; Falcon 1973; Robinson and May 1974; Vita Finzi 1973), as well as the expedition carried out by the Royal Geographical Society/Imperial College of Science and Technology on the Makran coast of Iran (Falcon 1975; Vita Finzi 1975).

Data collected from offshore hydrographic and geological work as well as from the alluvial stratigraphy show that the Musandam peninsula is tilting towards the north-east, the drowned valleys of the peninsula indicating that locally the area is sinking towards north-east, although the Oman mountains have been uplifted. Vita Finzi (1973) reported that the vertical displacement appears to have exceeded 60 m during the last 10,000 years. No raised beaches have been observed in the peninsula, but there are alluvial deposits which are well exposed and provide the basis for the identification of a generalized sequence called the Makhus formation. According to Vita Finzi, these deposits were formed about 35,000 years ago equating with the last marine regression. The deposits have been traced inside the Gulf as cemented wadi gravels which have been covered by the Flandrian transgression. It seems also that the Makhus alluvial deposits were deposited at the same time as that of the Tehran alluvials of northern and central Iran dated archaeologically between 7000 and 36,000 years BP (Vita Finzi 1969). Alluvial deposits similar to the Makhus deposits of Musandam have, according to Vita Finzi (1978), been traced as far south as Ras al Khaimah. Brice (1978) suggests that both Vita Finzi (1978) and Stevens (1978) reach broad agreement about the timing of the late glacial phase in Arabia which was associated with large-scale formation of fans and outwash zones in piedmont areas. These deposits were carried further downstream and deposited as deltas into the rising water of the Gulf during the wetter conditions of post-glacial times.

During the Würm glaciation the sea level was about −100 or −120 m which meant that the Gulf (where contemporary maximum depths do not exceed −100 m or −110 m near the Strait of Hormuz) would have then been in the main a dry basin depression. Coinciding with the maximum retreat of sea level, river channels and wadi degradation became active and those facing the coastal areas of the Gulf would have experienced a corresponding lowering of their base level. Kassler (1973) reported the occurrence of a network of submarine channels which represent drowned river valleys, these appearing most clearly in the steps between submarine terraces. According to his interpretation these river valleys were formed when the climate was wetter than at present.

Evidence of climatic changes during post-glacial times in the Gulf has been obtained in different ways, such as in the fluctuations in sediment loads brought to deltas and changes of sea level which affected the river regimes as well as the occurrence of submarine terraces at different levels. Melguen (1973), who investigated the sedimentary history of the Rud Hilla delta on the Iranian coast of the Gulf, reported that there has been an increase in the rate of sedimentation since about 6000 years BP; there is also evidence of differentiation between the Pleistocene and Holocene sediments.

Several environmental reasons for this have been suggested by Melguen (1973), among which is the increase in the amount of rainfall affecting Iran during the Holocene period. This increase caused periodic violent runoff accompanied by an increased supply of terrigenous materials, this increasing the rate of erosion and sedimentation and causing a growth in the rate of progradation of the head of Gulf delta towards the central part of the Gulf.

The changes in sea level in the progression to the present level, since the beginning of the Flandrian transgression of about 20,000 to 17,000 years ago, have also left their clear marks as submarine platforms formed during standstills of rising sea level. Houbolt (1957) recognized four submarine terrace-like levels to the north and east of the Qatar peninsula, terming them: (1) the near-shore terrace, at a depth of 3–9 fathoms (5.5–16.5 m); (2) the first offshore terrace at a depth of 11–17 fathoms (20–31 m); (3) the second offshore terrace at a depth of 17–28 fathoms (31–51 m); and (4) the Central Persian Gulf, which is the deepest terrace at about 30–40 fathoms (55–73 m). All these terraces have slight dips towards the axis of the Gulf deeps and are separated from each other by clear zones of steep gradient towards the offset Gulf axis. According to Houbolt these four terraces or levels represent five stages of post-glacial rise of sea level. By comparing the value of the deepest terrace (40 fathoms or 73 m) with estimates which have been made of the amount of recession of sea level during the Pleistocene, he came to the conclusion that the area around the Qatar peninsula and most parts of the Arabian Gulf were above sea level during periods of maximum glaciation and assumed that these terraces were formed by eustatic changes of sea level during the post-glacial period. Radiocarbon dating of sabkha sediments in Abu Dhabi has revealed that the sabkha was formed during the past 7000 years when a marine transgression inundated the coastal area of Abu Dhabi, reworking the coastal sandstones, and reaching a level of about 1 m above present sea level some time before 4000 years BP (Evans *et al.* 1969). Taylor and Illing (1969) noted that there are old strandlines in Qatar, represented by cemented sands which stand at heights of 1.5 – 2.5 m above sea level. The ages of the dated samples are 3930 ± 130, 4200 ± 200 and 4340 ± 180 years BP. Felber *et al.* (1978) have confirmed marine transgression along the eastern coast of Saudi Arabia in the Middle Holocene. C_{14} dating of shells from a well-developed terrace near Jubail and near Ras Tannurah has given ages respectively of 1090 ± 80 and 3380 ± 180, 3990 ± 90 and 4670 ± 190 BP.

In a study of the late Pleistocene/early Holocene transgression in the Gulf presented by Sarnthein (1972), he maintained that during the last glacial period the Shatt al Arab reached the shelf margin in the Gulf of Oman which today stands at – 110 m, holding that throughout the Holocene the Gulf was not affected severely by tectonic movements. He supported this point of view by referring to several drowned aeolian dunes at different depths representing standstill transgressions. The surfaces are at – 64 to – 61 m, – 53 to – 40 m and about – 30 m. Kassler (1973) seems to agree with Sarnthein about the post-glacial submarine platforms which represent standstills during the rising of sea level. He identified six levels which are mainly found in the Arabian part of the Gulf:

(1) – 62 fathoms (– 120 m); (2) – 55 fathoms (– 100 m); (3) – 36 to – 44 fathoms (– 66 to – 80 m); (4) – 16 to – 20 fathoms (– 29 to – 37 m); (5) – 10 fathoms (– 18 m); (6) – 5 fathoms (– 9 m).

Kassler suggested also that both the north-east of the Gulf of Oman and the Arabian Gulf are tectonically stable today. Al-Asfour (1978) also recorded evidence of a higher sea level from the coast of Kuwait during the last 6000 years and has suggested a detailed correlation of late Pleistocene and Holocene sea level changes in the Arabian Gulf.

Evidence from a study of sea/land changes in Kuwait

Apart from the Wadi Al-Batin depression, the Jal-az-Zor escarpment and Ahmadi ridge, Kuwait generally has a gently graded, low-relief surface with altitudes ranging from sea level in the east to about 300 m in the south-west part of the country; northern and western parts of Kuwait are extensively covered by sand and gravels. The present work has been concerned with morphological evidence of climatic and sea/land changes north of Kuwait Bay in the areas of Kathma, Ghidhai, Mudairah and Al-Bahra (Fig. 10.2). The most prominent feature in the area is the Jal-az-Zor escarpment which extends for a distance of about 65 km, with a maximum height of about 145 m. The escarpment is dissected by several wadis which are most evident in the area between Kathma and Mudairah. Several terraces and terrace fragments covered by marine and terrestrial deposits can be observed.

The methods of investigation involved levelling, sediment analysis and radiocarbon dating. Levelling was carried out to determine the number and

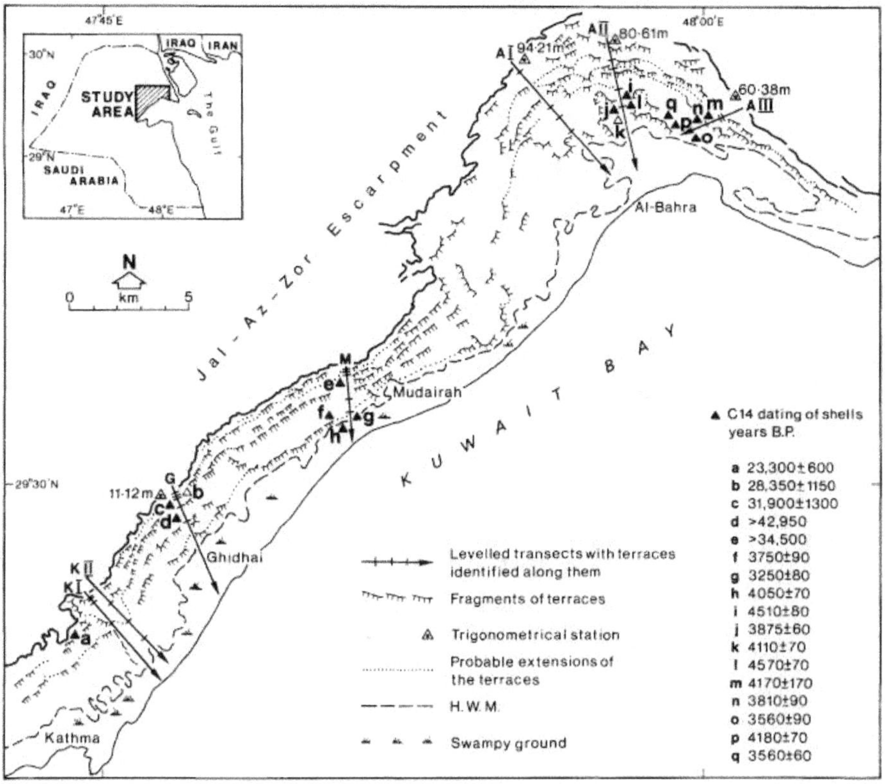

Fig. 10.2 Tentative genetic correlation of terraces north of Kuwait Bay.

height of the terraces in the area, to provide essential data for the understanding of the structural stability and instability of the research area. Table 10.1, which has been drawn from the result of the levelling, indicates several points. (1) The number of the levelled terraces is six and these are distributed in the area from Kathma to Al-Bahra. (2) The mean heights of the terraces range between a maximum of about 93 m a.s.l. recorded in Ghidhai and a minimum of about 6 m for the lowest terrace in Al-Bahra area. (3) The terraces do not follow constant heights, but gradually decrease in elevation from west to east. (4) Accompanied with the general eastwards decrease of the terrace elevations, there are indications supported by the differences in height that the area between Ghidhai and Al-Bahra has been exposed to some local structural instability.

The results of sediment analysis, which involved both particle size analysis as well as the interpretation of scanning electron microscopy, show that there are materials formed in three kinds of sedimentary environments prevailing in

Table 10.1 Tentative genetic correlation of the Kuwait terraces (all heights in metres above Kuwait datum)

	Location						
Terrace number	*Kathma I*	*Kathma II*	*Ghidhai*	*Mudairah**	*Al-Bahra I**	*Al-Bahra II*	*Al-Bahra III*
1	90.60	81.53	92.62	71.12	79.85		
2	77.92	69.65	79.27	59.57	56.10	63.72	33.11
3	67.38		65.00	52.30	40.39	47.82	18.58
4	50.76	40.55	52.94	31.23	31.79?	39.45	11.72
5		10.70?	33.60		22.32?	29.86	
6			13.00	11.61	16.10	14.43	5.99

* The distinction between Mudairah and Al-Bahra I is a separation of two main sub-areas (see text).

the area, these being fluvial, marine and aeolian. Although it is difficult to isolate each of the sedimentary environments because there is some mixing over the whole area, the results can be summarized as follows:

(1) The gravelly deposits in general are concentrated on the top of the escarpment but are also found on some parts of the first, second and third terraces.
(2) The materials of the marine environment, as well as being found on the present coastline where they have the characteristics of adjacent sediments to the north of the Gulf, have been found on the second, third, fourth, fifth and on the lowest sixth terrace in Al-Bahra area.
(3) The aeolian sedimentary environment is, in fact, not possible to isolate from the other two environments, as aeolian material is deposited continuously.

Radiocarbon dating technique was applied to shell samples, and gave two groups of ages. The first is for the samples collected from elevations about 40 m, and the results suggest Pleistocene age dating between >42,950 years BP and 23,300 ± 600 years BP, with no simple direct relationship between age and height; for example the highest dated sample which comes from Ghidhai is at an elevation of 63.3 m and has an age of 28,350 ± 1150 years BP. The second group of datings suggests that the lowest terraces in Mudairah and Al-Bahra are Holocene in age; the dates range between 4570 ± 70 and 3,250 ± 80 years BP. The mean height of these latter terraces varies from about 12 m in Mudairah, 14 m in the middle of Al-Bahra to about 6 m east of Al-Bahra. Figure 10.2, which is compiled from the combined results of the field survey and information constructed by examining aerial photographs as well as topographic maps, shows that throughout the area there is evidence of several breaks of slope representing remnants of terraces. It is also possible to hypothesize that the fragments of the lowest terraces observed between the transects of Al-Bahra and Mudairah are remnants of one continuous terrace which extends from Al-Bahra in the east at least as far as Mudairah. This may extend further west to Ghidhai (aerial photographic evidence) and even extend to Kathma, but this latter cannot be definitely proven.

Discussion

The problem of interpretation becomes complicated when both localized and regional crustal movements as well as eustatic changes of sea level have to be included in any multi-factor explanation of the development of the landscape and terraces north of Kuwait Bay. The limited amount of data available from the fieldwork and literature about Kuwait and adjacent areas indicate that: (1) The area north of Kuwait Bay is tectonically unstable and this has affected the general topography of the area. The Al-Bahra anticline has been associated with the Kuwait arch which extends to the south of Kuwait and is possibly still active. (2) A general dip in the elevation of the terraces towards the north-east has been noted above (Table 10.1), and this dip may accord with the general subsidence referred to by Lees and Falcon (1952). (3) Recent species of shells occur at higher levels than would be expected either in the absence of positive movement of land or of a general falling of sea level. It is clear that there are several levels of terraces extending along the foot of the escarpment which reflect changes of sea level during the Pleistocene period.

The combination of all the evidence suggests that the terraces were formed by two main agents of erosion and deposition, these agents being marine and freshwater action. The present dissected staircase topography, the deposits covering them and the results of C_{14} datings indicate that the terraces have been influenced by sea and land movements since at least the beginning of the Pleistocene. It is possible therefore to suggest that the terraces were first cut by marine action during high stages of sea level. As the sea dropped to a lower level, stream erosion would have been increased and streams would be deeply incised, with heavier rains washing down the fluvial deposits from the top of the escarpment to the lower levels. If the sea level rose again, stream erosion would decrease and streams would tend to aggrade their loads. These changes occurred through either eustatic drop of sea level or local and/or regional movements, or both. The probability that the upper terraces were partly formed by freshwater action is strengthened when one accepts all the evidence for past widespread fluvial action in the area. The alluvial agent of freshwater deposition no longer operates under present climatic conditions, but the northern and western parts of Kuwait are covered by fluviatile materials and are dissected by several drainage systems which originated during some of the pluvial periods of the Pleistocene. A major remnant of these river systems is the Wadi Al-Batin in the west of the country together with, possibly, the Wadi Al-Musannat in the south-west. Other morphological evidence (such as wadi drainage systems and fluvial gravels) of climatic changes in Arabia have been reported by Holm (1960), Glennie (1970) and Stevens (1978).

Although the height range of the Kuwait terraces (apart from the lowest terrace in Mudairah and Al-Bahra) in general fits with the Mediterranean sequence of Pleistocene high stages of sea level (Déperet 1918–22), correlation between them is not completely possible. Among the reasons for this are the height variations within the study area as well as the fact that the uniformity of the Mediterranean shorelines described by Déperet has been questioned by Baulig (1935) and by Castany and Ottmann (1957). The raised beach sequence of the Makran coast, 80–90, 60, 30 and 15 m above sea level (Falcon 1947; Butzer 1958), also agrees with the Kuwait terraces, but one is faced with the problem of structural instability in both areas. The fifth terrace traced in

Kathma, Ghidhai and Al-Bahra can be tentatively correlated with the evidence reported from Iraq (Voute 1957; Mitchell 1958; Larsen 1975). It also conforms with the Qatar raised beaches at +25 to 30 m reported by Kapel (1967) to have an age of >39,800 years. As far as the lowest terrace in Mudairah and Al-Bahra is concerned, the radiocarbon dating gave ages of Holocene 4570 ± 70 and 3560 ± 60 years BP, indicating a higher stage of sea level during the last 6000 years which may be correlated with the older Peron terrace of Australia (Fairbridge 1961).

Conclusion

The picture which appears from these studies is one of continued disagreement concerning the relative importance of the dominant processes affecting landform development in the Arabian Gulf. Nevertheless, it is possible to conclude that during different periods of the Pleistocene and Holocene the two main forces, eustatic sea level changes and structural instability (including isostatic changes), were variously balanced in different areas of the Gulf. One is therefore faced with a regionally varied and complex geomorphological history which can only be elucidated by further detailed studies. The author will shortly be publishing preliminary results of one such study of Bubiyan Island, Kuwait.

References

AL-ASFOUR, T. (1978) 'The marine terraces of the Bay of Kuwait', in Brice, W. C. (ed.) *The Environmental History of the Near and Middle East: Since the Last Ice Age*, London, Academic Press, 245–54.

BAULIG, H. (1935) *The Changing Sea Level*, Institute of British Geographers, London, No. 3.

BEKE, C. T. (1835) 'On the geological evidence of the advance of the land at the head of the Persian Gulf', *Philosophical Magazine*, 7, 40–6.

BLANDFORD, W. T. (1872) 'Note on Makran and Musandam on the East Coast of Arabia', *Records of the Geological Survey of India*, 5, 75–7.

BRICE, W. C. (ed.) (1978) *The Environmental History of the Near and Middle East: Since the Last Ice Age*, London, Academic Press.

BUTZER, K. W. (1958) 'Quaternary stratigraphy and climate in the Near East', *Bonner Geographische Abhandlungen*, 24, 1–57.

CASTANY, G. and OTTMANN, F. (1957) 'Le Quaternaire marine de la Méditerranée Occidentale', *Revue de Géographie Physique et Géologie Dynamique*, Ser. 2, 1, 46–55.

CHAPMAN, R. W. (1971) 'Climatic changes and the evolution of landforms in the Eastern Province of Saudi Arabia', *Bulletin of the Geological Society of America*, 82, 2713–28.

CORNELIUS, P. F. S., FALCON, N. L., SOUTH, D. and VITA FINZI, C. (1973) 'The

Musandam Expedition 1971–72, Scientific Results: Part I', *Geographical Journal*, 139, 400–25.

DE MORGAN, J. (1900) *Délégation en Perse, Mémoires*, Tome 1.

DEPERET, C. (1918–22) 'Essai de coordination chronologique des temps Quaternaires', *Compt. Rend. Acad. Sci. Paris*, 116:480, 636, 884, 118:868; 120:159; 121:212; 124:1502, 1595.

EMERY, K. O. (1956) 'Sediments and water of the Persian Gulf', *Bulletin of the American Association of Petroleum Geologists*, 46, 2354–83.

EVANS, G. (1970) 'Coastal and nearshore sedimentation: a comparison of classical and carbonate deposition', *Proceedings of the Geological Association*, London, 81, 493–508.

EVANS, G., KENDAL, C. G. St. C. and SKIPWITH, Sir PATRICK, A. D. E. (1964) 'Origin of the coastal flats, the sabkha of the Trucial Coast, Persian Gulf', *Nature*, 202, No. 4934, 579–600.

EVANS, G., SCHMIDT, V., BUSH, P. and NELSON, H. (1969) 'Stratigraphy and geologic history of the sabkha, Abu Dhabi, Persian Gulf', *Sedimentology*, 12, 145–59.

FAIRBRIDGE, R. W. (1961) 'Eustatic changes in sea level', in Ahrens, L. H. *et al.* (eds) *Physics and Chemistry of the Earth*, vol. 4, Oxford, Pergamon Press, 99–185.

FALCON, N. L. (1947) 'Raised beaches and terraces of the Iranian Makran coast', *Geographical Journal*, 109, 149–51.

—— (1973) 'The Musandam (Northern Oman) Expedition 1971/72', *Geographical Journal*, 139, 1–19.

—— (1975) 'From Musandam to the Iranian Makran', *Geographical Journal*, 141, 55–8.

FELBER, H., HOTZL, H., MAURIN, V., MOSER, H., RAUERT, W. and ZOTL, J. G. (1978) 'Sea level fluctuations during the Quarternary period', in Saad, S., Al-Sayari and Zotl, J. G. (eds) *Quaternary Period in Saudi Arabia*, Springer-Verlag, 50–7.

FUCHS, W., GATTINGER, T. E. and HOLZER, H. F. (1968) *Explanatory Text to the Synoptic Geologic Map of Kuwait*, Geologic Survey of Austria.

GLENNIE, K. W. (1970) *Desert Sedimentary Environment*, Amsterdam, Elsevier.

HARRISON, J. V. (1941) 'Coastal Makran', *Geographical Journal*, 97, 1–17.

HENSON, F. R. S. (1951) 'Observation of the geology and petroleum occurrences of the Middle East', *Third World Petroleum Congress*, The Hague, Sec. 1, 118–40.

HOLM, D. A. (1960) 'Desert geomorphology in the Arabian peninsula', *Science*, 132, 1369–79.

HOLMES, A. (1944) *Principles of Physical Geology*, London, Nelson.

HOUBOLT, J. J. H. C. (1957) *Surface Sediments of the Persian Gulf near the Qatar Peninsula*, The Hague, Mouton & Co.

KAPEL, H. (1967) *Atlas of the Stone-Age Cultures of Qatar*, Denmark, Aahrus University Press.

KASSLER, P. (1973) 'The structural and geomorphic evolution of the Persian Gulf', in Purser, B. H. (ed.) op. cit., 11–32.

KENDALL, C. G. St. C., and SKIPWITH, Sir PATRICK, A. D. E (1969) 'Holocene shallow water carbonate and evaporate sediments of Khor al Basam, Abu Dhabi, Southwest Persian Gulf', *Bulletin of the American Association of Petroleum Geologists*, 53, 841–69.

LARSEN, C. E. (1975) 'The Mesopotamian delta region: a reconsideration of Lees and Falcon', *Journal of the American Oriental Society*, 95, 43–57.

LARSEN, C. E. and EVANS, G. (1978) 'The Holocene geological history of the Tigris–Euphrates–Karun delta', in Brice, W. C. (ed.) op. cit., 227–44.

LEES, G. M. and FALCON, N. L. (1952) 'The geographical history of the Mesopotamian plains', *Geographical Journal*, 118, 24–39.

LEES, G. M. and RICHARDSON, F. D. S. (1940) 'The geology of the oil-field belt of South-west Iran and Iraq', *Geological Magazine*, 78, 227–52.

LOFTUS, W. K. (1857) *Travel and Research in Chaldaea and Susiana*, London.

MELGUEN, M. (1973) 'Correspondence analysis for recognition of facies in homogeneous sediments off an Iranian river mouth', in Purser, B. H. (ed.) op. cit., 99–121.

MINA, P., RAZAGHNIA, M. T. and PARAN, Y. (1967) 'Geological and Geophysical Studies and Exploratory Drilling of the Iranian Continental Shelf – Persian Gulf', *Seventh World Petroleum Congress*, 872–901.

MITCHELL, R. C. (1958) 'Instability of the Mesopotamian plains', *Bulletin de la Société de Géographie d'Egypte*, 31, 127–40.

PILGRIM, H. G. E. (1908) 'The geology of the Persian Gulf and the adjoining portions of Persia and Arabia', *Memoirs of the Geological Survey of India*, 34 (4), 1–177.

PILKEY, O. H. and NOBLE, D. (1966) 'Carbonate and clay minerology of the Persian Gulf', *Deep Sea Research*, 13, 1–16.

POWERS, R. W., RAMIREZ, L. F., REDMOND, C. D. and ELBERG, E. L. Jr. (1966) *Geology of the Arabian Peninsula: Sedimentary Geology of Saudi Arabia*, US Geological Survey, Professional Papers, 560–D.

PURSER, B. H. (ed.) (1973) *The Persian Gulf: Holocene Carbonate Sedimentation and Diagenesis in a Shallow Epicontinental Sea*, Springer-Verlag.

PURSER, B. H. and SEIBOLD, E. (1973) 'The principal environmental factors influencing Holocene sedimentation and diagenesis in the Persian Gulf', in Purser, B. H. (ed.) op. cit., 1–9.

RAWLINSON, H. C. (1857) 'Notes on the ancient geography of Mohamrah and the Vicinity', *Journal of the Royal Geographical Society*, 27, 185–90.

ROBINSON, G. P. G. and MAY, H. P. (1974) 'The Musandam Expedition: Scientific Results, Part II', *Geographical Journal*, 140, 94–104.

SALEH, A. A. AL-M. (1975) 'Pleistocene and Holocene Oolitic sediments in the Al-Khiran Area, Kuwait', unpublished MSc thesis, Univ. Kuwait, Kuwait.

SARNTHEIN, M. (1972) 'Sediments and history of the postglacial transgression in the Persian Gulf and north-west Gulf of Oman', *Marine Geology*, 12, 245–66.

SHINN, Z. A. (1973) 'Carbonate coastal accretion in an area of longshore transport, N. E. Qatar, Persian Gulf', in Purser, B. H. (ed.) op. cit., 179–91.

STEVENS, J. H. (1978) 'Post-pluvial changes in the soils of the Arabian Peninsula', in Brice, W. C. (ed.) op. cit., 263–74.

TAYLOR, J. C. M. and ILLING, L. V. (1969) 'Holocene intertidal, calcium carbonate-cementation, Qatar, Persian Gulf', *Sedimentology*, 12, 69–107.

VITA FINZI, C. (1969) 'Late Quaternary alluvial chronology of Iran', *Geographische Rundschau*, 58, 951–73.

—— (1973) 'Late Quaternary subsidence', in Cornelius *et al.*, op. cit., 414–21.

—— (1975) 'Quaternary deposits in the Iranian Makran', *Geographical Journal*, 141, 415–20.

—— (1978) 'Recent alluvial history in the catchment of the Arabo-Persian Gulf', in Brice, W. C. (ed.) op. cit., 255–61.

VOUTE, C. (1957) 'A prehistoric find near Razzaza (Karbala Liwa): its significance for the morphological and geological history of the Abu Dibbis depression and surrounding area', *Sumer*, 33, 1–14.

WOOLLEY, Sir C. L. (1938) *Ur of the Chaldees*, Harmondsworth, Pelican Books.

CHAPTER ELEVEN

Fisheries of the Arabian peninsula

WILLIAM J. DONALDSON

> It might well be that for each expert sent abroad to modernize and industrialize there should be at least one *anti-expert* seriously conveying concepts of de-industrialization and de-modernization – to make for some kind of balance.
>
> (Galtung 1977: 20)

It is by no means remarkable that the inhabitants of a seaboard should eat fish. For the author of the *Periplus of the Erythraean Sea* (Schoff 1912) to call the coastal dwellers of the Hijaz and Yemen specifically Ichthyophagi, the Fish-Eaters, in the first century AD suggests, therefore, that fish and fishing formed an outstandingly important aspect of their life and livelihood. The coastal communities of the lower Arabian Gulf and the Makran, too, were called by the same name by classical authors (Zwemer 1902; Brunt 1976). Marco Polo (Latham 1958) and later Niebuhr (Heron 1792) were also impressed by the importance of fish to the inhabitants of the Hadhramaut and Gulf of Oman.

With the nineteenth century, European travellers to Arabia became more frequent and often the records of their journeys contain references to fishing. These are summed up substantially by Lorimer (1908–15) and Miles (1919), and together with later reports and studies, e.g. Bertram (1948), they indicate that fishing in the Arabian peninsula was not only important but had changed remarkably little from classical times well into the present century. Fishing units were small family concerns. Equipment consisted of hand-spun cotton nets and lines, date-frond traps, and ropes hand-made from date or coconut fibre. Except for the larger beach seine craft, fishing boats were small, holding up to five or six men only. Power was invariably by sail, oar or paddle. Fishing communities, permanent or seasonal, were found on most stretches of the coast of the peninsula, though by far the greatest concentrations seem always to have been on the southern and south-eastern coasts.

This pattern has formed the basis of Arabian fisheries – the traditional, or artisanal, fisheries – for many centuries. Indeed, with important but recent innovations over the last three decades, including motor power, larger craft and nylon nets and ropes, these artisanal fisheries were the only form of fishery in the peninsula until the 1950s.

The industrial sector

Industrial, or commercial, fishing first made its appearance in the peninsula on the Arabian Gulf side in the late 1950s. Since 1959 Bahrain and Kuwait have had large shrimp-fishing companies which catch and pack shrimps by modern commercial processes. As the 1960s progressed, Gulf shrimp fishing grew and companies were set up in the eastern province of Saudi Arabia and in Qatar, while Iranian, Soviet and, from 1967, Iraqi commercial shrimping vessels were also fishing in Gulf waters. Kuwait, however, with its four shrimping companies retained the largest part of the total fleet, with seventy of the total of about 130 vessels in 1967 (Boerema and Job 1968).

More recently in 1976, the Ras al Khaimah government and a Norwegian fishing company have set up a large fish-meal project, the first of three planned for the United Arab Emirates.

In addition, several countries of the region have from the 1960s entered into concession agreements with second parties for the catching of finfish. These include South Yemen (with Japan, USSR and, from November 1977, Iraq), and more recently Oman (with Japan and subsequently, from early 1978, Korea). Generally such agreements have included the delivery of part of the catch of finfish to the grantor country to supplement its supply for home consumption. Capital-intensive shore preservation and marketing facilities have in such cases usually been installed in the grantor country to cope with this supply.

Such projects as these differ radically from the artisanal fisheries. First, their technology is of a much more advanced order, their capital-to-labour ratio is high, and the basic unit tends to be a large public or private company rather than a family. Secondly, the primary object of almost all the large-scale commercial enterprises so far set up in the Arabian peninsula has been to produce a refined product for export, and any supply of finfish for home consumption has been a by-product. With the artisanal fisheries on the other hand local consumption has always been the primary aim, although where production has been consistently in excess of local needs export of cured products has resulted, notably from South Yemen and Oman.

Experience has already shown, however, that to set up and operate successfully an industrial fisheries enterprise in the Arabian peninsula is a more difficult task than might at first sight be supposed. The difficulties that have beset the shrimp fishing industry in the 1970s are illustrative of this.

From 1959 the commercial shrimp catch rose steadily to a peak in the 1968–9 season, estimated by Ellis (1975) at 16,500 tonnes live weight. At the same time the shrimping fleet had risen to around 130 vessels in 1967–8. The 1969–70 season unfortunately saw a large drop in the catch and the 1968–9 peak has never been approached since. Nevertheless, expansion of the fleet continued for several years, though by 1972 it was clear to the largest companies that they had over-extended. The three largest Kuwaiti concerns amalgamated to form a single public company, the United Fisheries Company (UFC), with 47 per cent government participation, and reduced considerably the size of their fleet. Even so, in 1977 manpower difficulties meant that only one-third of the reduced UFC fleet was active at any time, while the catch of the company was reportedly only some 400 tonnes compared with several thousand in the boom period.

The precise reasons for the dramatic decline in shrimp catches after 1969 are not entirely clear. Although harbour development and land reclamation in Kuwait, Bahrain and elsewhere may locally have been partly responsible by destroying spawning grounds, Ellis (1975) is firmly of the opinion that considerable over-exploitation, especially in Kuwaiti and Saudi waters, was the main reason, and that a decrease in the fishing effort might well have produced a larger total catch. Whatever the causes, the Gulf commercial shrimping industry underwent spectacular growth in the 1960s, and since 1970 has suffered as spectacular a decline. Retrenchment and wastage of capital investment has been the result.

In the Red Sea too, commercial fishing projects have met with difficulties and indeed none has so far proved to be a successful long-term concern. A Saudi fishing company set up in the late 1960s to catch finfish with trawls and lines is now defunct, presumably because of insufficiently large catches. Both Japanese and Korean commercial fishing projects in the area of the Farasan Banks, thought to be the most promising part of Saudi Red Sea waters, have similarly failed.

In Omani waters the catches of the Japanese commercial concession agreement which took effect from April 1975 proved disappointingly small, and the contract was terminated in late 1977 and replaced with one involving the Koreans. Initial indications are that the Koreans, too, are not meeting their targets.

With such concession agreements any risk there may be in the venture lies squarely on the shoulders of the grantee company. On the other hand, in joint projects such as the UAE fish-meal ventures the government is heavily committed. In June 1977 the Ras al Khaimah fish-meal project was opened as a joint venture with a Norwegian company. White and Barwani (1970) had predicted that a catch per season of at least 180,000 tons of sardines alone was possible from the lower Gulf waters in addition to large quantities of anchovies and other fish, and had concluded that a fish-meal industry was feasible. In this light, the gearing of the Ras al Khaimah project to an annual catch of 125,000 tonnes live weight from seven vessels sounds reasonable. However, the basis of White and Barwani's estimates appears never to have been explained or tested in detail. By late 1977 it had become obvious that their estimates were much too optimistic and that the actual catch rate of the scheme in the first year of operation would be only about one-third of the desired capacity. As a result, there have been rumours that the Ras al Khaimah plant will be forced to close.

Before the Ras al Khaimah problems became clear two other emirates of the UAE had already embarked on plans to set up even more ambitious joint fish-meal projects. Ajman was assured by a Peruvian fisheries corporation that stocks in the lower Gulf could support a fish-meal factory served by twelve craft with an annual capacity of an astounding 400,000 tonnes live weight, and construction is under way at the time of writing. The third such project in the UAE is planned for Kalba on the Gulf of Oman coast of Sharjah. On the experience of the Ras al Khaimah venture, it would now seem very unlikely that the Ajman and Kalba projects will ever be completed, still less fully productive.

At first sight, the lesson to be learned from past difficulties and failures of industrial fisheries ventures is that more intensive research is needed into fish stocks, movements and spawning habits to arrive at a realistic assessment of what are maximum sustainable yields. More information is certainly desirable, and such is the nature of fisheries that there will always remain yet more to be discovered, measured and assessed. Indeed, scarcely has a report been written or a conference convened over the last thirty years on the fisheries of the peninsula which has not called for more data.

On the other hand, thanks to the efforts of international research projects, notably the International Indian Ocean Expedition and the programmes of the Indian Ocean Fisheries Commission, as well as numerous national and independent surveys, very much more is known now of the potential of the fisheries of the peninsula than ever before, and arguably enough is known to enable effective, decisive action to be taken to develop the region's fisheries. The assumption that yet more information will solve commercial fisheries' problems is by no means certain. The key to a successful fisheries policy, at least in the short to medium term, may well lie not with the commercial sector, but elsewhere.

A re-examination of problems and potentialities

Despite the efforts of the various industrial fisheries ventures, it is the artisanal fisheries which still provide most, by far, of the fish consumed in the countries of the peninsula, not only in rural areas but in all the large coastal towns as well. In view of the difficulties encountered by industrial projects, this is likely to remain the case for at least the next decade and probably in fact for much longer.

Further, there are in the peninsula as a whole considerably more fishermen active in the artisanal fisheries than there ever have been or ever will be in the industrial sector.

Moreover, there is an increasing volume of evidence from other parts of the developing world that, even where an industrial fisheries project does succeed in increasing production, the lot of most of the population including the artisanal fishermen is not necessarily thereby improved, *even when this is the primary aim of the project.* (See for example Galtung's (?1977) convincing indictment of the Indo-Norwegian project in Kerala, India.)

These three observations can perhaps be taken as a starting point for re-examining the problems and potentialities of fisheries in the peninsula. The problems first of all can be interpreted in relation to three separate areal distributions: the fishery resource, the fishing population and the demand for fish. If all three distributions coincided there would be less room for difficulty. The fact that they do not coincide can be interpreted as the root of the problem.

Considering first the fisheries resource, there is general agreement that the richest areas are the south-eastern coasts of the Arabian peninsula where the western Arabian Sea upwelling occurs, and specifically the coasts of the Hadhramaut, Dhufar and south-east Oman. Lesser but still fruitful grounds are the Gulf of Oman and the southern third of the Red Sea. The rest of the Red Sea and the Arabian Gulf are considered to have a much lower production potential.

The distribution of the active fishing population in the recent past to a large extent mirrored the availability of fish. That is to say, the largest concentrations of fishermen have traditionally been on the coasts of the Yemen, the Hadhramaut, Dhufar and the Batina coast of Oman. Fishing communities have without doubt existed on the Hijaz and Gulf coasts since prehistoric times (see, for example, De Cardi (1976) for evidence from the Gulf), but it is unlikely that they have ever been as important as on the southern and south-eastern coasts of the peninsula either in absolute numbers or as a proportion of the population.

The areal distribution of the demand for fish on the other hand shows a completely different picture. It is firstly much more nodal than the artisanal supply, being concentrated in urban centres. Secondly, it is greatest in the urban centres of the Arabian Gulf and the Hijaz, the very areas where potential supply is least. The net effect therefore is that the richest areas are under-utilized, while the poorer areas run the risk of being overfished.

These circumstances have given rise to two main results, which though interrelated are too often viewed independently as problems in their own right. First of all, demand in the urban centres of the Gulf and the Hijaz has for long outstripped supply. Secondly, there has occurred over the last three decades or more a pendular migration of fishermen from the Yemen to the Hijaz coast and from Oman to the Gulf.

Since fish supply to the main urban centres of the Gulf and western Saudi Arabia has for long been insufficient to meet demand, prices there have tended to be relatively high. Several governmental and municipal authorities in the region have tried to ameliorate the situation by attempting to increase or regulate the fish supply to the markets or control artificially the price of fish, but everywhere have either failed completely or been only marginally successful. In Bahrain, for example, the government attempts to supplement the artisanal supply by the provision of deep-frozen fish at low prices from the state-owned Bahrain Fisheries Company. In Kuwait too commercially caught frozen fish, in this case imported from Pakistan and since 1978 Oman, is supplied and sells at lower prices. In both cases, however, consumer resistance to frozen fish in the first place and the fact that the supply is only a small fraction of the total demand make these measures only very marginally effective. Kuwait, again, and Abu Dhabi have attempted in recent years to control prices directly by imposing legal maxima. In Kuwait, although nominal control has applied to retail prices since 1974, it has never been adhered to since wholesale auction prices cannot be controlled in the same way. In Abu Dhabi the control applied to the prices which could be paid to fishermen, but had to be abandoned after only a few weeks since the fishermen very understandably stopped delivering fish to Abu Dhabi and unloaded instead at Dubai and other markets where the control did not apply.

This last example leads on to the second effect of the distribution of resources, manpower and demand, namely the mobility of the fishermen within the region. Tempted by much higher earnings per unit effort than they can earn in their home settlements, fishermen of the Yemen and Oman have for several decades been attracted respectively to the Hijaz and Gulf coasts for periodic or seasonal fishing. Recent fieldwork by the present author indicates that about one-third of the 3000 adult fishermen of the Batina coast of Oman go to fish seasonally in the Gulf states, especially Bahrain.

This has been accompanied in the Gulf states and in the urban centres of the Hijaz coast by a relentless move of the nationals away from fishing to more profitable and less strenuous jobs on land provided directly or indirectly by the oil industry. The result has been that the supplying of fresh fish to these countries has over the last twenty years passed almost wholly into the hands of foreign nationals. Thus, in Kuwait in 1967 the artisanal fishermen were 60 per cent Iranian and 40 per cent Omani (Boerema and Job 1968). Today almost the whole fishing workforce is Iranian, even though some of the vessels remain Kuwaiti. In Bahrain, Qatar, Abu Dhabi and Dubai the fishing force is almost wholly from the northern states of the UAE, from the Sultanate of Oman and to a lesser extent from Iran, while in Jeddah and other towns on the Saudi Red Sea coast the active full-time fishermen are North Yemenis almost to a man. Most nationals of these oil states who do still retain a link with fishing do so only as a part-time or leisure activity, or as non-seagoing owners of fishing vessels with foreign crews.

In these circumstances the collection of data by these countries on their fisheries, and still more the rational control and development of them, is extremely difficult and demands a degree of regional co-operation not so far seen.

On the other hand, a re-examination of the artisanal fisheries also brings to light potentialities which are more hopeful than the above discussion of problems suggests could be the case, but which have not been stressed sufficiently in recent writings on Arabian fisheries.

Firstly, the coasts of the Hadhramaut and Oman have for centuries caught and cured enormous quantities of fish of several kinds, but in particular sardines, anchovies, sharks and various tuna and tuna-like fish. Until the last couple of decades, the large quantities surplus to local requirements formed the basis of an important export trade to India and East Africa. The trade has declined for a number of reasons, among them the increasing demand for higher-quality products than were capable of being produced on the coasts of southern Arabia and Oman at the time. However, what the formerly large export trade does indicate is that an artisanal fishery (which was entirely non-motorized) is capable of producing very large quantities of fish given conducive demand conditions and a plentiful fish resource. Further, it has already been noted above that it is indeed the artisanal fishery still which supplies the large bulk of the fresh fish consumed in the peninsula, though the best grounds are under-utilized.

Secondly, the Yemens, Oman and the northern UAE still have sizeable native fishing populations though many work for much or most of the time elsewhere, as has been seen. Accurate statistics are unavailable for all the countries concerned (where figures do exist they are usually inflated), but the number of fishermen native to these four countries at present may be placed at 18,000 to 20,000. In absolute terms this may appear small, but as a proportion of the non-urban coastal population it is substantial. Moreover, recent research of the present writer in Oman and the northern UAE tends to indicate that the artisanal fishermen form strong socially cohesive communities.

Both economically and socially, therefore, the artisanal fishermen form an important element in the countries concerned, though their potential is at present only partly realized.

The third encouraging aspect of the artisanal fisheries, which is also perhaps the most immediately important, is that as artisanal fisheries they are generally well equipped. A high percentage – 80 per cent or more – of the fishing vessels of all sizes, from the 25 m *sambūks* of the Red Sea and upper Gulf to the ubiquitous dugout *houri*, are motorized, while nylon netting and ropes and galvanized wire for traps are now universal. With more powerful diesel engines, larger fishing craft have been locally developed and, in the Red Sea and upper Gulf at least, the sterns of the larger vessels have begun to be adapted so as to accommodate more easily purse seine and trawl gear.

Grant-loan schemes for equipment have been operating since 1971 in the UAE and since 1974 in Saudi Arabia. It was, however, wholly on the initiative of the fishermen themselves that all the above innovations were introduced in the first place, and their spread throughout the peninsula has been almost wholly at their own expense.

In short, the frequent criticisms levelled at the artisanal fishermen of the peninsula and their equipment and methods, describing them as 'inefficient' and 'relatively unproductive' (as for example in *Middle East Economic Digest*, 23 March 1978), indicate an ignorance of the true situation. Compared with most of the industrial fisheries enterprises which have become mere costly failures, the artisanal fishermen have been, are now and are likely to continue to be both highly productive and very efficient in their use of capital and labour. In contrast to the industrial fisheries, the artisanal fishermen are flexible and able and willing to adopt and adapt innovations which are appropriate for their physical and social environment.

If increased production and improved living standards are to be the main objectives of fisheries development policy in the peninsula, then it would appear to the present writer that there is more hope of success in attaining them (and at far less financial cost) by assisting the artisanal fisheries and by creating conditions in which they can flourish, rather than by devoting nearly all the available attention and resources to commercial ventures. As with most complex problems, it is true that the most appropriate solution will probably not be produced by an 'either–or' answer. In future years commercial fisheries may well grow out of the present artisanal framework as a gradual and logical development. Indeed, large-scale industrial-type ventures may well prove eventually to be the only suitable way of exploiting to the full the undoubtedly large potential of the south-eastern coasts. However, a modern, highly technological industrial fishery cannot be bought ready made and self-operating off the shelf. The artisanal fishermen, on the other hand, already form a ready-made fishery and have proved themselves a viable proposition.

Note

This article is based on field data collected since 1974 in the Sultanate of Oman and other countries of the Arabian peninsula as part of the University of Durham Oman Research Project. The author wishes to acknowledge the Centre for Middle Eastern and Islamic Studies of the University of Durham for a travel grant which enabled him to revisit the peninsula in December 1977–January 1978. The opinions expressed are however the responsibility of the author.

Postscript

Since this article was written, the Ras al Khaimah fish-meal plant has closed down (1978) and both the Bahrain Fishing Company and the Qatar National Fishing Company have met with financial difficulties (1979).

References

BERTRAM, G. C. L. (1948) *The Fisheries of the Sultanate of Muscat and Oman*, Report to the Sultan of Muscat and Oman.

BOEREMA, L. K. and JOB, T. J. (1968) 'Report on a visit to Iran, Iraq, Kuwait, Saudi Arabia, Bahrain, Qatar and the Trucial States in order to collect fisheries data relevant to assessment of the fisheries resources in the Gulf', *Fisheries Travel Report and Aide Memoire No. 130*, Rome, FAO Department of Fisheries.

BRUNT, P. A. (1976) (trans. and ed.) Arrian's *Indica*, Loeb Classical Library, Cambridge, Mass., Harvard University Press; and London, Heinemann.

DE CARDI, B. (1976) 'Ras al Khaimah: further archaeological discoveries', *Antiquity*, 50, 216–22.

ELLIS, R. W. (1975) 'An analysis of the state of the shrimp stocks in the Gulf between Iran and the Arabian Peninsula', *Indian Ocean Fisheries Commission, Fourth Session, Mombassa, Kenya*, 21–5 July 1975, Rome, FAO.

GALTUNG, J. (?1977) *Development from Above and the Blue Revolution: the Indo-Norwegian Project in Kerala*, Oslo, University of Oslo, International Peace Research Institute Papers, No. 52.

HERON, R. (1792) (trans. and ed.) *Travels through Arabia and Other Countries in the East, by C. Niebuhr*, 2 vols, Edinburgh; reprinted (n.d.) Beirut, Librairie du Liban.

LATHAM, R. (1958) (trans. and ed.) *The Travels of Marco Polo*, Harmondsworth, Penguin.

LORIMER, J. G. (1908–15) *Gazetteer of the Persian Gulf, Oman and Central Arabia*, Calcutta, Superintendent, Government Printing Office. Reprinted 1970, Farnborough, Gregg International Publishers Ltd.

MILES, S. B. (1919) *The Countries and Tribes of the Persian Gulf*, 2nd edn, 1966, London, Cass.

SCHOFF, W. H. (1912) (trans. and ed.) *The Periplus of the Erythraean Sea*, New York, Longman, Green & Co.

WHITE, A. W. and BARWANI, M. A. (1970) *A Survey of the Trucial States Fishery Resource with Reference to the Sultanate of Oman*, Dubai, Trucial States Council.

ZWEMER, S. M. (1902) 'Three journeys in Northern Oman', *Geographical Journal*, 19, 54–64.

CHAPTER TWELVE

A rural community development project in Oman

RODERIC W. DUTTON

This paper attempts a definition of rural community development and assesses the changing need for it in Oman; continues with an analysis of the approach made by the Durham University Khabura Development Project to help meet this need; tables the project's specific objectives; examines the programme so far; and closes with an optimistic look into the future.

A definition of terms

A sanguine approach to establishing a development project might be to ask the people concerned what they want and then create a system which allows them to get it. For many, in the short term this would be called progress, but others would look ahead, taking a longer-term view, and think of what the community needed in order to supply tomorrow's wants. Perhaps, then, a developed community is one where an equilibrium has been established between wants and needs. Such an equilibrium could of course be set at any one of the many levels, depending on the aspirations of the members of the community. Thus the bedouin tribal society of southern Arabia in the late 1940s, as described by Thesiger (1977), could be termed as fully 'developed' as the great industrial societies of western Europe.

A society has the right to attempt to satisfy its wants but also has an obligation to cater for the needs of future years and future generations. Finding and maintaining the right balance between these rights and obligations in turn implies control. Currie (1978) maintains that a key to development is 'better control over or dominance of the physical, social, political, economic and demographic environment', but it is control as a means of achieving the equilibrium that is important; control by itself can be negative and destructive. The right control in any given traditional society is usually attained by empirical response 'refined and sharpened by countless years of experience' (Schultz 1964). Control can be maintained by self-subjugation to a given code of conduct, by the direction and organization of a responsive leadership, or by a general realization that to lose control would bring disaster. Examples of all three types could be found in Oman. The main problem with the first two types is that they can become inflexible in adjusting the equilibrium to meet changing circumstances; it is of the essence that the equilibrium should be dynamic and not static. The main problem with the last form of control is that a sudden removal of the constraint can result in complete loss of equilibrium and a painful process of readjustment. An example of the latter is the introduction of pumpwells to villages whose water supply had been entirely dependent on *aflaj*, the complex system of underground gallery water extraction, transportation and distribution, known as qanat in Iran (Wilkinson 1977). The *falaj* system, in which it was accepted that the 'principle of sharing is of greater practical value to everyone than the idea of competitiveness' (Bowen-Jones 1975) was faced with the prospect of hydrological, economic and social decay once individualism was made possible by the importation of pumpwell equipment (University of Durham 1978c).

This approach to a definition of development perhaps underlines a weakness in that argument which states that 'above all' of the human factors essential for development is 'their desire for national betterment and their willingness to make necessary sacrifices to attain it' (Mountjoy 1971). It is all too easy for this latter approach to cater for 'material betterment' wants now while the 'necessary sacrifices' are in fact the future needs.

If we extrapolate from this argument we can then premise that a 'community' can most easily be defined as a human aggregation which has developed into something more. Community development is, therefore, not development of a community but development to create a community. The main features of a true community are: that its members accept that they are mutually dependent, have obligations to the future and act accordingly; that they perceive that in order to remain a community they must live in harmony with their human and physical environment and respond to changes in that environment; and that they perform complementary activities. An indigenous rural community lacking extensive contacts with the outside world will be more or less self-reliant. Material requirements will be derived from local natural resources and processed by local artisans for local usage. The community will then have independence as one of its valuable assets.

With the world getting ever smaller, and because of the enormous economic and social divergence of different world cultures that can now be thrown together, the greatest danger to an indigenous community lies in being overwhelmed by change from without. The consequences of the discovery and exportation of oil were potentially overwhelming to the rural communities of a poor and traditional country such as Oman during the 1960s when oil exploitation began.

The present need for rural community development

The regional growth of the oil industry, first in neighbouring countries and finally in Oman from 1967, entailed massive urban expansion. Rural Oman was affected by these developments primarily as a source of relatively plentiful and cheap labour. By the early 1970s, when Oman's own oil industry was rapidly expanding and neighbouring Abu Dhabi had become one of the world's major oil-exporting countries, the impact of the oil boom on the life of rural Oman was becoming apparent even to the casual observer. At the same time the new Omani government, under Sultan Qaboos, realized the need not only to aid rural Oman but for a much more detailed understanding of what was happening. It was at this time that Professor Bowen-Jones of the University of Durham was asked by Shell International to design and direct an interdisciplinary survey programme examining a cross-section of rural Oman from Khabura on the Batina coast to and beyond Ibri in the interior Dhahira.

These surveys, done in very considerable depth between 1972 and 1976, revealed *inter alia* the extent to which the traditional rural communities were affected by the impact of oil. Increasingly since the 1950s the young and active Omani men had been able to find employment in the Gulf states (Birks and Sinclair 1978). Since the late 1960s Omani oil has financed comparable jobs within Oman. Very few of these jobs have been created in rural Oman and equally few, in Oman or abroad, have built on the artisanal inheritance of the men involved; skilled and experienced rural producers have been turned into a raw and largely unskilled labour force. Most of the active men have left their villages though usually leaving their families behind and returning for holidays. The men left behind are usually the less enterprising, but because of scarcity of manpower at home and financial support from their relatives abroad, they can successfully demand a greater income for less productive work and therefore further diminish wealth creation. A much higher proportion of total village income originates in salaries earned outside the villages and is spent on goods produced overseas; rural Oman was and is in danger of becoming economically irrelevant. One new import, the pumpset, has been mentioned. Other imported goods, such as plastic sandals, woollen rugs, cement for concrete blocks, timber, plywood, brushes, baskets, rope, beakers, water pots and storage jars all replaced goods that previously had been produced at home. This had crushed Omani craft industries and also, in the case of crafts such as rope and mat making, reduced the opportunities for the old and disabled to be active and self-respecting members of the village.

Meanwhile the government increasingly takes responsibility for a wide range of village activities. In many ways this is beneficial and necessary – e.g. the establishment of schools, extension centres, hospitals and clinics – but it has three weaknesses from the point of view of development as above defined. First, national organization can be too complex and remote for it to be fully responsive to local requirements – this is particularly true of teaching skills. Secondly, it diminishes people's local sense of responsibility for their own future – this can be true of heavily subsidized services such as tractor cultivation, subsidized foodstuffs and even some price controls. Thirdly, it diminishes the self-reliance of the rural community by weakening that measure of control over the development equilibrium which is due to fear of disaster.

In brief, therefore, the University of Durham surveys revealed that the indigenous rural communities were in disarray: gone was the feeling of mutual dependence; gone was the sense of obligation to the future; gone was the concept of living in harmony with the environment; and going was the independence and the skilled use of local resources that was associated with it. Not only was the symbiosis with the environment weakened but the idea that equilibrium control was necessary was tacitly challenged in a period when everything that anybody 'wanted' arrived on an apparently endless stream of oil wealth.

Does this disarray of the rural communities matter? The answer to this question is yes, for two reasons. First, the stream of oil wealth is not endless. Oman itself is not a particularly oil-rich country and even the wealthiest of its immediate neighbours will only be rich in oil for a few more decades. Therefore Omanis must look to a future when they are again economically dependent on the produce of the land and the sea and on their own skills. In particular, the skills needed for work in their homeland must be developed, not allowed to disappear.

Secondly, as rural Oman appears more irrelevant to the national economy the social and spiritual life inextricably associated with rural economies also breaks down. The consequences of this are multiple: one of the worst, rural – urban drift, is predictable because it has characterized most Middle-Eastern countries undergoing rapid change during this century (Clarke and Fisher 1972). In most cases this population movement is associated with a wide range of undesirable consequences as well as some undoubted benefits.

There has appeared therefore an undoubted need for rural community development in Oman and it is in this context that the Durham University Khabura Development Project grew out of the 1972–6 survey programme.

The Durham University Khabura Development Project

The approach

In 1974 Bowen-Jones was invited by Petroleum Development (Oman) Ltd (PDO) to design and direct a rural village community development project (Fig. 12.1), this to be based on the findings of the Durham survey programme. It is the experience gained in this project, established at Khabura, and under the direction of the author since 1979, which is examined below.

The fundamental concept behind the project was expressed thus: 'the general objective is to achieve a self-energizing movement towards higher living standards' (Bowen-Jones 1974), such higher living standards to include both social and economic desiderata, this within a given Omani village community. In order to achieve this general strategic objective, certain specific and relevant activities were developed; the list of these continues to expand.

The project aimed to assist people to become: more independent (with an increasing use of local resources); more mutually dependent (with a reliance on their complementary activities); and more aware of future needs and the obligation, therefore, to live in harmony with their environment. At the same time there had to be realistic acceptance of the new, oil-based economy which had given people higher aspirations and made them potentially part of a wider national community. Oil had helped create the problems defined above, but it also created an opportunity to set the equilibrium of economic activity at a higher level so that some of the new aspirations could be satisfied. But if more was to be taken out of the system, more would have to be put into it to achieve effective and stable growth. In a region with virtually no virgin resources this could only imply the application of more labour, harder work, a wider range of remunerative skills or more productive labour.

But were these aims realistic in Oman in the mid-1970s? Certainly, objectives that simply required more labour and harder work would have been unrealistic. The traditional organization of rural activities in Oman was dependent on long hours of hard manual work in a harsh and enervating climate, an unattractive prospect given other opportunities. But just because production had been so dependent on manual labour it was regarded as possible to replace some labour requirements with improved techniques, without becoming over-reliant on high-cost alien technology. Therefore movement towards more productive labour was of prime importance to project strategy.

Because of emigration to work, the number of active local producers was diminished, but, if those that remained showed that they were responsive to demonstrable opportunities for change, there was a point of departure. Such 'enterprising peasants' have been found in Africa (Hill 1970; Tiffen 1976) and

Fig. 12.1 Air photograph, Khabura village, northern Oman, showing Community Development Centre (about 1:5000).

in other parts of the world (Schultz 1964). They could also be found in Oman. For example, by the mid-1970s nearly all the farmers who had previously lifted water using an ox and trip-bucket had replaced this with a diesel pumpset. During the mid-1970s, they responded to an expanding market for alfalfa in al Ayn by increasing their alfalfa acreage. Meanwhile the fishermen had replaced their cotton nets with nylon nets and they were using new types of fish traps and fitting their boats with inboard and outboard engines (University of Durham 1978a; Donaldson 1979). The fishermen had perhaps adjusted to the new situation better than other groups, but because their numbers had fallen they were landing fewer fish than before. Craftsmen, meanwhile, had made no changes at all and were being steadily squeezed out of their jobs by economic forces quite beyond their control. But in general it could be said that the people were not averse to new ideas; some were even initiating change and discovering empirically whether or not it was economically sound. Also the traditionally hospitable nature of the Omanis made them open and responsive to the influence of outsiders.

In a community development project the work has to be multi-faceted. The central importance of this is stressed by Hutchinson (1966) who compared various schemes and found that the successful had in common a rise in living standards 'basically from the increased productivity of the community, largely in things they used themselves'. This of course requires a fuller use of local resources and the development of skills to process them. It responds to the need for greater community independence based on mutual dependence. It also allows one to be aware of and to limit the 'disequilibria and observable lags in adjustments' that Schultz (1972a) noted will occur under dynamic conditions of economic growth; growth in one sector will not necessarily have spin-off effects in others.

A multi-faceted approach has other more mundane advantages, important from the project viewpoint. First, it allows more realistically for the fact that the understanding of the local potentials and problems will be imperfect, particularly at the outset. In a relatively wide-based programme the failure of some less critical activities has to be accepted and this in turn helps relieve some of the pressure to 'get everything right every time'. No project can achieve that ideal but if the central strategy is correct and the base-survey data sound, and key activities going well, it can continue to carry conviction locally. Finally, a wider programme allows the project to adjust its activities and its approach to changing circumstances; the work does not take place in isolation or in a static social and economic environment.

Fundamental to the approach is accepting that a 'proposed change must stand in some organic relationship to what the people are already doing' (Schumacher 1973). Nearly all the enterprises of rural Oman are on a small scale: an owner-operator working by himself, with his family, or with a very limited amount of paid help. It is these small enterprises which have to be reached if productivity is to rise. This approach is less spectacular and more difficult than the creation of a large new enterprise, and the achievement of results is slow. But as it involves a far greater proportion of the rural population, each working for himself within the village group, the results for the community and for growth appear more certain in the long run. Working at a small scale does not, of course, exclude an active interest in assisting small enterprises to grow or in helping large local enterprises in appropriate ways.

Working from the known has other advantages. Radical departures are hazardous in a sector such as agriculture, where success depends on a complex mix of physical, economic and social factors, and a one-step-ahead approach also gives the opportunity of making minimum changes to overcome major bottlenecks and to evaluate the full potential of the local production system. It may well be that the minimum-modified system is the best, relative to local conditions, or even in Oman frequently the best in absolute terms.

But how could the project meet the challenge encompassed by the development strategy given above? First, there is the foundation provided by the accumulated detailed understanding of the region gained by the survey programme; continuity was given by the directors; Durham University itself supplied extensive experience of modern Middle-Eastern affairs and an entrée to other professional expertise.

Secondly, in Oman the project had the broad sympathy of the Omani government and the logistic support of the oil company, PDO. Thirdly, the project called not only on the specialist skills but also on a degree of commitment and a sense of service from its staff. Moreover, living and working within the village of Khabura, staff were made forcibly aware of people as individuals and not as aggregates. People, not land and water, are the key to development (Schultz 1972b) or in Schumacher's words (1973) 'people are the primary source of wealth'.

The project has also had time. Both the Oman government and PDO wanted results, but neither made unrealistic demands for instant achievement. Time was essential if the aim was to reach a large number of people most of whom had no education. People unused to formal learning assimilate ideas slowly and are only likely to accept them as they become fully proven in practice. Apart from this, in any form of agriculture progress is partly governed by the seasons or by livestock cycles, and can only move at their pace. The main achievement might then be not the establishment of a new technique but accepting that change can be beneficial and that ideas coming from the project are very probably good. As Tiffen (1976) says, community development and growth ('a modernized economy') is dependent on 'cumulative smallscale changes' – such changes demand time.

The approach thus outlined stands in contrast to but is compatible with the central-planning approach to development. The latter has an essential place in all spheres of activity, but is most effective where dependence on the local physical environment and human element is smallest, e.g. in the building of roads, ports, schools, etc. It is at its weakest where success depends upon the accurate understanding of many small-scale physical and human variables and their interrelationships – and also upon quick responses to changes in these – as in community development and the transformation of existing agricultural systems (Dutton 1976).

Project design and points of departure

The project was located at Khabura because: (a) within the area for which the survey programme provided essential information it presented the fewest exogenous problems; (b) since it was reasonably representative of conditions along some 270 km of the Batina, replication, within that region at least, would have a good chance of success; (c) possessing a wide range of indigenous activities including fishing, agriculture, some craft industries and a busy market (University of Durham 1978b and 1978c), Khabura offered a wide range of development potential.

Agriculture, directly and indirectly, involves more people in Khabura than any other single sector and therefore it became the focus for project concern with production. Within the agricultural sector, the activity which has the greatest number of 'upstream' and 'downstream' linkages – from irrigation technology and fodder production to diet and saleable products – is that concerned with livestock. Given the need to concentrate relatively limited resources, within defined time schedules at a few key points, the development of appropriate livestock farming systems became central to project work.

Since the relatively youthful official agricultural services were under enormous pressure and since the scientific information base was still limited, project activities had to include practical demonstration, field testing of innovations and the monitoring and recording of a wide range of technical and human responses. The project operations had therefore to be based on a small farm, typical in size, location, terrain, soil and water of local farm-holdings.

The project had to provide not only a conceptual bridge to the outside world but also a work base at and through which non-farm activities could be developed. It had also to possess sufficient logistical and working capability to be worth incorporating, in time, into an Omani institutional framework.

Practice and progress

'Appropriate development' in the chosen sector of livestock keeping was based on the author's work and that of others in the survey (University of

Durham 1979a,b). What did we know? First, goats and sheep were by far the most important stock numerically, and the most suited to ecological conditions at both contemporary and foreseeable scales of farming. Most village households owned two or three head, larger flocks and herds of fifty or more being confined to the arid ranges of plain and highland and owned by the shawawi (family-group pastoralists). Though goats and sheep are traditionally important for celebration slaughtering and relatively infrequently consumed animal products, rearing practices were unsatisfactory. Villagers' animals, dependent on browsing on non-irrigated village land and on minimal supplementary feeding on cultivated fodders, suffer from the vagaries of unpredictable and very low rainfall. Malnourished, they have little defence against disease, give low milk yields and carry little flesh; conception and reproduction are difficult and haphazard and have low success rates.

The requirement here was to change attitudes from stock keeping to stock farming. This has implied the development of a farming system which could include a livestock unit of up to 150 head – large enough to give a significant income to a family enterprise and to support veterinary and other services. In turn, such a system implies that the well-being of the livestock must be less directly dependent on the natural environment and that quality and production standards rise.

In addition to the development of livestock farming, with all that this implied, other points of departure could be identified. Among these were domestic-scale craft industries such as wool spinning and weaving, beekeeping and honey production, the utilization of renewable energy sources (windpower and solar energy) and improvements in artisanal fishery. These and other analogous efforts are regarded as important because they extend the project into non-farming but still rural activities and networks.

On 2 ha of village land the project got under way in 1976. On 1.5 ha a modified traditional system of flood-basin irrigation is used, a simple but important modification being the regular alignment of larger than normal irrigation basins. This saves land, allows easier access and permits the use of machinery without the regular destruction of perimeter ridges. In turn, less labour is used in cultivation and basin construction and the incentive to shorten the previously overlong life cycle of perennial crops is increased. Experiments with cheap and durable linings for water channels, important for water conservation and soil management, are leading to trials of date-fibre-reinforced cement forms. With the help of the Intermediate Technology Development Group (ITDG), London, it is hoped to produce portable, lightweight channel sections which are cheap, resistant to decay and damage, and which are suitable for easy local manufacture. Incorporating locally available raw materials, they could form the basis of small-scale local industry replicable elsewhere.

Alfalfa is already well known, and local varieties (which out-perform exotics) give excellent yields with reasonable husbandry improvements to provide one basic fodder for stall-feeding. Rhodes grass has been introduced as a high-yielding, salt- and drought-tolerant alternative. Perennial and with stoloniferous growth which colonizes bare ground and suppresses weeds, it can also be grazed intensively with consequent saving of the effort, time and cost necessitated by fodder cutting.

Irrigated cultivation traditionally relies on the heavy use of manual labour, supplemented by the hire at a subsidized rate from the Ministry of Agriculture of a limited number of tractor-mounted cultivators and drivers. This latter service cannot keep up with demand and conventional large four-wheel tractors are unsuited to the majority of operations – cultivating between datepalms or in small patches of land between standing crops. Much time is spent in unproductive travel between centres and farms. Project production trials of a two-wheel rotavator have demonstrated its suitability for Omani conditions on enlarged irrigation basins (10 m by 20 m). Khabura farmers appreciate its advantages, in tilth making and in manoeuvrability, and show this by hiring it. In association with the government and manufacturer it is planned to expand its use and, hopefully, to lay the basis of a local machinery co-operative.

Using local breeds of sheep and both local, imported Anglo-Nubian and cross-bred goats, the project has demonstrated that it is locally feasible to maintain 100–20 head of stall-fed and/or grazed livestock on 1.5 ha of irrigated feed. The aim of producing saleable stock, of a quality (in all respects) which is demonstrably higher than local standards and yet well within local capabilities, has been achieved and the emulation spin-off is occurring. Moreover, the necessary innovations do not require total specialization or any infeasible scale of operation.

One downstream linkage is with spinning and weaving. Weaving, a labour-intensive craft, had declined in Khabura to a single-market operation. The shawawi still utilize a small number of rug-blankets, woven on demand from yarn spun, when required, by the shawawi women. Most wool was thrown away and productive craft employment was nearing extinction. The Ministry of Agriculture now funds a joint project, which even at this early stage has produced an excellent response. Wool and hair from the project farm and from local flocks are now regularly spun by women in their own homes and sales of yarn and woven goods are increasing and finding new markets for the improved-quality goods of attractive types. Another joint project is beginning to increase the quality and quantity of honey produced by beekeepers, locally and elsewhere in Oman, from stronger swarms in improved hives.

Collaboration with Omani and other organizations is extensive and normal. The windpump, shortly to be erected on the outskirts of Khabura to supply domestic water, was designed by ITDG, and constructed by PDO and a Euro-Omani firm. A solar still installed for water heating is a joint ITDG and commercial product.

A research programme associated with the honey production work involves co-operation with the flora surveys of the Royal Botanic Gardens, Edinburgh, and the artisanal fisheries development by the Omani Department of Fisheries has been able to utilize the results of research on fisheries carried out from Khabura by Donaldson (see Chapter 11).

What of the future?

Given the approach, the background survey work, the design and implementation of the development project, the collaborative ventures and a realistic evaluation of progress so far, it is possible, without complacency or over-optimism, to sketch two future scenarios.

The first concerns a hypothetical local family. It has been attracted to developing livestock production as part of a small mixed farm unit, partly because the system adopted does not require too much skill or effort yet produces results. The animals mostly feed themselves by grazing and also manure the land. In addition, the combination of larger irrigation basins and rotavator cultivation makes good husbandry easy. The local machine co-operative, to which the farmer subscribes, carries a good stock of spares and has a skilled mechanic in charge. The high stocking rate ensures that there are always animals for sale and improved stock nutrition supplies surplus milk which is drunk by the family or sold. The children help with the irrigation, made easy and reliable with the fibre/cement channels bought from the nearby workshop. Part of the water is raised by a windpump (which saves increasingly costly fuel) manufactured by a company based in the region. The women spin and weave the wool and hair from the sheep and goats and they cure the animal skins for sale to the leathersmith.

There is still one son employed in Abu Dhabi but the other, now working full time on the farm, has developed a particular interest in beekeeping. The bees are pollinating the limes and the alfalfa and, through local honey sales, are netting a useful income. He takes the honey and other goods for sale to the local market in his donkey cart, purchased, together with a range of implements for use with the donkey, from a second local workshop. Father and son are contemplating expanding into commercial vegetable production and thinking that they may have to buy a bit more land to make the enterprise worthwhile.

The second scenario, on a national scale, sees other community development centres, dependent initially on the experience gained at Khabura, established in other parts of Oman. Their function as demonstration farms within the overall policy for livestock improvement adopted by the government, is to link government breeding stations and small farmers. The demonstration farms take improved stock from the breeding stations, fit them to economically viable production systems incorporating the more successful aspects of the work described above and encourage local farmers to emulate their example. Oman is less dependent on imports of animals and animal products. As the average quality of the stock has risen the better farmers are concentrating on producing higher-quality animals commanding a premium in the market. The development centres are creating a socio-economic environment in which production and purchasing co-operatives are flourishing and are the nuclei around which rural Oman, within the national development programme, is re-creating itself.

These partial views of the future may be sanguine but they are based on cautious evaluations of what it is possible to extrapolate from established trends. This is not the only path of rural progress but much of the Khabura experience is fundamentally relevant to rural community development elsewhere in the Middle East and the world.

References

BIRKS, J. S. and SINCLAIR, C. A. (1978) 'The Sultanate of Oman: economic development, the domestic labour market and international migration', *Migration for Employment Project*, ILO.

BOWEN-JONES, H. (1974) 'Memorandum on a possible rural development project in Oman' (unpublished).

—— (1975) *Rural Community Development in Oman*, Conference on Rural Development, Min. of Social Affairs and Labour, Oman, and UNESCO.

CLARKE, J. I. and FISHER, W. B. (1972) (eds) *Populations of the Middle East and North Africa. A Geographical Approach*, London, ULP.

CURRIE, L. (1978) 'The objectives of development', *World Development*, 6, 1–10.

DONALDSON, W. J. (1979) 'Fishing and fish marketing in Northern Oman', unpublished Ph D thesis, University of Durham.

DUTTON, R. W. (1976) 'Farming in the Lower Medjerda Valley', in Stone, R. A. and Simmons, J. (eds) *Change in Tunisia*, New York, State Univ. of NY Press, 3–24.

HILL, P. (1970) *Studies in Rural Capitalism in West Africa*, London, CUP.

HUTCHINSON, Sir J. (1966) 'Introduction', *The Transformation of Rural Communities*, World Land Use Survey, Occ. Papers No.7.

MOUNTJOY, A. B. (1971) *Industrialisation and Underdeveloped Countries*, London, Hutchinson.

SCHULTZ, T. W. (1964) *Transforming Traditional Agriculture*, New Haven, Conn., Yale Univ. Press.

—— (1972a) 'Production opportunities in Asian agriculture: an economist's agenda', in Johnson, W. L. and Kamerschen, D. R. (eds) *Readings in Economic Development*, Cincinnati, South Western Publishing Co.

—— (1972b) 'Investment in human capital in poor countries', in *Readings in Economic Development*, op. cit.

SCHUMACHER, E. F. (1973) *Small is Beautiful*, London, Blond & Briggs.

THESIGER, W. (1977) *Arabian Sands*, London, A. Lane, Harmondsworth (new imprint).

TIFFEN, M. (1976) *The Enterprising Peasant: Economic Development in Gombe Emirate, North-Eastern State, Nigeria, 1900–1968*, London, HMSO.

UNIVERSITY OF DURHAM (1978a) *Fishing and Fish Marketing*, Research and Development Surveys in Northern Oman: Final Report, vol. 5, Bowen-Jones, H. (ed.).

—— (1978b) *Marketing*, Research and Development Surveys in Northern Oman: Final Report, vol. 6, Bowen-Jones, H. (ed.).

—— (1978c) *Water*, Research and Development Surveys in Northern Oman: Final Report, vol. 2, Bowen-Jones, H. (ed.).

—— (1979a) *Farming*, Research and Development Surveys in Northern Oman: Final Report, vol. 4, Bowen-Jones, H. (ed.).

—— (1979b) *The People*, Research and Development Surveys in Northern Oman: Final Report, vol. 7, Bowen-Jones, H. (ed.).

WILKINSON, J. C. (1977) *Water and Tribal Settlement in South-East Arabia*, Oxford, Clarendon.

CHAPTER THIRTEEN

Agricultural development in Turkey

JOHN C. DEWDNEY

Nearly sixty years have now elapsed since the establishment of the Turkish Republic in 1923 and throughout this period the prime objective of development planning has been the transformation of a traditional agrarian economy and society into a modern industrial state. Considerable progress has been achieved, especially during the last twenty years, in the industrialization of the Turkish economy. Nevertheless, the agricultural sector continues to play a vital role, while at the same time presenting some of the most intractable problems in the country's economic development.

The role of agriculture in the Turkish economy

In terms of employment, agriculture remains a dominant sector. At the 1975 census, farming employed about two-thirds of the labour force, including some 55 per cent of all male workers. Since non-agricultural employment is heavily concentrated in a few districts, the agricultural sector is the main support of the great majority of the population over extensive areas of the country. In terms of the national income, on the other hand, agriculture appears much less important, contributing only about a quarter of the Gross National Product compared with about 22 per cent from extractive and manufacturing industry and 53 per cent from other activities. One result is a major disparity in incomes between the agricultural and non-agricultural sectors – the average income of a farm worker is little more than 40 per cent of the average for all workers – a gap which has widened considerably in recent years. In addition, there are marked regional variations in farmers' incomes with large numbers well below the national average.

It is also important to note the significance of agriculture in Turkey's foreign trade. Although the proportion has declined in recent years, agricultural commodities, especially cotton, tobacco, fruit and nuts, still account for about three-quarters of all exports and are likely to continue their dominance in the foreseeable future. Both as a source of foreign currency and, more directly, as a source of raw materials, agriculture remains a major support for the industrialization process.

The role of agriculture must also be viewed against the current demographic situation. The post-war period has seen a marked demographic upsurge with an annual rate of population growth persistently above 2½ per cent and as yet little evidence of any slowing down of this growth. Whereas between 1927 and 1945 the population of Turkey rose by some 38 per cent, from 13.6 million to 18.8 million, since the latter date it has more than doubled, topping 40 million in 1975. Current estimates suggest a total around 60 million by the end of the 1980s. While much of this growth has been absorbed in the growth of towns, the rural element is still large (23.4 million or 58 per cent in 1975) and continues to expand by at least 250,000 a year. It would appear that neither agricultural development nor industrialization has provided sufficient work for the expanding population; unemployment and underemployment, both urban and rural, are a growing problem reflected in the large numbers of Turkish workers who have sought jobs in western Europe.

Thus there are obvious reasons why agricultural development has been and remains so vital to the development of the Turkish economy as a whole – to supply basic foodstuffs for a rapidly growing population, to give employment to the still growing rural element, to lessen the gap between agricultural and non-agricultural incomes, to provide a major source of overseas income while there is a chronic balance-of-payments deficit and to support industrialization. Given its scale and importance, it is perhaps surprising that in recent five-year plans agriculture has been alloted only 12 per cent of planned capital investment.

Physical constraints on agricultural development

In land resources, Turkey would appear to be among the most well-endowed countries of the Middle East. While her total land area (781,000 sq km) is considerably smaller than that of Algeria, Egypt, Iran, Libya or Saudi Arabia, all these have large stretches of desert or semi-desert territory which greatly reduce the size of their habitable area. In the case of Turkey, such climatically negative zones are virtually absent and her ecumene is the largest in the region. According to official figures no less than 83 per cent of the land surface of Turkey is utilized in one way or another; 36 per cent is cultivated, 21 per cent used for grazing and 26 per cent is classed as forest. For a total population of 42 million (1978), this gives each inhabitant about 1.5 ha of utilized land, including 0.56 ha of cropland, while for the rural population the equivalent figures are 2.8 and 1.1 ha respectively; relatively favourable balances when compared with most other Middle-Eastern states.

However, the situation is a good deal less favourable than these crude figures would suggest, for there are severe physical constraints which restrict the agricultural potential of much of the country; these include aspects of altitude, relief and climate. Like her Northern Tier neighbour Iran, Turkey is an elevated country: some 55 per cent of the land surface stands between 500 m and 1500 m above sea level and a further 22 per cent is above 1500 m; the median altitude is about 1130 m. In addition to its altitude the land is characterized by a predominance of steep slopes. According to Oakes (1957), 'level or gently sloping land' with gradients below 3 per cent constitutes only 9.8 per cent of the total and 'moderately sloping land' (3.8 per cent slope) another 6.4 per cent, leaving well over 80 per cent classed as 'rough and broken' or 'rough and mountainous' terrain. Plains are thus very limited in their extent and lowland plains even more so. The most extensive areas of level or gently sloping land are in the elevated interior of Anatolia at heights around 1000 m and in the south-east at 500–700 m. Lowland plains are confined to the coastal periphery and even there are by no means continuous; they are most extensive in Trakya (Turkish Thrace) and around the Sea of Marmara and in the down-faulted troughs of the Gediz, Menderes and other rivers flowing to the Aegean; on the Mediterranean coast they are confined to the widely separated Adana and Antalya plains; they are particularly restricted along the Black Sea, occurring only on the deltas of the major rivers (Kizilirmak and Yeşilirmak), where there are drainage problems. All these lowland areas and especially those along the Mediterranean and the Black Sea are cut off from the interior by formidable mountain barriers.

It is in the interior, because of its altitude, that the restrictive effects of climate on agriculture are most obvious. Here, the climatic regime is distinctly continental and winter cold restricts both the length of the growing season and the possible range of crops; the number of months with mean temperatures below 5 °C increases eastwards from three at Konya to five at Erzurum. At the same time, rainfall in the interior is rather limited and its effectiveness reduced by high summer evaporation rates. Annual precipitation exceeds 500 mm in the east, a region where there is little level or gently sloping land, whereas on the plains of the centre and west it is 300–50 mm, barely adequate for most types of rainfed agriculture.

The coastal zones have a more plentiful precipitation and escape the effects of winter cold, though killing frosts occur occasionally along the Aegean and Mediterranean coasts, both of which have to contend with summer drought. Only on the Black Sea coast, with its moderate temperatures and plentiful, year-round rainfall, is there no moisture problem (save that of occasional excess), but it is this coast which has the narrowest coastal plain.

Thus, over Turkey as a whole a combination of positively favourable conditions of altitude, slope, temperature and moisture availability is a somewhat rare occurrence.

Agricultural development since the Second World War

Despite these physical difficulties and in response to the challenge of population growth, there has been considerable agricultural growth over the past thirty years, as the figures given in Tables 13.1 and 13.2 clearly illustrate. According to Aresvik (1975), the average annual increase in the value of agricultural production since the late 1940s has been in the region of 2½–3 per cent, roughly in line with population growth, though well below that of total GNP, where the annual increase averaged 5.5 per cent. In volume of production, output of most commodities has over the thirty-year period as a whole increased more rapidly than population numbers.

Growth of agricultural output since 1945 has involved both expansion of the land area used for farming and intensification of production, two processes which were, in the case of Turkey, carried out sequentially rather than concurrently. Expansion was dominant in the late 1940s and early 1950s; intensification in the ensuing twenty years.

Expansion of the arable area was an extremely rapid process. Between 1945 and 1960 it rose by no less than 84 per cent (from 12.7 to 23.3 million ha), the great bulk of this growth occurring between 1948 and 1955. Since 1960, in striking contrast, it has risen only by a further 5 per cent. The cultivation of the additional land was made possible by a single technological innovation, namely the introduction of the tractor and the progressive substitution of mechanical for animal-powered cultivation. In the mid-1940s there were barely 1000 tractors in the whole country; by 1960 there were 42,000, working about 15 per cent of the arable area, and today there are 250,000, working 75 per cent. This, like many other developments, was facilitated by the advent of massive foreign, chiefly United States, financial assistance, which in turn was linked to Turkey's adherence to the western alliance exemplified by her membership of NATO and CENTO.

The proportion of land under the plough rose from 16 to 31 per cent, the latter figure being well above the 20 per cent suggested by Oakes as being suitable for arable farming in the light of the physical conditions already discussed. But no great changes in agricultural organization or cropping systems accompanied this massive expansion of the arable area. The proportion of the sown area devoted to cereals remained virtually unchanged at 84–5 per cent, as did the fallow element at about one-third of the arable.

Table 13.1 Crop areas and livestock numbers in Turkey 1945–75

	1945	*1950*	*1955*	*1960*	*1965*	*1970*	*1975*
Total land area (mill. ha)	78.14	78.14	78.14	78.14	78.14	78.14	78.14
Arable area (mill. ha)	12.66	14.54	20.99	23.26	23.56	24.3	24.41
% of total land area	16.2	18.6	26.9	29.8	30.2	31.1	31.2
Fallow (mill. ha)	4.62	4.67	6.79	7.96	8.26	8.71	8.18
% of arable area	36.5	32.1	32.4	34.2	35.1	35.8	33.5
Sown area (mill. ha)	8.04	9.87	14.2	15.3	15.3	15.59	16.23
All cereals (mill. ha)	6.89	8.24	12.08	12.95	12.96	13.24	13.61
% of sown area	85.7	83.5	85.1	84.6	84.7	86.5	83.9
Wheat (mill. ha)	3.74	4.48	7.06	7.7	7.9	8.6	9.25
% of sown area	46.5	45.4	49.7	50.3	51.6	56.2	57.0
Barley (mill. ha)	1.63	1.9	2.64	2.84	2.77	2.59	2.6
% of sown area	20.3	19.3	18.6	18.6	18.1	16.9	16.0
Rye (mill. ha)	0.4	0.49	0.64	0.67	0.73	0.65	0.57
% of sown area	5.0	5.0	4.5	4.4	4.8	4.3	3.5
Maize (mill. ha)	0.51	0.59	0.71	0.7	0.65	0.65	0.6
% of sown area	6.3	6.0	5.0	4.6	4.3	4.3	3.7
Pulses (mill. ha)	0.34	0.43	0.52	0.57	0.55	0.53	0.57
% of sown area	4.2	4.4	3.7	3.7	3.6	3.5	3.5
All industrial crops (mill. ha)	0.44	0.72	0.98	1.11	1.13	1.03	1.17
% of sown area	5.5	7.3	6.9	7.3	7.4	6.7	7.2
Tobacco (mill. ha)	0.09	0.13	0.17	0.19	0.22	0.33	0.23
% of sown area	1.1	1.3	1.2	1.2	1.4	2.2	1.4
Sugar beet (mill. ha)	0.05	0.05	0.1	0.2	0.16	0.12	0.21
% of sown area	0.6	0.5	0.7	1.3	1.1	0.8	1.3
Cotton (mill. ha)	0.23	0.45	0.63	0.62	0.68	0.53	0.67
% of sown area	2.9	4.6	4.4	4.1	4.4	3.5	4.1
Sunflower (mill. ha)	0.02	0.11	0.15	0.14	0.16	0.36	0.42
% of sown area	0.3	1.1	1.1	0.9	1.1	2.4	2.6
Potatoes (mill. ha)	0.05	0.08	0.11	0.16	0.15	0.15	0.18
% of sown area	0.6	0.8	0.8	1.1	1.0	1.0	1.1
Vineyards, gardens and orchards (mill. ha)	*	1.46	1.81	2.06	2.27	3.0	3.85
% of total land area	*	1.9	2.3	2.6	2.9	3.8	4.9
All livestock (mill.)	53.08	55.61	62.71	76.11	72.06	73.03	77.61
Cattle (mill.)	9.81	10.12	11.06	12.44	13.2	12.76	13.75
Sheep (mill.)	23.39	23.08	26.44	34.46	33.38	36.47	41.37
Goats (mill.)	16.25	18.46	21.03	24.63	20.8	19.48	18.76

Sources: *Summary of Agricultural Statistics, 1941–1962*, Publication No. 447, State Institute of Statistics, Ankara, 1963; *Summary of Agricultural Statistics, 1975*, Publication No. 781, State Institute of Statistics, Ankara, 1976.

* Not available.

Table 13.2 Agricultural production in Turkey 1941–75 (annual averages in millions of tons).

	1941–5	*1946–50*	*1951–5*	*1956–60*	*1961–5*	*1966–70*	*1971–5*
All cereals	6.43	7.06	11.86	14.14	14.74	16.43	18.82
Wheat	3.32	3.63	6.37	7.91	8.45	9.92	12.29
Barley	1.58	1.72	2.98	3.43	3.45	3.63	3.72
Rye	0.33	0.40	0.62	0.68	0.73	0.80	0.73
Maize	0.63	0.63	0.84	0.91	0.95	1.01	1.13
Pulses	0.31	0.33	0.45	0.56	0.59	0.59	0.64
Tobacco	0.06	0.09	0.10	0.12	0.13	0.16	0.18
Sugar beet	0.58	0.72	1.31	2.84	3.40	4.44	5.92
Cotton lint	0.06	0.08	0.15	0.17	0.27	0.40	0.53
Cotton seed	0.11	0.14	0.29	0.30	0.45	0.64	0.85
Sunflower	*	0.06	0.12	0.11	0.11	0.27	0.50
Potatoes	0.27	0.41	0.93	1.33	1.57	1.83	2.25
Grapes	1.15	1.36	1.83	2.72	3.08	3.56	3.45
Olives	0.21	0.23	0.32	0.41	0.54	0.63	0.62
Milk	1.78	2.89	3.24	3.95	4.09	4.37	4.57
Meat**	0.07	0.08	0.12	0.15	0.17	0.20	0.21

Sources: *Summary of Agricultural Statistics, 1941–1962*, Publication No. 447, State Institute of Statistics, Ankara, 1963; *Summary of Agricultural Statistics, 1975*, Publication No. 781, State Institute of Statistics, Ankara, 1976.

* Not available.

** Meat from municipal and state slaughterhouses only.

At the same time there was an equivalent and in some cases even greater expansion in other forms of cropping, though the proportion of non-cereal cultivation remained low. Cotton was a special and significant case. The demand for raw cotton on the world market and in Turkey's expanding textile industry sparked off a cotton boom and the area devoted to the crop more than doubled between 1945 and 1955 (since when it has risen by less than 10 per cent).

The growth of agricultural output in the late 1940s and early 1950s was well in advance of population growth – there was even a brief period (*c*. 1950–6) when Turkey was a large-scale exporter of cereals – while the increased output of cotton, tobacco, fruit and nuts helped both the trade balance and the industrialization drive. But the picture was generally one of growth without much innovation, the extension of traditional farming systems on to new land with the aid of mechanization, a process common at this time in other parts of the Middle East (e.g. Syria) and elsewhere (e.g. the Soviet Union's 'Virgin Lands' scheme). This process, which by its very nature was a 'once for all' affair, was by no means without its dangers. The growing population was fed, and for a brief period a surplus was created, mainly by the spread of dry-farmed cereals on to land much of which was marginal in terms of precipitation and/or slope, thus exacerbating the long-standing problem of soil erosion. Other pressures were created: the demand for cotton, for example, often led to the use of some of the best alluvial soils of the coastlands for that purpose, relegating food crops to more steeply sloping areas, and uncoordinated attempts by individual farmers to irrigate the crop led sometimes to waterlogging and soil deterioration and a subsequent decline in cotton yields.

The expansion of the arable land had side-effects on other forms of land use. While it took cereal growing into more and more marginal terrain it also consumed large areas, often the best, of grazing land. Between 1945 and 1960 the area of 'meadow and pasture' was reduced by nearly one-third. Since the numbers of livestock rose by nearly 50 per cent over the same period, the problem of over-grazing, already serious, was made much worse. This, in turn, affected the 'forest' area, which was further degraded by uncontrolled grazing, especially of goats. Thus, by 1960, despite Turkey's apparently quite favourable man/land ratio, several of the classic symptoms of pressure on land resources were apparent, essentially as a result of the extensive nature of the dominant farming systems.

The 1960s saw increased government involvement in economic planning, epitomized by a series of five-year plans, the first of which covered the period 1963–7. Since the economy remained a mixture of state and private enterprise, with the latter overwhelmingly predominant in the agricultural sector (83 per cent of all holdings are owner-occupied), these plans were a set of desirable objectives rather than targets to be met and interest centres on what was considered desirable. The plan documents showed official awareness of current and potential agricultural problems in the light of continuing population growth and the vital role of agriculture in the economy. They made it clear that there was little or no potential for further expansion of the cultivated area and that further rises in production could be achieved only from more intensive use of the existing cultivated land. The extent of the arable area should be stabilized, or even reduced by the return of marginal land to grazing, while the forest area should be increased at the expense of some of the poorer grazing. Yields from a stabilized arable area should be increased by extension of irrigation, by the increased use of fertilizers and by the introduction of new strains of cereals and other crops. Intensification was also seen to be essential on the livestock side – livestock numbers should be stabilized (that of goats reduced) and higher yields of meat, milk and wool achieved by the establishment of better breeds and by improved feeding.

The matter of stock-feeding provides an insight into the difficulties involved in changing traditional farming systems. Throughout Turkey, livestock are fed mainly on natural grazing and on fallow land. It is true that in recent years the availability of other types of feed has greatly increased as a result of increased production of cereals, sugar beet and cotton, but crops grown specifically for fodder are still something of a rarity. The several five-year plans and other sources repeatedly comment on this anomaly, pointing out the advantages to both livestock and arable farming which could be achieved from the replacement of fallow by leguminous fodders in cereal-growing areas. Yet of all the crops featuring in official agricultural statistics, the group 'pulses', which includes leguminous fodders, has shown one of the slowest expansion rates. In the past thirty years, the area devoted to such crops has barely doubled and today stands at around 600,000 ha, a mere 2.5 per cent of the arable area, as against some 8 million ha (33.5 per cent) fallow.

Current trends and future developments

In the period since 1960, during which both the size of the arable area and its allocation among the various crops have shown relatively little change, total agricultural output has continued to expand. However, there has been a marked imbalance among the various sectors of agriculture. The production of basic foodstuffs has barely kept pace with population growth and within this sector the supply of livestock products has failed to meet the growth in demand resulting from population increase and rising per capita income. The most rapid increases in output have been achieved in industrial crops and high-value crops destined largely for export. While these trends have continued to assist both the industrialization programme and the balance-of-payments situation, the provision of an assured food supply and a change in emphasis from cereals towards livestock products in the diet of the population remain major concerns.

There would thus appear to be a case for giving higher priority to agriculture as a whole in the allocation of investment capital, and for measures to stimulate a more rapid growth in the output of basic foodstuffs, especially livestock products. Much has already been achieved, for example by the introduction of new varieties of cereals and other crops and by the extension of irrigation; much more could be done, particularly as regards irrigation. At present some 6 or 7 per cent of Turkey's cropland is irrigated; according to Aresvik (1975: 195), this could be raised as high as 30 per cent, a long-term aim which would require very large-scale capital investment. So far, modern irrigation technology has been applied mainly to high-value crops, but much could also be gained by irrigating large areas at present under rainfed cereal cultivation. Development in this field should go hand in hand with improvements on the livestock side, including the provision of fodder by the establishment of cereal–legume rotations as widely as possible.

Little or no mention has so far been made of regional problems and lack of

Table 13.3 Regional variations in Turkish agriculture, 1975

	*Regions**							
	Turkey	*1*	*2*	*3*	*4*	*5*	*6*	*7*
Total area (mill. ha)	78.14	29.94	7.41	11.34	4.43	9.94	8.16	6.89
% of total area	100.0	38.3	9.5	14.5	5.7	12.7	10.4	8.8
Sown area (mill. ha)	16.23	7.4	0.74	1.77	1.4	1.89	1.81	1.23
% of total sown	100.0	45.6	4.6	10.9	8.6	11.6	11.1	7.6
Crops: percentage of sown area								
All cereals	83.9	92.6	91.9	91.9	67.1	60.0	62.1	85.2
Wheat	57.0	63.7	57.0	72.0	44.6	33.7	44.7	32.8
Barley	16.0	8.7	28.0	16.8	4.4	15.0	11.2	12.2
Rye	3.5	2.7	5.1	0.7	2.0	2.6	1.0	2.4
Maize	3.7	0.4	1.4	0.2	8.6	4.1	0.4	28.5
Pulses	3.5	3.2	2.1	3.7	1.8	7.0	5.7	2.9
All industrial crops	7.2	2.2	2.5	4.4	3.8	28.6	29.3	7.2
Tobacco	1.4	2.4		0.5	0.9	13.8	0.2	5.5
Sugar beet	1.3	0.7	1.4	0.1	2.2	0.7	0.3	1.4
Cotton	4.1	0.1	1.0	3.3	0.4	13.5	28.6	0
Sunflower	2.6	0.2		0.4	25.5	3.1	1.9	1.0
Potatoes	1.1	1.2	3.3	0.2	0.8	0.8	0.6	3.1
Production: percentage of total								
All cereals	100.0	41.6	2.8	8.9	12.8	12.2	12.9	8.9
Wheat	100.0	41.7	2.7	10.7	14.3	11.1	15.0	4.5
Barley	100.0	50.8	3.9	8.4	3.9	15.5	11.6	6.0
Rye	100.0	69.7	4.1	2.5	6.2	10.2	2.8	4.5
Maize	100.0	12.3	1.2	0.4	18.6	12.4	1.3	53.8
Pulses	100.0	34.5	1.9	10.0	4.4	23.9	18.7	6.6
Tobacco	100.0	9.5		5.8	6.1	66.9	1.0	10.8
Sugar beet	100.0	56.6	3.7	0.4	21.6	7.1	2.9	7.7
Cotton	100.0	0.9	6.5	0.3	0.4	37.6	54.2	0
Sunflower	100.0	7.9	0.1	0.1	77.6	10.3	0.5	3.6
Potatoes	100.0	49.2	11.5	1.4	6.9	12.2	5.2	13.5
Fruit	100.0	25.2	1.3	0.9	8.4	30.3	27.7	6.1
Grapes	100.0	28.7	0.5	16.1	8.1	24.5	21.9	0.5
Nuts	100.0	17.4	1.7	8.8	11.8	7.1	5.2	47.9
Vegetables	100.0	21.5	1.2	5.5	16.6	24.0	21.9	9.3
Milk	100.0	30.4	12.7	15.0	6.4	12.9	8.6	14.1
Meat	100.0	26.1	8.8	5.0	30.7	13.5	9.0	7.0

Source: *Agricultural Structure and Production, 1975*, State Institute of Statistics, Ankara, 1976.
* (See Fig. 13.1) 1: Central Anatolia; 2: north-east; 3: south-east; 4: Marmara; 5: Aegean; 6: Mediterranean; 7: Black Sea.

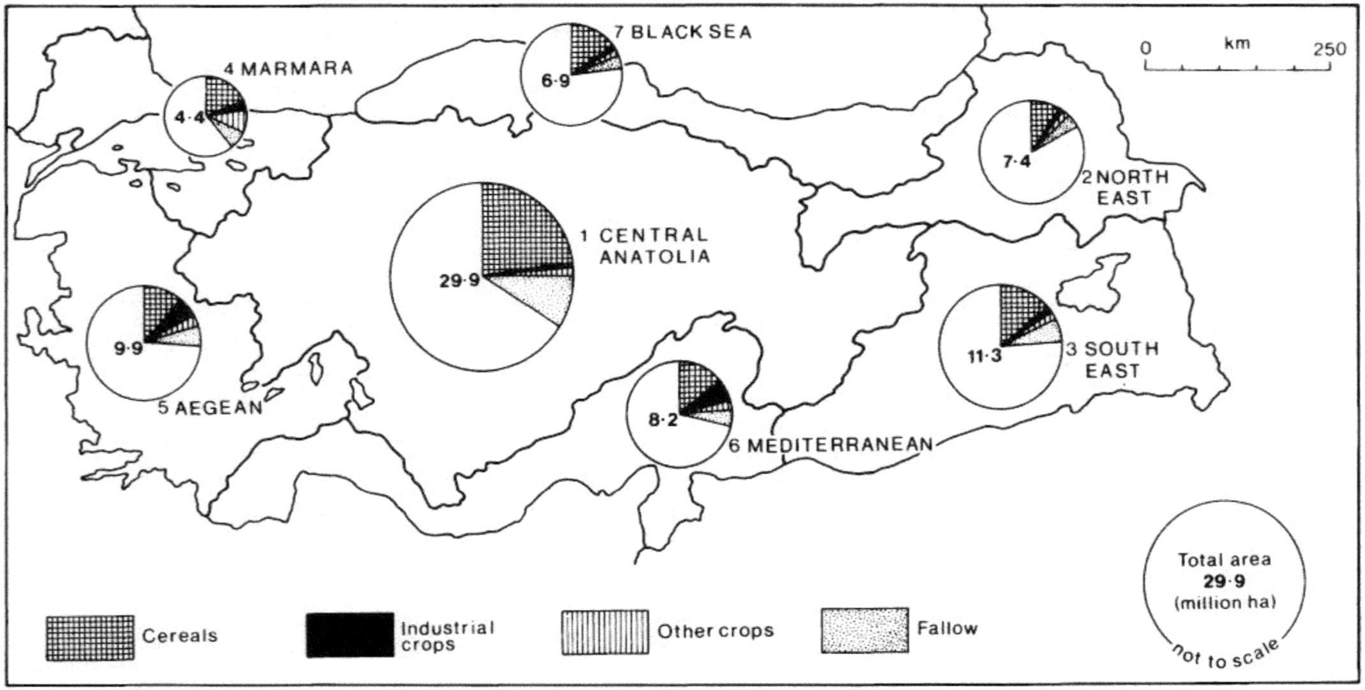

Fig. 13.1 Agricultural regions and land use in Turkey, 1975.

space prevents a full treatment. It must, however, be emphasized that there are major regional disparities in agricultural productivity and thus in the incomes and living standards of the rural population (Table 13.3). The basic contrast is that between a relatively prosperous coastal periphery which produces the great bulk of the high-value crops and a less prosperous interior devoted mainly to extensive cereal and livestock production. This dichotomy is to a large degree imposed by the physical environment, but has been intensified by agricultural development over the past twenty years, which has been most successful in the coastal zones, as well as by industrialization, which has also occurred mainly in these peripheral areas. The lessening of regional disparities is another major objective requiring more attention to the agricultural problems of the interior and modernization of the highly traditional cereal and livestock sectors.

The various developments suggested above should enable Turkey to cope with her population growth in the short term; the country's agricultural resources, if fully and efficiently utilized, could support much larger numbers. Problems will, however, intensify if population growth is not checked. Agricultural and indeed all forms of economic development must be accompanied by more forceful family-planning programmes aimed at the eventual stabilization of population numbers.

Bibliography and references

AKTAN, R. (1966) 'Agricultural problems of Turkey', *Mediterranea*, 12, 266–75.

ARESVIK, O. (1975) *The Agricultural Development of Turkey*, New York, Praeger.

COHN, E. J. (1970) *Turkish Economic, Political and Social Change*, New York, Praeger.

DEWDNEY, J. C. (1972) 'Turkey: recent population trends', in Clarke, J. I. and Fisher, W. B. (eds) *Populations of the Middle East and North Africa*, London, ULP.

—— (1976) 'Agricultural problems and regional development in Turkey', in Hale, W. M. (ed.) *Aspects of Modern Turkey*, London, Bowker.

HERSHLAG, Z. Y. (1968) *Turkey: The Challenge of Growth*, Leiden, Brill.

OAKES, H. (1957) *The Soils of Turkey*, Ankara, Ministry of Agriculture, Soil Conservation of Farm Irrigation Division.

PALMER, Z. E. (1966) *Agriculture in Turkey: Long Term Projections of Supply and Demand*, Istanbul, Roberts College.

STATE INSTITUTE OF STATISTICS (annual) *Agricultural Structure and Production*, Ankara.

—— (annual) *Summary of Agricultural Statistics*, Ankara.

STATE PLANNING ORGANIZATION (1962) *First Five-Year Development Plan, 1963–67*, Ankara.

—— (1967) *Second Five-Year Development Plan, 1968–72*, Ankara.

—— (1972) *Third Five-Year Development Plan, 1973–77*, Ankara.

CHAPTER FOURTEEN

A fieldscape in transition: the case of a Turkish orchard

BRIAN W. BEELEY

In contrast with the macro-analysis of Turkish agriculture in Chapter 13, this chapter is devoted to a micro-analysis of one orchard located in the village of Çakırlar near Antalya in southern Turkey.

This orchard is not typical of most of Turkey's agriculture. It is fully irrigable. It produces citrus fruit and has brought great profit to the farmer who has built it up. Changes in the orchard over the past two decades illustrate the transition from the subsistence-orientated family enterprise of the past towards the commercial unit of the present. Neither is the producer in this case typical of Turkish farmers, being one of the most innovative men in his village of some 250 households. This coincidence of exceptional land and an outstanding man is an extreme case in the realignment of the man – land relationship, but its study can be a useful counterweight to concern with averages measured over wide regions or broad sectors of the economy. The orchard plan thus presents an unashamedly non-random fieldscape where visible changes reflect adjustments in the rural scene. Clearly, such a 'micro' case cannot be a sensitive gauge of change in Turkey's 40,000 villages, but it may provide both planner and farmer with an indicator of possibilities for the future.

The expansion of citrus in this selected coastal-plain village, sited on alluvium some 50 m above sea level near the Mediterranean port of Antalya, began after 1945 and accelerated in the 1960s as villagers realized the potential income from oranges and saw that orchard management demanded less work than did the ground crops which had taken most of their attention hitherto. The oldest existing orchards in the village date from the early years of this century, when they were seen as providing fruit for individual families and for some exchange within the community. They were back-garden orchards in the village centre, with a profusion of different varieties of orange and many non-citrus fruits. Between 1955 and 1960, the area of village lands under orchard

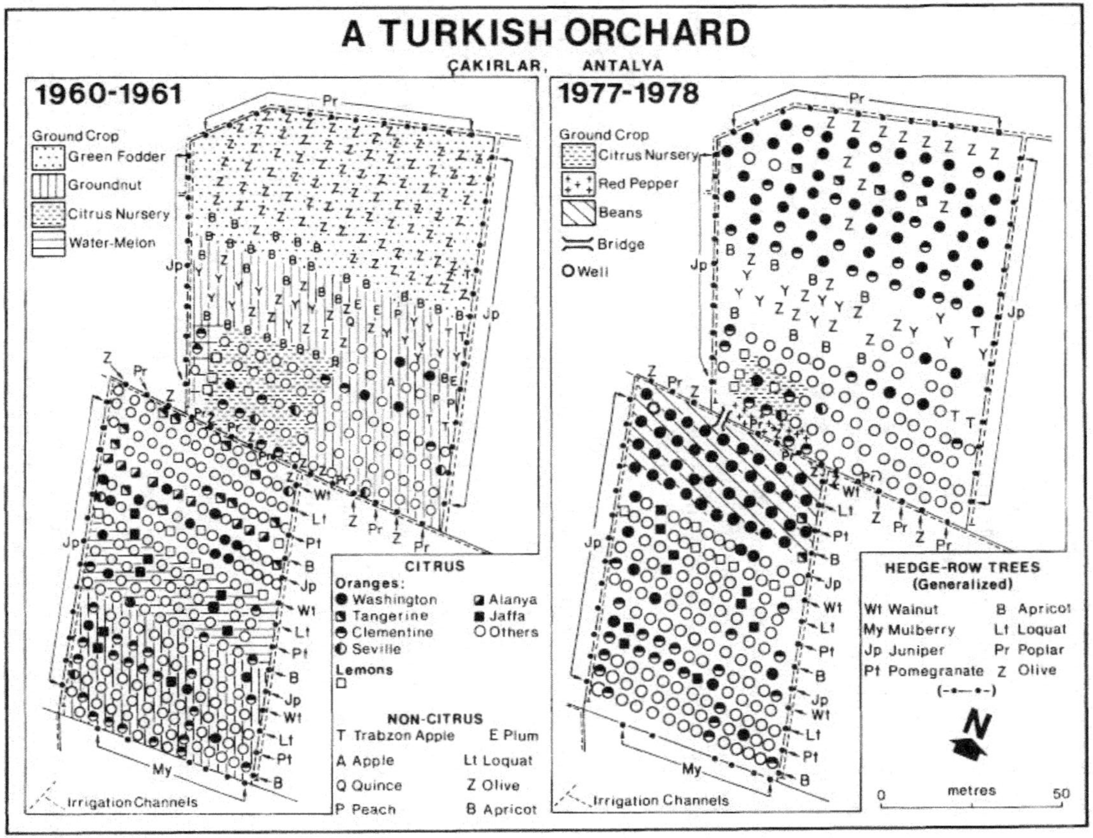

Fig. 14.1 A Turkish orchard, Çakirlar, Antalya.

more than doubled. In 1960, more than a third of the 480 trees in the two orchard plots were non-citrus – mainly olive, though eight other fruits were represented (Fig. 14.1). The older trees were spaced more closely together than have been subsequent plantings which have usually been put 5–6 m apart in response to advice from extension agents. By 1977–8, most of the non-citrus had disappeared; even the once-popular olive was reduced to only a score of trees, though some of these were cleared on account of age. Meanwhile, ninety-three orange trees were planted. The large and popular 'Washington' variety accounted for seventy-eight of these, especially in the northern section of the smaller orchard field where the tree density was reduced by about one-third and where green beans were planted to take temporary advantage of the time during which the trees would remain relatively small. Four lemon and some eighty older and less profitable orange trees were removed during the seventeen-year period, with the remaining lemons (not regarded here as a commercial crop) being retained for home consumption. The net result of the changes in the orchard between 1960 and 1978 was a 23 per cent reduction in the total of trees, though many empty spaces remained available to take new orange trees from the small nursery or from beyond the village.

In 1977–8, as in 1960–1, the perimeter of the orchard was marked by several varieties of tree, notably poplar and juniper. The junipers stand along the north–south trending field boundaries as efficient wind-breaks, though one section was removed in the intervening period to provide wood to build a house in the upland area (*yayla*) to which this transhumant villager retreats during the summer. The few productive hedgerow trees, such as walnut and pomegranate, remain but seem nowadays to be tolerated rather than cultivated. In 1960–1, much of the ground beneath the trees in the orchard was put to groundnut, water-melon and green fodder. Most of this ground cultivation ceased by 1965 because the trees were said to be getting too large to allow sufficient light to penetrate. In 1977–8 ground crops included only the beans beneath the new 'Washington' orange trees, a small patch of red pepper (for home consumption) and a residual citrus nursery. The intercultivation of ground crops during the growing period of orchard trees continues to be standard practice in intensive farming operations of this type in Mediterranean Turkey.

A number of other changes in the orchard during the seventeen years are worth noting. In 1971 the farmer built a well in his orchard to give access to water under his personal control. He continues to use the public supply of irrigation water fed to the orchard (and to the other 1200 field plots in this village) through the open channels along field perimeters. The upkeep of this system is a responsibility of the village council, as is the allocation of water supplies on an agreed basis. Shortly before his well was completed, the village arranged for the irrigation channel to be concreted, to prevent loss by seepage, as far as the point of entry into this orchard. The distribution, though frequently the cause of dispute, is as well attuned to the needs of individual cultivators as it is to the availability of water fed into the system at the upper end of the 5 km stretch of village lands by streams from the mountains to the west. Well water for irrigation, on the other hand, is a novelty, though wells for domestic supply have long been available in the village centre and have been supplemented by piped water during the period reviewed here. If the number of wells sunk in the fields and orchards increases substantially, there could be serious implications for the overall supply as well as a need for new regulatory procedures. Other improvements have included a small 'bridge' which this farmer has constructed between the two parts of his orchard. The bridge consists of a pipe, bought for the purpose, topped with concrete. This endeavour can be linked with the fact that this farmer was, in the late 1940s, the first in the village to attempt any kind of bridging of a field irrigation channel – at that time it was to permit his horse-drawn cart (another village 'first' for the period) to carry substantial loads along rough tracks.

The innovative endeavours of this farmer and others like him in this prosperous part of Turkey are clearly above average for the country. But what is the average in innovation? Is such an average more important in planning and developmental terms than the exceptional case where a new advance first appears or where an individual is the first within his group to respond to some new stimulus such as a hard-surface road or a credit scheme? A broadly based 1962 survey of rural attitudes in Turkey found little variation between regions in villagers' 'propensity to innovate' except for the easternmost part of the country, though it asked questions about willingness to innovate rather than achieved innovation (Frey 1966). The critical element may be the variations from the mean rather than average levels *per se*: the existence of risk-takers and innovators in even small numbers may make possible socio-economic re-alignments which might otherwise be slow to occur. If, however, there were few rural risk-takers, and the 1962 survey accurately portrayed a lack of variation in propensity to innovate over most of Turkey, then the observed variation in achieved development level remains to be explained. Presumably such differences reflect factors, such as roads and credit systems, which are outside the psychological characteristics of individuals and which are, significantly, sensitive to manipulation by government and planners just as traditionalism and fatalism may not be (Mitchell 1971: 212–15).

A study of a tangerine-growing village on Turkey's Aegean coast shows many of the features exemplified in the orchard in this case study. In the Aegean village, commercial production of tangerines first expanded immediately prior to the Second World War as cultivation techniques were copied from outside. During the 1950s an extensive Turkish road-construction programme stimulated the tangerine business and the production of vegetable cash crops (Starr 1978: 27–31). Early produce for the Ankara and Istanbul markets – and even for western Europe – was stressed as subsistence fields were converted to take advantage of the new opportunities. By the mid-1960s, one-third of the households in this village were producing cash crops, with those villagers who were able to make an early move from subsistence into cash crop production gaining an important advantage over those who followed later. Those first to make the move were generally the bigger entrepreneurs in the village, able to make agreements with truckers (or even to purchase vehicles) and to contact middlemen connected with the shipment of produce to distant markets. The farmer who missed the early 'take-off' into commercial production failed to make the same horizon-broadening contacts and, indeed, frequently became dependent on the better-established village exporters for shipment of his more limited marketable surpluses. Over all, this Aegean study suggests that the villagers have not yet shown readiness for collective action. In short, there is in their commercial cash-cropping activities a legacy of the independence of their subsistent past. This legacy and the distinction between early and late-starting cash croppers both obtain for the Mediterranean case study.

The Aegean case and the 1962 survey of rural Turkey attempt to classify villagers; it is also relevant to try to classify villages in their response to developmental stimuli. One attempt distinguishes urban- and rural-directed villages by the stage of their transition from subsistence agriculture to market-oriented operations yielding cash incomes (Kolars 1967: 64–77). Urban-directed villages depend on towns for most of their food supply and employment. Rural-directed settlements, however, show different levels of penetration in the money economy. Market-ignoring villages lack crop surpluses for sale, while market-recognizing villages have occasional surpluses of subsistence crops for sale. Finally, market-seeking villages show continuing expansion of the area under cash crops. This case-study orchard village is clearly one of these – as is the example from the Aegean coastlands. Explicitly as well as implicitly, it is decided to abandon individual household (or, at least, village) subsistence in favour of a commerce-based mode of life, perceived as 'urban' and 'modern'. Indeed, a major topic of interest for the orchard farmer in this case study, and for his fellow villagers, at the end of the period under review, was the prospect of the village being designated a town (*belediye*) in the Turkish administrative hierarchy. How far do market-recognizing villages emerge because of 'eccentric' individual innovators and risk-takers rather than as a result of some outside stimulus? Conversely, how many non-innovators can be carried by a market-seeking community? Clearly, no attempt can be made to deal adequately with such questions on the basis of one man and his orchard. But, at least, such an atypical case may lead to key questions being asked.

Traditionalism, or the alleged resistance to change in the developing world, is often seen as fatalism, for which three sub-types have been suggested (Niehoff and Anderson 1966) based respectively on anticipation of the supernatural, on reliance on government to ameliorate local circumstances and on a negative attitude to development projects derived from awareness of previous failures. If the first two types of fatalism are decreasing, project negativism clearly is not. It is created by change agents themselves and, as such, may be open to redress by them. In the case of this orchard, the farmer responsible for the changes has undertaken each initiative in the belief that the risk and the effort would be his. In evaluating a particular choice, he responds to controls within his community and to stimuli from outside. The former include his material status at any given time (notably capital and manpower) and the attitudes of relatives and neighbours. The outside stimuli range from the specific efforts of the agricultural extension service to trends within the economy at large.

A distinction between decision-making under conditions of commitment and of evaluation is relevant here (Kolars 1965–8: 37–46). When an initial decision is made, such as one to plant a number of orange trees, the farmer is committed to a series of subsequent actions. Moreover, such a sequence of anticipated events is likely to spread over many years in the case of a fruit grower, as against the shorter-term commitment to grains or ground crops. Evaluative decisions are those which are made prior to the start of a new sequence of commitment. In this orchard case, it is clear that the farmer has sometimes evaluated his progress and modified his plans during a commitment sequence. For example, during the late 1960s, groundnuts were rapidly displacing the orange in the estimation of local growers and there was a sharp slow-down in the rate of citrus planting. Yet, by 1977–8 the orange (notably the tangerine) had regained its prime status and non-citrus as well as less profitable types of orange were being progressively removed to make way for it. The farmer still keeps a small quantity of many crops for home consumption, as if he is ultimately reluctant to commit himself to full monocultural plantation cash cropping, however much he might profit thereby.

Thus the continuing profusion of types of trees in the orchard may indicate four things. It is the work of a farmer who, until very recently, was close to the subsistence level, providing something of everything for the support of family and immediate circle. Secondly stands the farmer's perception of the unreliability of the market: with numerous varieties, he feels to some degree protected against fluctuating prices and demand. A third, operational, factor seems to be that a range of different trees and crops or other commitments enables the individual producer to use his own labour, or that of his immediate relatives, as efficiently as possible with his various concerns making demands at different times. Fourthly, orchards require less labour than do most ground crops and are thus attractive to the producer reluctant to pay rising wages to hired field workers or whose sons are leaving full-time farming. This orchard farmer has already lost one of his two sons to a government job in the town – even as the parental generation 'hits the jackpot' in profits.

The new market-seeking order brings involvement in a regional and national system, whereas the subsistent peasant of the past regarded the outside as unnecessary or even undesirable. Links between village, province and state include infrastructural developments such as roads, credit schemes and improved telephone and postal systems, as well as the efforts of national and regional planners, extension agents and teachers to encourage farmers to produce new crops or to produce them in 'better' ways.

One of the most influential outside stimuli is price, especially where recently near-subsistent farmers may fail to take account of their real production costs and see only the gross receipt on sale to middleman or market. In the county (*ilçe*) where the case-study orchard is located, the number of orange trees doubled between 1945 and 1960–1. Only five years later the total was five times the 1945 figure. Since then the increase in numbers has been less rapid but prices have continued to climb, particularly between 1970 and 1978 when there was a fivefold rise in wholesale rates for oranges. But even these increases were only half the level for tangerine and mandarin oranges, while lemon prices climbed most of all, though the increase in the number of trees in the county has been the least among the citrus fruits considered here.

Another financial stimulus has been agricultural credit at very low interest rates. The citrus grower and others with a high potential agricultural asset to offer were quick to take advantage of cheap credit. At the same time, the authorities became better able to influence the pattern of investment until the individual producer's profit gave him greater freedom in making investment decisions. Lately the appearance of large numbers of tractors, trucks and private cars, along with many dozen huge greenhouses and a range of expansion in domestic building, suggests that a citrus village such as this has reached a stage where growth can be financed from profits; though such growth may give way to over-investment in tractors and equipment as mechanization, perceived as enhancing the independence of the once-subsistent producer, becomes an end rather than a means thereto. Yet even here the authorities can influence the pattern of activity in the village through import duties, other taxes and price-support. Hence the farmer who invests in a tractor acquires a commitment to a series of decision choices over an extended period, just as he acquires a new vulnerability to constraints and stimuli from outside his enterprise and his immediate community.

An outside stimulus directly involved in rural decision-making is the extension service of the Ministry of Agriculture, though there is a shortage of qualified personnel. Those available lack adequate transport and clerical and technical support and perform many regulatory and information-gathering duties of little direct benefit to farmers. Such bleak comments are, however, once again about averages: the coastal-plain villages round Antalya were and are relatively advantaged. Extension agents are frequent visitors to the village of the case-study orchard, and to others in the group of alluvial valley settlements of which it is part. Communications with offices, agencies and suppliers in the nearby provincial town were good and are now better. Occasionally, extension agents set up experimental plantings in this area or arrange demonstrations from which some villagers are quick to copy. Most importantly, relations between farmers and Ministry personnel became easier than they are over much of the country, where infrequent visitors from whatever government office continue to be regarded with a circumspection not conducive to the rapid exchange of ideas and confidence. In short, the innovative farmer responsible for this case-study orchard had the additional advantage of access to agricultural support services. Visible evidence includes the regular plantation layout (a major break with past practices) and the increased spacings between trees which villagers adapted to their own inclination to intercultivate ground crops during the early growing years of orchard trees. Also evident are new techniques for grafting, spraying, pruning and harvesting. Many of these came from brochures accessible to literate villagers in an 'advanced' village such as this or from radio broadcasts for farmers, as well as from face-to-face discussions with advisers.

The Ministry of Education is another branch of government which acts as a stimulus to change in the village, where teachers might explain a brochure during an evening session in a coffee house or give an opinion on some new scheme or crop variety being promoted by the extension service. Meanwhile, the transistor and, more recently, mains electricity have opened the air waves to all villagers, literate or not. An important difference between this village and others closer to the average for the country is that here this fruit grower and his neighbours not only have access to personal advice as a supplement to broadcast, published or hearsay information, but they are ready to question what they hear or to ask officialdom for goods and services. Such villagers have thus passed a take-off point in decision-making.

But attractive prices, credit, extension support and even personal initiative would have meant little without infrastructural improvements such as roads and transportation. In the village where the sample orchard is located, the horse-drawn wooden cart was introduced as recently as 1940 when pack animals were still being used to move the small amounts of produce taken for sale in the Antalya market. At that time there was no all-season roadway to the town. As late as 1960–1 the village was cut off by floodwater long enough for frost to inflict serious damage to half the citrus crop. Concerned villagers persuaded the provincial authorities to link them, by 1977–8, to the regional and national markets by a hard-surface road. This removed a major constraint on village exports and citrus producers were able to take advantage of retail prices in Ankara 50 per cent above local levels. Again, the accident of location is a further advantage, putting the producer who has developed this case-study orchard into a favoured position as compared with his countrymen in other places and other types of productive enterprise. The orchard and its village lie in a part of Turkey which has attracted much public and private investment and so already has a relatively well-developed regional infrastructure of roads and other communications and commercial and administrative support.

Despite his advantages it is not certain that this citrus farmer can maintain the growth momentum of the 1960–78 period. He remains reluctant to abandon the independence of the traditional family enterprise, reluctant to hire labour or to combine with other producers. Meanwhile, costs of imported equipment, fuel and fertilizers seem set to escalate against a background of continuing uncertainty in the Turkish economy.

References

FREY, F. W. (1966) *Regional Variations in Rural Turkey*, Cambridge, Mass., Massachusetts Institute of Technology, Center for International Studies (Rural Development Research Project, Report No. 4).

KOLARS, J. F. (1965–8) 'Decision and commitment in Turkish agriculture', *Review of the Geographical Institute of the University of Istanbul* (International Edition), 11, 37–46.

—— (1967) 'Types of rural development', in Shorter, F. C. (ed.) *Four Studies on the Economic Development of Turkey*, London, Frank Cass, 63–87.

MITCHELL, W. A. (1971) 'Çayırhan: development and modernization in a Turkish village', *The Professional Geographer*, 23, 212–15.

NIEHOFF, A. H. and ANDERSON, J. C. (1966) 'Peasant fatalism and socio-economic innovation', *Human Organization*, 25, 273–83.

STARR, J. (1978) *Dispute and Settlement in Rural Turkey: An Ethnography of Law*, Leiden, Brill (Social, Economic and Political Studies of the Middle East, vol. 22).

CHAPTER FIFTEEN

The settlement of Armenian refugees in Syria and Lebanon, 1915–39

THOMAS H. GREENSHIELDS

While the importance of the ethnic mosaic pattern in the Middle East has long been recognized, the significance of ethnicity as a social force in the twentieth century is now questioned. At the urban level in particular, several writers have suggested a tendency to the disintegration of ethnic clusters as part of a movement towards a new social organization based on socio-economic class structure (Adam 1974; Baer 1964; Churchill 1967; Clark and Costello 1973; De Planhol 1959). Implicit in this argument is the assumption that these ethnic clusters were themselves originally established independently of economic status, that is that they were a reflection primarily of the social organization of ethnic groups and their social relations with their host society. Recent studies in the geography of ethnic groups outside the Middle East, however, have suggested that ethnic population patterns may be largely determined by the economic status of the ethnic group; that ethnic concentration may be a by-product of the concentration of persons of the same economic status (Lee 1973). Such an explanation upsets the simple dichotomy between ethnicity and economic status invoked above. It demands a reinvestigation of the significance of ethnicity and its relationship to economic status in determining the distribution of minority groups in the Middle East. This paper presents some observations on this relationship based on the writer's work on the settlement of Armenian refugees in Syria and Lebanon between 1915 and 1939.

By force and intimidation, about 80,000 Armenian refugees arrived in French-mandated Syria and Lebanon from Turkey in 1920, 1921, 1922–4 and 1929. They came principally to the coastal towns, especially Beirut, and to Alexandretta and Aleppo, the first two large towns on the routes from the north, though some of the 1929 arrivals came directly to the growing settlements of north-east Syria. The 1921 arrivals were partly dispersed to the interior by the French High Commission in order to relieve congestion in the arrival points and to spread them according to the economic absorptive capacity of the country. For similar reasons, and possibly also to avoid offending Turkish susceptibilities, the French authorities dispersed more refugees from Alexandretta in 1922 and again dispersed some of the 1922–4 arrivals from Aleppo to Beirut and Damascus. This government-inspired dispersal was largely responsible for such movement of refugees as did take place from their arrival points during the period. Subsequently, despite the attempts by the League of Nations Nansen Office and various philanthropic organizations to settle the Armenians on the land, the pattern already established, with its strong relationship to arrival points, its overwhelming concentration of Armenians in urban rather than rural settlements and in particular their concentration in the principal cities of Aleppo, Beirut, Damascus and Alexandretta, persisted to a large extent for the rest of the period (Fig. 15.1). In the cities the Armenians settled partly within the existing

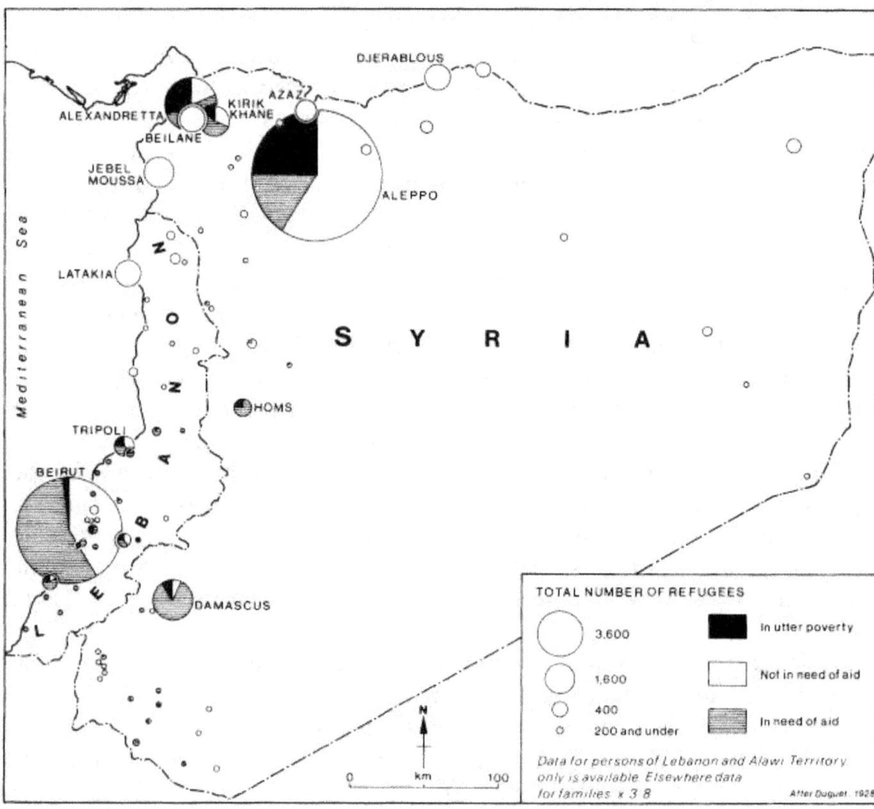

Fig. 15.1 Distribution of refugees in Syria and Lebanon (1928).

fabric, partly within 'camps' or shanty-towns where they remained concentrated on the outskirts. Later, the Armenians in the camps were largely transferred to new Armenian quarters on the city margins, where, however, they still retained their original concentration.

This lack of dispersal and parallel over-concentration were clearly encouraged by the operation of social constraints. Within Syria and Lebanon the Armenians endeavoured to preserve their own national identity. They constructed their own schools, their own churches and ran their own charitable, cultural and athletic societies. In their desire to preserve their own culture they were frankly non-assimilatory, conserving 'une solidarité ethnique sans rivale au monde' (Weulersse 1940). 'The Armenians', wrote Marshall Fox of the Society of Friends, 'as "the Scotch [*sic*] of the Near East" remind one often of the Scotch in their clannishness, thrift and an independence bordering upon arrogance'. Old Armenian communities were reconstituted anew in Syria and Lebanon, a tendency demonstrated most noticeably in the formation of Compatriotic Unions in the towns. These Unions corresponded to more or less spatially distinct communities, within the camps of Aleppo and Beirut, based on town or region of origin. Subsequently the Unions played an important role in the transfer of the Armenians from the camps to new quarters on the outskirts of the towns, such that the community reconstitution apparent in the camps was re-established in the new quarters, receiving its clearest expression in Beirut, where were created for example the quarters of Nor-Adana, Nor-Marache, Nor-Sis and Nor-Hadjin.

In clustering together, the Armenians were not solely concerned to preserve their own culture and community structure. After their traumatic experiences in Turkey they were also genuinely concerned about their own security. Thus, for example, they opposed early French attempts to disperse them on the land unless this should be in large groups. When the Nansen Office settlement scheme commenced, they argued against settlement in the interior and in favour of settlement in allegedly more secure coastal regions. They particularly favoured the Sanjak of Alexandretta, which contained an indigenous Armenian population in Alexandretta, Antioch, the Jabal Moussa and the Jabal Aqra (Qassab). By contrast, settlement in the interior was rejected, not only in the Euphrates region and the southern Hauran but also in the more immediate vicinity of Aleppo. Only in 1929 did the Armenians offer support for Nansen Office plans of extensive resettlement in the Euphrates region and Palmyra, a volte-face which undoubtedly reflected the better conditions of security then prevailing in the eastern regions of Syria but which may also have been related to political considerations. The scheme in any case came to nought, as shortly afterwards the Office was obliged to concentrate virtually exclusively on urban resettlement.

The Armenians' need for security, related partly to their horrific experiences since 1915, was increased by the coolness, sometimes open hostility, of the indigenous Arab population, an attitude which was based on ethnic, political and economic grounds. The criterion for ethnic distinction was not simply religion but linguistic, cultural or 'national' identity. While Syria and Lebanon were together composed of a mosaic of ethnic groups, the overwhelming majority of the population of both territories was Arabic speaking. The strength of Arabism as a cultural-political force among the confessional groups of Syria and Lebanon at this time is difficult to evaluate, but in so far as Syrian nationalism was identified with Arabism the Armenians inevitably ran the risk of being treated as foreigners. The only way that the Armenians could avoid the resulting antagonism was to shed their own national allegiance to which they were so attached. Arab hostility was of course encouraged by the Armenians' own non-assimilatory tendencies and by their preference for concentrating in compact groups. Such ethnic antagonism was not the only cause of friction between Armenians and Arabs but its importance was fundamental. The ethnic label provided the basis for the identification of inequalities and the perpetuation of other grievances and fears. One such fear was political. The confessional mosaic of the Syrian and Lebanese population provided the French authorities with the opportunity to use the Armenians to their own political advantage. When it was felt that this was happening, for example during the Druze Revolt (1925–7) or in certain settlement schemes, hostility was particularly intense. Perhaps most important, however, was the feeling that the impoverished Armenians were lowering the wages of the indigenous labour force by their 'unfair' competition in accepting lower wages themselves. The most striking manifestation of Arab–Armenian hostility occurred during the Druze Revolt when the Armenian refugee camp at Kadem, south of Damascus, was attacked by rebels. The attack precipitated the movement of thousands of Armenian refugees from Damascus to Beirut. This movement was not however typical. Generally Arab–Armenian hostility was more subtle in its influence on Armenian settlement, merely reinforcing the Armenians' need for security and therefore for concentration and segregation.

In the ethnic clustering of the Armenians there were two contrasting social dimensions; the inter-aid and continuity offered by the Compatriotic Unions and the national pride and insecurity felt by all Armenians. While in the Compatriotic Unions original regional or urban identity divided one Armenian from another, insecurity was a property common to all Armenians, irrespective of origin. The extent to which these two aspects of ethnicity were interdependent cannot be assessed without detailed sociological research, but it does seem likely that the insecurity felt by the Armenians collectively would have helped to maintain the regional sub-groups, or indeed that the principal vehicle by which the Armenians sought the security they all required was the regionally exclusive Compatriotic Union. The relationship between ethnic group and sub-group is a fascinating question worthy of more research.

The existence of intra-community sub-groups has been observed in many bidonville situations, a fact which suggests a relationship between the Armenian settlement and the 'normal' processes of settlement of rural–urban migrants under rapid urbanization. In the bidonville situation ethnic sub-groups have been recognized as transient features by some writers, characteristic of rural life and disappearing with increasing urbanization (Adam 1974; Naciri 1963; Prenant 1968; Sebag 1958). An alternative view, and that preferred by the writer, would see these sub-groups as a vital element in the process of migrant adjustment, contributing in time to a new urban synthesis in which the ethnic sub-groups persevere as long as they have this useful urban role to play (Karpat 1976). Ethnic social organization is then seen not as a rural import but as the product of the interaction between the aspirations of the immigrant group and the demands of the social environment. In the Armenian case they imported their own insecurity, nationalism, culture and community structure, but within the host society they encountered both hostility and the problems of adjusting to a new life. Their concentration and refusal to disperse was partly a response to the interaction of these social forces.

The identification of a relationship between Armenian settlement and the bidonville situation suggests a correspondingly strong economic base to Armenian concentration and segregation and it is evident that severe economic constraints were also operating to maintain concentration and inhibit dispersal. The Armenians had arrived in an economy which simply was not able to support them in the cities where employment in industry was actually decreasing. They were obliged, therefore, to accept employment in jobs of low economic status where they were highly vulnerable to economic crises, and they seem to have been unable to afford the expense of movement and reinstallation on the land where, given the capital, the real opportunities lay. Within the cities, as an economically weak population, they were obliged to find accommodation where they could in rented khans and houses. Those unable or unwilling to do this settled *en masse* in camps outside the towns, which developed into shanty-towns where living conditions were unsatisfactory and even dangerous. The Armenians who thus settled in the camps appear to have formed the poorer part of the refugee population, that is those unable to rent accommodation in the town and who enjoyed the freedom from rent and tax which their situation in the camps as *de facto* squatters initially gave them.

Schemes to settle the refugees on the land met with little success. The most ambitious, the Nansen Office scheme, was itself increasingly retarded and finally extinguished by the adoption of alternative schemes of urban resettlement, themselves largely dictated by the development of housing crises in the principal centres of Armenian concentration. Without dispersal the Armenians in the camps came under increasing pressure to move, either from the landowners who desired to evict from their property those refugees unable to pay rent, or from municipal improvement schemes. There was also a fear on the part of the Mandatory authorities that the squalid conditions under which the Armenians were living in the camps would encourage the growth of communism. Thus the great bulk of the refugee population in the camps of Aleppo and Beirut was transferred to new quarters on the outskirts of the towns, while new quarters were established also in Damascus and Alexandretta. In the transfer the economic status of the refugees was not transformed, for there was no real economic progress made in the country to permit this. The transfer was no real solution to the problem and the deprivation which the Armenians had previously experienced in terms of the squalid living conditions of the camps was now expressed in terms of distance from the town centre and lack of urban amenities. While the transfer process radically altered the position of the Armenians in the cities, its effect on the regional distribution was, of course, to perpetuate the status quo that is to maintain the overwhelming concentration of the Armenians in the principal cities.

The appreciation of these economic constraints on settlement again draws attention to the comparison with the bidonvilles of North Africa, which ironically were also attracting attention for the first time from French scholars in the 1930s (Baron, Huot and Paye 1936, 1937). In almost every respect the situation of the Armenians in the camps was a classic bidonville situation, with the same problems of insecure or unregulated tenure, dismal living conditions and forced resettlement which have been observed in the Middle East and elsewhere in so many situations (Adam 1949–50; Bourgey and Phares 1973; Delisle n.d.; Descloitres *et al.* 1961; Harrison 1967; Montagne 1951). The concentration of the Armenians in the cities indeed made their migration to Syria and Lebanon, for a substantial but indeterminate number, a rural-urban migration. Viewed in this light, the settlement experience of the Armenians was essentially a function of their low economic status within an economy whose capacity to support its members was weak even without their presence. It is a matter of individual perception and political definition as to whether their camps became shanty-towns and part of the urban scene and when they ceased to be 'refugees' and came to be regarded and regard themselves as Syrian or Lebanese Armenians. The point is that, whatever the uniqueness of their titular status as 'Armenians' or 'refugees', from the moment of their arrival in Syria and Lebanon the Armenians were inextricably linked to, and part of, the economic system of the receiving states and subject to the constraints imposed by that system.

This is not, however, to replace our first explanation of concentration and segregation in terms of ethnicity with an explanation based purely on economic status, for the ethnic component in the settlement process was demonstrably real. Instead it is necessary to construct a new model embracing the mutual interdependence of ethnic and economic constraints. Both constraints, by acting in the same direction to inhibit dispersal and reinforce concentration, were mutually reinforcing. Thus, increasing concentration helped to foster Arab hostility to the Armenians not only as a compact ethnic group, but also because, being concentrated, the Armenians had that much greater effect on the local economy. As already observed, Arab hostility perpetuated Armenian insecurity and consequent concentration. Similarly, increasing concentration perpetuated the imbalance between the number of Armenians and the number of economic opportunities and hence the operation of economic constraints. Ultimately it was this situation which led to the demolition of the camps and the transfer of the Armenians to the new quarters, a process which not only solidified the existing social structure and concentration, but also diverted funds from proposed plans of dispersal. In brief, while both economic and social constraints acted in favour of concentration rather than dispersal, concentration itself reinforced both constraints. Thus ethnic and economic constraints were interdependent and their principal effect was to maintain a self-perpetuating process of concentration and segregation.

This discussion is of course highly simplistic. It neglects intra-Armenian differences in settlement behaviour and ignores political constraints on settlement which, though also related intimately to social and economic constraints, were less important. It does however shed some light on the significance of ethnicity as a social force in the Middle East and on its relationship to economic status as a determinant of settlement behaviour. It demonstrates, first, the continued operation of ethnicity as a social force in settlement behaviour in the Middle East in the early twentieth century. In the case of such a 'national' minority as the Armenians this is not surprising. De Planhol (1959), for example, has drawn a distinction between 'national' minorities, which appear to be irreducible, and old religious minorities which, he believes, tend either to transform themselves into 'national' minorities or disappear entirely. The discussion demonstrates, secondly, the need, when investigating the settlement behaviour of an ethnic group, to identify also the sub-groups operating within the larger community. In the Armenian case ethnic constraints on settlement appear to have acted at two significant levels, that is at the level of the Armenian community as a whole, and at the level of the Compatriotic Unions. This relationship between group and sub-group behaviour requires further investigation. The third and most important point to emerge from the paper, however, is the danger of attempting to force a conflict between ethnicity and class status as determinants of settlement behaviour. Both constraints may operate simultaneously without mutual opposition. To an extent this paper endorses the argument in favour of an economic base to ethnic segregation, suggested in contemporary studies of non-Middle-Eastern situations, as there was in the Armenian camps clear correspondence between ethnic concentration and economic status. It does not follow in the Armenian case, however, that class status was the sole or dominant determinant of ethnic concentration. Rather ethnicity and economic status were mutually reinforcing interdependent determinants of settlement behaviour. The idea of a dichotomy between 'ethnic' population patterns and 'class-based' patterns may be rejected. Since the acceptance of the breakdown of ethnicity as a social force has hitherto rested on the assumption of such a dichotomy between 'traditional' ethnic settlement patterns and 'modern' class-based patterns, its rejection may have significance not only for the study of contemporary settlement processes but also for the understanding of so-called 'traditional' structures.

References

ADAM, A. (1949–50) 'Le bidonville de Ben Msik à Casablanca', *Annales de l'Institut d'Etudes Orientales d'Alger*, 8, 61–199.

—— (1974) 'Urbanisation et changement culturel au Maghreb' in, Duchac, R. *et al.*, *Villes et Sociétés au Maghreb – Etudes sur l'Urbanisation*, Aix-en-Provence, Centre de Recherches et d'Etudes sur les Sociétés Méditerranéennes.

BAER, G. (1964) *Population and Society in the Arab East*, London, Routledge & Kegan Paul.

BARON, R., HUOT, Ltnt., and PAYE, L. (1936) 'Conditions d'habitation des émigrants indigènes à Rabat', *Revue Africaine*, 79, 875–98.

—— (1937) 'Logements et loyers des travailleurs indigènes à Rabat-Salé', *Revue Africaine*, 81, 723–42.

BOURGEY, A. and PHARES, J. (1973) 'Les bidonvilles de l'agglomération de Beyrouth', *Revue de Géographie de Lyon*, 48, 107–39.

CHURCHILL, C. (1967) 'An American sociologist's view of seven Arab cities', *Middle East Economic Papers*, 14, 13–39.

CLARK, B. D. and COSTELLO, V. (1973) 'The urban system and social patterns in Iranian cities', *Transactions Institute of British Geographers*, 59, 99–128.

DELISLE, S. (n.d.) 'Le prolétariat marocain de Port-Lyautey', *Cahiers de l'Afrique et l'Asie*, 1, 109–228.

DE PLANHOL, X. (1959) *The World of Islam*, New York, Cornell Univ. Press.

DESCLOITRES, R. *et al.* (1961) *L'Algérie des Bidonvilles*, Paris, Mouton.

HARRISON, R. S. (1967) 'Migrants in the city of Tripoli', *Geographical Review*, 57, 397–423.

KARPAT, K. (1976) *Rural Migration and Urbanisation in Turkey: The Gecekondu*, London, CUP.

LEE, T. R. (1973) 'Ethnic and social class factors in residential segregation', *Environment and Planning*, 5, 477–90.

MONTAGNE, R. (1951) 'Naissance du prolétariat Marocain', *Cahiers de l'Afrique et l'Asie*, 3.

NACIRI, M. (1963) 'Salé: étude de géographie urbaine', *Revue de Géographie Marocaine*, 3–4, 13–82.

PRENANT, A. (1968) 'Rapports villes-campagnes dans le Maghreb: l'exemple de l'Algérie', *Revue Tunisienne des Sciences Sociales*, 5, 191–216.

SEBAG, P. (1958) 'Le bidonville de Bourgel', *Cahiers de Tunisie*, 6, 267–309.

WEULERSSE, J. (1940) *Le Pays des Alaouites*, Tours, Arrault.

CHAPTER SIXTEEN

Labour mobility and manpower planning in Tunisia

ALLAN M. FINDLAY

Social mobility is the most potent force in labour redistribution. Irrespective of local employment opportunities, most young people from rural areas of Tunisia aspire to live in the city. They have come to view village life as drab, rejecting the mores and lifestyle of their forebears. Urban residence, albeit in a squalid *gourbiville*, has become socially more prestigious than village life. The penetration of modern transport networks even to the remotest regions has facilitated rural–urban migration, while the presence of relatives and former villagers in squatter settlements around the large cities has promoted an atmosphere conducive to further in-migration. As long as spatial mobility remains the main way by which social mobility may be achieved, massive labour redistribution from the village to the city is likely to persist.

Within Tunisia, participation in the modern urban milieu is only possible in a restricted number of locations: in Tunis, Sousse and Sfax. Other cities certainly draw migrants, but have a less magnetic cosmopolitan atmosphere. Resultant patterns of internal labour migration from the urban void of central and north-western Tunisia (Miossec and Signoles 1977) to the economic foci of the eastern littoral accentuate rather than diminish problems for the manpower planner. Continued population redistribution would appear to add further congestion in the housing and employment markets in zones of in-migration and to accelerate economic depression and social disruption in the hearths of out-migration.

Since independence in 1956 the Tunisian government has had considerable success in encouraging economic growth. Peak levels of economic expansion occurred between 1970 and 1974 when average growth in GNP of 9.7 per cent per annum was achieved. Preoccupation with rapid expansion has been at the expense of considering alternative spatial strategies for development to the highly polarized pattern of activities inherited from the French colonial period. Attempts at decentralization in the 1960s were limited to capital-intensive projects, such as the chemical-processing industries located at Gabes. Despite awareness of the spatial imbalance in employment prospects and selective depopulation of the less developed regions of Tunisia, no effective policy emerged to distribute the labour-absorptive textile factories which were established in Tunisia during the late 1960s and early 1970s. While no development strategy concerning the regional provision of new jobs was enforced, the promotion of a nationwide education system added to the numbers of school leavers seeking office and service jobs in every region of Tunisia. Education and a receptiveness to the mass media have accentuated the desire among the younger cohorts of the rural population, not merely for non-agricultural employment, but for participation in city life. Patterns of labour migration in Tunisia suggest that aspirations for the social advancement associated with becoming an urbanite transcend desires for the limited improvement in occupational status which might accompany employment in a rural factory.

Spatial dimensions of the labour market

Labour migration is necessary for two reasons. First, patterns of job creation and demand are ill matched in Tunisia. Second, opportunities for social and occupational mobility are neither equivalent nor ubiquitous.

Population distribution influences patterns both of labour supply and of consumer demand, whether it be for shelter, sustenance, manufactured goods or welfare services. The largest concentration of population, and consequently the largest labour market, is located in the conurbation of Tunis, which at the time of the 1975 census had a population of 944,000 persons. By contrast, in the vast and largely arid territory of the inland gouvernorats (Fig. 16.1a), there are only four towns of over 20,000 inhabitants and rural areas are sparsely populated.

The pattern of Tunisian population does not correspond directly with patterns of labour supply. The male activity rate for the population aged 15–59 years was 85.5 per cent in 1975, but many members of this 'active' labour force were in fact recorded by the latest census as unemployed and many more were seriously underemployed. For religious and cultural reasons, female participation in the labour market was much lower than that of the male population. Female emancipation has been one of the key objectives of the post-independence political regime, yet only in the Sahel, and most notably in the gouvernorat of Monastir, have significant employment opportunities become available for women in textile and shoe factories (Fig. 16.1b).

Labour qualifications act as a second form of spatial bias. Workers in the interior gouvernorats have received less professional training than in the Sahel and the north-east and this has resulted in considerable spatial variation in the

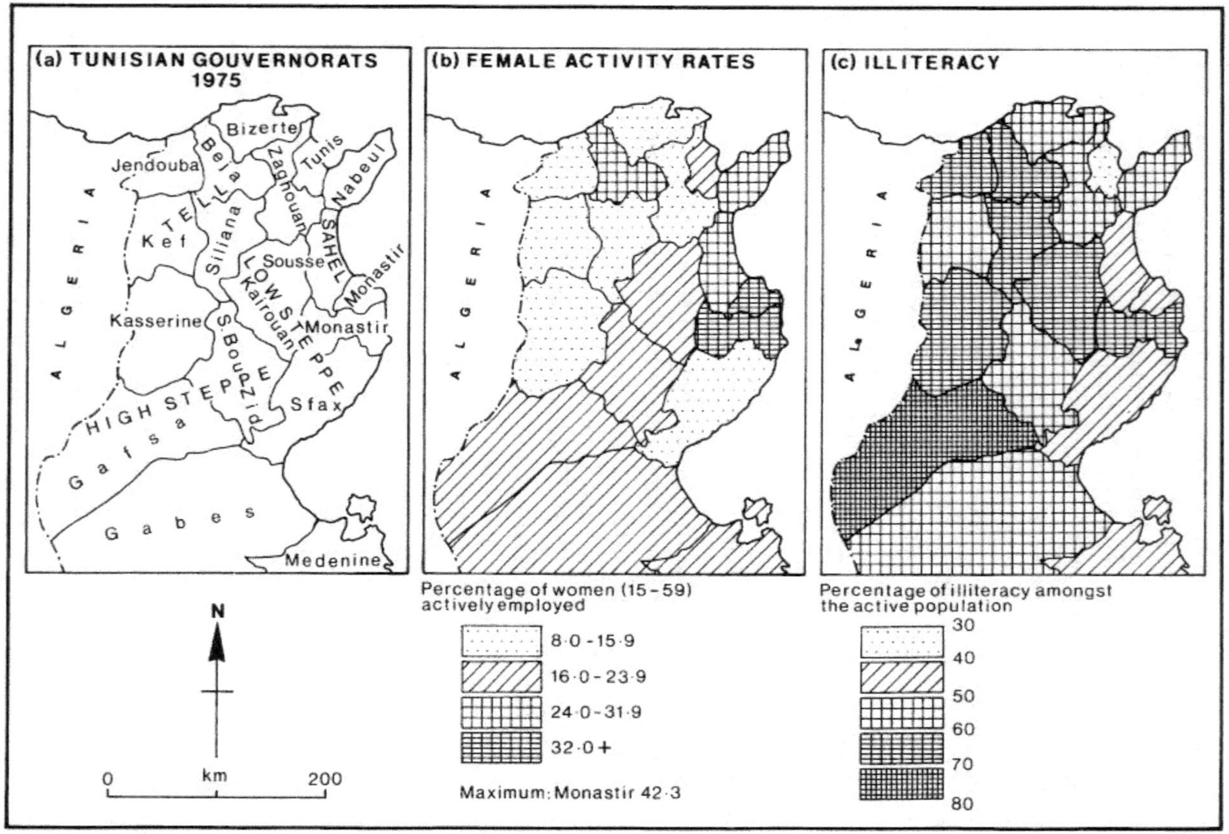

Fig. 16.1 Tunisian gouvernorats, female activity rates and illiteracy, 1975.

skill levels of the labour force. Similarly Figure 16.1c indicates that levels of adult illiteracy are also lower on the eastern littoral where the colonial education system was most influential. In spite of the apparently lower quality of labour supply in the interior and north-western region, labour productivity in manufacturing activities is only marginally below the national average. Value added per employee in the small number of food processing units, such as the Beja sugar factories, is moderately high (1834 dinars p.a.), comparing favourably with the productivity of industrial employees in the Sahel (1698 dinars p.a.). Chemical industries in Gabes and around Tunis assist in giving these gouvernorats the highest levels of productivity, with 3102 dinars and 2830 dinars added in value per employee per annum (Rep. Tunisienne 1976). Value added per employee is also high in light engineering and metal industries throughout the north-east (1349 dinars p.a.).

Patterns of contemporary labour demand mirror the distribution of economic activities established by the French. The Tunisian economy, once based almost exclusively on the export of olives and phosphates, continues to be highly dependent on exports. Although diversification has occurred in the range of activities now earning foreign exchange, France continues to be the dominant trading partner. Susceptibility to swings in the French economy affects the Tunisian labour market not only directly through a reduction in demand for export commodities (often as the result of prohibitive tariff barriers, as in the case of textiles), but also indirectly through reduced demand for migrant workers. Along with other Maghreb states, Tunisia might justly claim that French restrictions on immigration, following the economic recession of 1973, virtually amounted to the export of French unemployment.

Employment in non-agricultural activities is almost entirely confined to the four largest coastal cities of Tunis, Sfax, Sousse and Bizerte. In 1975, 56 per cent of all male employment in the industrial sector was concentrated in the gouvernorat of Tunis (Rép. Tunisienne 1975a). By contrast, the gouvernorats of central and north-western Tunisia accounted for less than 8 per cent of all industrial employment. The small numbers employed in secondary activities in these rural gouvernorats can be closely equated with employment in the construction industry. Investment of migrant remittances in new or improved housing has been one of the chief forces stimulating building activities in these areas. The vicissitudes of international migration policies substantially influence the availability of employment opportunities in the construction sector in gouvernorats such as Medenine and Jendouba.

Throughout Tunisia there is an interesting correlation between the size of the secondary and tertiary sectors, reflecting the interdependence of development in manufacturing and service employment (Fig. 16.2). Tunis has the highest level of involvement in the tertiary sector, resulting from its dominant role as administrative and commercial capital. Rapid growth of manufactur-

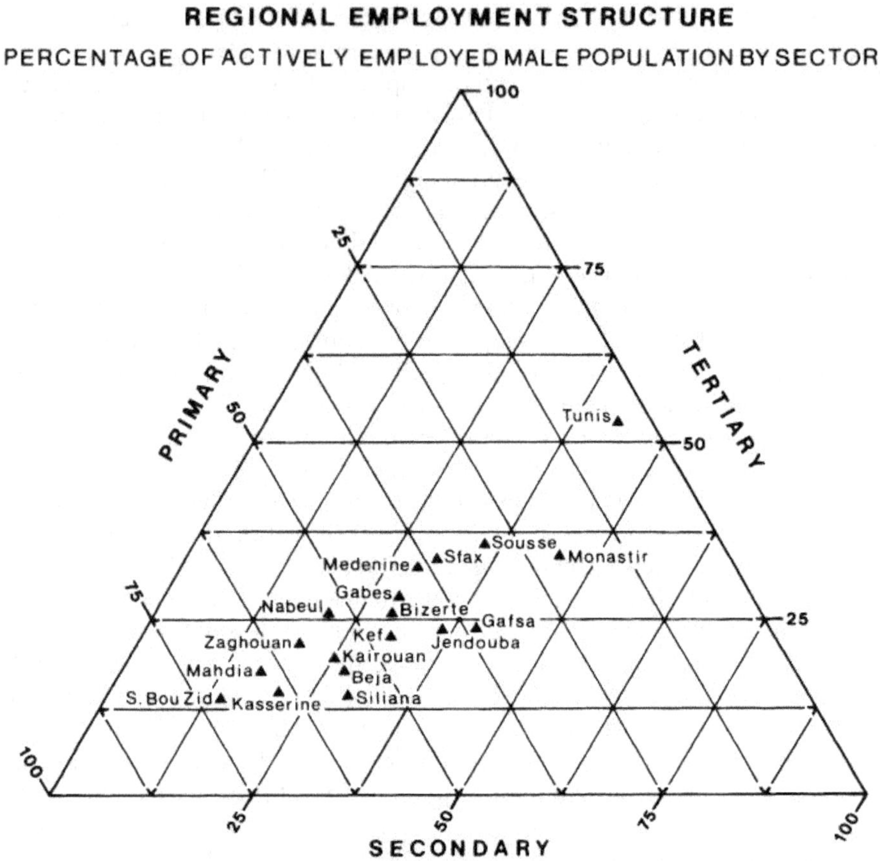

Fig. 16.2 Regional employment structure in Tunisia by gouvernorat, 1975.

ing opportunities, as well as the proliferation of office jobs in the headquarters of national organizations, has encouraged the expansion of the 'dependent informal' sector. Increasing numbers of street merchants, pedlars and bootblacks subsist by providing a wide variety of cheap services to the more fortunate populace involved in modern sector activities. Growth of the 'dependent informal' sector has stimulated further expansion in the capacity of the city to employ migrants. Each migrant family contributes additional demands for city services.

Recent trends in agricultural, industrial and service employment have augmented rather than reduced regional disequilibrium. The apparently intractable problems of rural development have been neither overcome by socialist co-operatives, initiated in the period 1964–9, nor resolved by the subsequent return to economic liberalism. Although indices of agricultural productivity improved substantially between 1970 and 1975, this arose largely from mechanization and technical advance in already prosperous agricultural zones such as Cap Bon. Land reform and amelioration of rural infrastructure has been extremely slow in the Tell and the steppelands. Government programmes for widespread village electrification, sewage disposal and piped water supply were started only in 1974 (Lawless 1976:10). Opportunities for further employment in agriculture remained poor, both within the traditional systems of central Tunisia and in the increasingly mechanized modern sector. In some areas the outflow of labour from agriculture has even exceeded the surplus of potential employees, suggesting that the desire to migrate has outpaced the necessity to find alternative employment.

One of the government's major strategies to boost economic growth has been the encouragement of foreign investment in labour-intensive factory units. It has been disappointing that a stronger initiative has not been taken in controlling the location of the 1600 new enterprises which have taken advantage of the government policy; 40 per cent of the jobs realized between 1973 and 1975 were located in Tunis and a further 22 per cent in the gouvernorats of Sousse and Monastir (API 1976:103), thus reinforcing existing industrial patterns (Bouthier *et al.* 1976).

Responses to imbalance

Quantitative and qualitative dislocation between labour supply and demand is partially reflected in the high levels of Tunisian unemployment. Table 16.1 shows that the national male unemployment rate was above 15 per cent at the time of both the 1966 and 1975 censuses. The highest levels of unemployment

Table 16.1 Male unemployment, 1966 and 1975

	1966		*1975*	
Age group	*Numbers (thousands)*	*Rate (%)*	*Numbers (thousands)*	*Rate (%)*
15–19	40.7	45.4	77.7	37.8
20–4	21.1	16.3	50.2	24.7
25–9	17.9	13.1	15.4	10.4
30–4	16.1	11.8	8.6	7.2
35–9	13.7	10.9	9.8	7.2
40–4	10.5	10.7	9.4	7.1
15–59	147.5	15.5	192.8	15.8

Source: Rép. Tunisienne 1977.

were to be found in the 15–24 age group, precisely the age band most prone to migratory movements.

The migration of unemployed persons in search of work prevents reciprocity of patterns of unemployment and labour demand. In 1975 rural unemployment rates were generally higher than those of urban areas, rising to over 20 per cent in central and north-west Tunisia and peaking at 33 per cent in the gouvernorat of Jendouba. On the other hand, the largest number of unemployed was to be found in Tunis.

High underemployment further reinforces labour-market congestion in the city of Tunis. A survey of in-migration and employment carried out in the conurbation in 1972–3 suggested that 17.8 per cent of migrants were seriously underemployed (working four days or less per week). A high percentage of workers hoped to find a better position, reflecting a sense of thwarted occupational mobility amongst the active population (Rép. Tunisienne 1974: 34 and 42). Although moderate occupational mobility existed in the immediate post-independence years as a result of the restructuring of the employment market, by the 1970s a more inflexible occupational hierarchy was already emerging. The trend towards lower occupational mobility has not reduced levels of spatial mobility. In 1975 319,000 persons, or over one-third of the population of Tunis, were in-migrants to the capital. The largest migrant flows from each gouvernorat were without exception destined for Tunis (Rép. Tunisienne 1975b). Rural districts of the Tell, the northern steppes and the gouvernorats of Medenine, Gabes and Mahdia appear to have been the chief zones of out-migration during the last decade (Koelstra and Tieleman 1977: 48). The internal redistribution of human resources has raised problems of social disruption in areas of departure and of re-integration in areas of in-migration.

Of even greater concern than the problems associated with restructuring regional economies to accommodate internal migration has been the upsurge of international migration, which has involved the total loss to the nation of valuable manpower. In 1975 there were some 174,000 Tunisians living in France and an estimated 50,000 in Libya. In 1977 alone 27,300 Tunisians departed for Libya, mostly from the southern and central gouvernorats. While merely reducing surplus labour supplies in some districts, the volume of emigration has severely drained the young and most able persons from the active labour force in the zones of highest emigration. Shortages of pickers were recorded at the time of the olive harvest and in the confectionery trade the government was forced to restrict the emigration of bakers during 1977 (Findlay 1978).

In areas of labour surplus, emigration to France and Libya has left a residual population of unskilled agricultural workers and elderly and dependent persons. Ironically one of the chief forces encouraging increased female employment in peripheral areas has been the emigration of the male workforce. Migrants have left their wives at home to maintain their agricultural holdings.

Despite a decade of heavy internal and international migration from the mountainous north-west, rates of unemployment amongst first-time job seekers remain the highest in Tunisia and news of the success of migrant workers has only increased local dissatisfaction with the limited employment opportunities of the regional labour market. In other areas of departure, such as Medenine gouvernorat, estimates of unemployment and underemployment are much lower, yet emigration has persisted, the local population finding job opportunities in Tunis, France and Libya preferable to local employment.

Migration, because it does not rectify the root causes of imbalance in the labour market, increases rather than redresses regional disparities.

Development and mobility

Worker migration and the development of new employment opportunities are not dichotomous options in manpower planning. Unsatisfactory job-creation schemes may encourage migration. Inversely, the viability of new economic projects may only be assured by a limited level of intra-regional labour redistribution.

The short-term policy of the Tunisian government towards rural unemployment has been to create work gangs to repair roads and improve agricultural structures. Attia (1966) considers that this form of temporary wage employment has done no more than transform the peasantry into a mass of day labourers. By accentuating local consciousness of underemployment and unemployment, it caused new values to emerge, and in particular a new perception of the utility of time. This in turn further stimulated migration.

The few experiments in longer-term development which have been undertaken in the interior of Tunisia have proved that industrial decentralization is no simple panacea to rural employment problems. Trabelsi (1975) has studied two infant industries – a tannery and a woollen mill – implanted in the steppe village of Hadjeb el Aioun. Amongst other problems, he found that the factory workers were not content with their recent upward mobility. Furnished with their newly acquired skills, many wished to migrate to still better jobs in Tunis and some 28 per cent wanted to work abroad.

The high turnover of employees and the constant need for training new workers adds to the costs of rural manufacturing. Not all projects succeed in utilizing local labour. For example, a factory set up at Kasserine was only able to draw 37 per cent of its workers from the locality (Boukraa 1968). Even in the industrial pole of Gabes a new chemical plant had to bring 12 per cent of its workers from the more favoured gouvernorat of Sfax, since the local labour pool was insufficiently skilled.

The change of work routine from the seasonal and daily rhythms of agriculture to the more rigid schedules of factory or office employment cannot be achieved rapidly. Even in the progressive village societies of the Sahel, such as Ksar Hellal, the introduction of factories has disrupted family life and heralded a revision of social values (Auerbach 1975). In summary, rural developments face one of two problems: either the local labour force is insufficiently skilled or lacks motivation to enter new factory employment, or else, if the workforce is capable of attaining the necessary skills, aspirations for social mobility amongst the local population exceed the opportunities which rural development can provide.

Migrants employed in even the most menial manual occupations find themselves held in high esteem when they return to rural areas from Tunis or foreign cities. In short, there is 'status inconsistency' between village and city (Abadan-Unat 1974). Absence of urban lifestyle rather than lack of non-agricultural employment is the critical factor in determining propensities to emigrate from rural Tunisia. The absence throughout most of Tunisia of urban infrastructure on a comparable and competitive scale with that of the cities of the north-east is the primary factor maintaining high levels of spatial mobility. As long as migration continues to be the easiest mode of achieving social advancement, the exodus of population from large rural areas is likely to persist regardless of local employment opportunities.

Conclusion

Labour mobility involves more than the spatial movement of workers. Migration from the less favoured peripheral regions towards Tunis and other urban cores has permitted social but not necessarily occupational mobility. Inversely, the development of rural industries in isolation from the urban environment has provided opportunities for occupational advancement, but has not offered an acceptable level of social mobility. Paradoxically, attempts at industrial decentralization undertaken independently from strategies for urban expansion have facilitated rather than halted rural–urban migration. Examination of the threefold nature of labour mobility – spatial, occupational and social – offers some resolution of this paradox and of associated problems of Tunisian manpower planning.

References

ABADAN-UNAT, N. (1974) 'Turkish external migration and social mobility', in Benedict, P. *et al.* (eds) *Turkey, Geographical and Social Perspectives*, Leiden, Brill.

AGENCE DE PROMOTION DES INVESTISSEMENTS (API) (1976) *Direction des études et du suivi, rapports*, Tunis.

ATTIA, H. (1966) 'Structures sociales et évolution en Tunisie Centrale', *Revue Tunisienne de Sciences Sociales*, 6, 1–41.

AUERBACH, S. (1975) 'Occupational options and adaptive strategies in a Tunisian town', unpublished PhD thesis, University of Illinois at Urbana Champagne.

BOUKRAA, R. (1968) *L'Entreprise industrielle en milieu rural*, Tunis, Publication du Bureau de Recherche Sociologique.

BOUTHIER, M., MIOSSEC, J. M. and SIGNOLES, P. (1976) 'Une approche de l'espace industriel Tunisien', paper presented to *IVe Colloque de Géographie Maghrebine*, Tunis, 15–20 November 1976.

FINDLAY, A. M. (1978) 'Tunisia', *International Migration Project Working Paper*, University of Durham.

KOELSTRA, R. W. and TIELEMAN, H. J. (1977) *Développement ou migration?* Hague, Remplod.

LAWLESS, R. I. (1976) 'Progress and problems of Maghreb agriculture', *Maghreb Review*, 1 (3), 6–11.

MIOSSEC, J. M. and SIGNOLES, P. (1977) 'Priorité de l'échelle nationale dans l'urbanisation et l'organisation spatiale de Tunisie', paper presented to *Table ronde sur l'urbanisation au Maghreb*, Tours, 17–19 November 1977.

REPUBLIQUE TUNISIENNE, INS (1974) *Enquête migration et emploi*, Fasc. 2, Tunis, Ministère du Plan.

—— (1975a) *Recensement des activités industrielles*, 7, Tableaux statistiques 1975, Tunis, Ministère du Plan.

—— (1975b) *Recensement général de la population et des logements*, Tunis, Ministère du Plan.

REPUBLIQUE TUNISIENNE, DAT (1976) *Eléments pour une décentralisation industrielle en Tunisie*, Tunis, Ministère de l'Equipement.

REPUBLIQUE TUNISIENNE (1977) *Le Chomage et le sous-emploi à la veille du Ve plan*, Tunis, Ministère du Plan.

TRABELSI, M. (1975) *L'Industrie et son impact sur le développement des zones rurales*, ronéotypé, Tunis, CERES.

CHAPTER SEVENTEEN

The industrialization of Libya

MOHAMMED EL MEHDAWI

Until recently the Libyan economy was clearly a deficient economy. Before the discovery and exportation of oil the prospects of economic and social development were discouraging to Libyans and foreigners alike (Higgins 1968). The country was classified as having one of the lowest standards of living in the world. The majority of the people led a simple life, dependent essentially on a subsistence level of food, clothing and housing. They had very limited knowledge of modern technology, about 90 per cent of the population being illiterate (Murabet 1964). More than 80 per cent depended for their livelihood on animal husbandry and agriculture, which were faced with extraordinary difficulties such as the inadequacy and unpredictability of rainfall, tribal ownership and tenure of land, lack of credit and primitive methods of production (Attiga 1973: 12).

Industry in Libya until the 1960s was limited in activities and production and this may be attributed to the extensive backwardness of the Libyan economy prior to the discovery and exploitation of oil. The main industry of Libya at present is, of course, the oil industry, but it is not within the scope of this chapter, which is concerned with manufacturing. The importance of oil for the manufacturing industry, however, lies in the fact that the government derives over 60 per cent of its annual revenue from that source. The value of exports of Libyan oil increased from about LD 196 million in 1964 to about LD 2375 million in 1974. Moreover, the activities of the oil companies have benefited the economy of the country through local purchase of materials, services and remuneration of employees at an annual average of over LD 60 million.

Manufacturing has been included in official policy in the development plans and has made significant progress in recent years, particularly during the 1970s. In fact, this sector has gained in terms of its contribution to the GDP, increasing from about LD 11.5 million in 1964 to LD 74.5 million in 1974

(Ministry of Planning and Development 1974). There is no doubt that the oil industry has had a marked effect in stimulating the growth of manufacturing.

Phase 1: before 1963

The industrial sector of Libya has passed through different stages of development. During the early years of this century industry was mainly in the nature of handicrafts. There was no opportunity for modern industry owing to the simple life of the people and the lack of capital, transportation, skilled labour and power supplies. Moreover, the Turkish government concerned itself only with collecting as much additional revenues as possible from its colonies. Traditional craft industries such as weaving, carpet making and rug weaving using hand looms, leatherwork, copperware, metalwork, woodwork and mat making were developed from local materials for home consumption. Production was carried on chiefly in bazaars, where retailing also took place.

During the Italian period (1911–43), some traditional industries suffered serious setbacks because of their inability to compete with more highly developed forms of European manufacture. The statistics show that in 1938 about 1900 Italians were engaged in various types of crafts (Bank of Libya 1970), most of them, such as photography, decoration and marbleware, having been only recently developed. However, after 1920 some small modern factories were introduced into Libya by the Italians; by 1938, 789 establishments making basic goods for consumption in the country were operating mainly in Tripoli, Benghazi, Derna and Misurata, 81 per cent of them being in the hands of Italians. The more important establishments were those producing building materials, metalwork and agricultural foodstuffs, particularly olive oil, macaroni, flour, soft drinks and tobacco.

Moreover, there was direct government investment in industry. Industries which engaged in processing agricultural and fishing products received about 191 million Italian lira. Consumer industries such as textiles, leatherwork, printing and paper received about 37 million lira. The introduction of such modern industries was due to colonial motives to assist in the settlement of Italian families in Libya.

Following the Second World War, Libya experienced a short period of political instability. A large number of the factories had been completely destroyed. The British Foreign Office estimated that the number of establishments which remained in operation was over 100 (FAO 1952: 72).

The Libyan industrial picture changed slightly during the 1950s. With UN assistance, two training centres were established in Tripoli and Benghazi. According to a 1956 census, there were 3121 industrial establishments in Libya, 87 per cent of which were small-size establishments employing less than ten workers. They were highly localized, Tripoli alone containing about 79 per cent of the total workers (IBRD 1960).

Phase 2: after 1963

In August 1963 the country's first five-year development plan was officially approved at an estimated cost of more than LD 169 million for the period 1963–8 (Lawless 1974). Since then considerable attention has been focused on developing the industrial sector. This concern is evidenced by the various government programmes: Industrial Research Centre, Real Estate Industrial Bank (REIB) and National Public Organization for Industrialization (NPOI) as well as tariff protection. But the government efforts before 1969 were different from those after that date. The most important changes were in the role of industry in the development plans. The share of the industrial sector in the plan was increased from 3 per cent in 1963–8 to 15 per cent in 1972–3, and to 19 per cent in 1976–80 (Ministry of Planning and Development 1976).

During 1965–76 capital investment in the public sector for manufacturing industries totalled about LD 606.2 million and averaged about 50.2 million per year (Table 17.1). After 1969 industrial development increased remarkably because the public sector began to receive more attention. With respect to direct investment in industry, the government before 1969 invested in about eight establishments as well as providing loans for private investors.

Table 17.1 The investments in the development of the Libyan industrial sector 1965–76 (in LD million)

Year	*Total amount planned*	*Actual value of investment*
1965–6	4.3	2.0
1966–7	5.8	4.7
1967–8	5.3	7.4
1968–9	7.7	5.0
1969–70	7.9	5.4
1970–1	20.9	15.0
1971–2	37.0	29.0
1972–3	68.1	65.1
1973	79.7	67.5
1974	133.7	135.4
1975	129.7	117.6
1976	199.4	152.1

Source: Secretary of Industry (1977) *The Achievements of the Revolution in the Industrial Sector*, Tripoli, 22.

However, in addition to the direct investment for industry one must also emphasize the activities of REIB and NPOI. Commercial banks seemed to be rather shy of offering loans to industries owing to the risk involved. During the period 1958–65 the share of industry derived from commercial banks was less than 13 per cent per year (Bank of Libya 1970). The REIB was therefore created in 1965 to help finance a programme of housing and the industrialization of the country. The bank was permitted to make loans to the industrial sector, to participate in the equity capital of industrial companies and to facilitate the import of machinery, raw material and technical information (el Mehdawi 1974).

As far as loans were concerned, until September 1970 the total value of the industrial loans granted by the bank amounted to about LD 11.7 million. During the period 1966–9 loans for over 60 per cent of the total individual investments were at a level exceeding LD 50,000, but after that period most of the loans were for small and medium-sized projects, based on a new policy which limited loans to LD 60,000. The loans were interest-free and chiefly for medium terms, more than half of these being for a period of three to seven years.

The major industry of food processing, notably tomato canning, olive oil processing, bakery and soft drinks, obtained more than 50 per cent of the total amount advanced by the bank; woodwork and metalwork received rather less. Moreover, it is clear from Table 17.2 that there are inequalities in the distribution of loans made by the bank to industry throughout the country. The main city regions of Tripoli and Benghazi have received about 65 per cent

Table 17.2 The distribution of the loans among Libyan muhafadat 1965–76

Muhafadat	*No. of loans*	*Total amount (in LD)*	*Percentage*
Tripoli	297	4,555,576	45.0
Benghazi	151	2,014,703	19.9
El-Zawia	121	1,585,133	15.7
Misurata	38	341,408	3.4
Homes	70	385,391	3.8
Derna	44	332,463	3.3
Gabal Akhdar	46	307,465	3.0
Gharian	50	333,000	3.3
Sebha	42	155,610	1.5
El-Kaleg	15	107,704	1.1
Total	874	10,118,453	100.0

Source: REIB (1977) *Statistical Information of the Bank's Activities 1965–76*, Tripoli, Share Press, 5.

of the total loans during 1965–76, while less than 2 per cent was spent in the desert muhafadat of El-Kaleg and Sebha.

In addition to the new policy, in 1970 the NPOI was established as the major organ for implementing the development plan for the public sector. The paid-up capital of the organization increased from LD 6 million in 1970–1 to about LD 230 million in 1973–5. This was of course due to the large sums of money available for investment in the industrial sector. The number of projects went up from five in the first year to about forty-three, employing about 8030 workers in 1977 (Secretary of Industry 1977). Moreover, a number of public-sector industrial establishments are either under construction or being actively considered. Up to 1980, it will deal with ninety-one projects with a total capital investment of LD 500 million. Most projects received more than LD one million, and one received more than LD 200 million. Foodstuffs, building materials, metalwork and engineering industries were the major industries to gain projects, receiving sixty-seven in all. This means that the NPOI's aim is to give priority to industries designed either to further the development of the country's natural resources or to manufacture items that have previously had to be imported.

Industrial structure

Employment and the number of establishments are used as variables throughout this section. They have been favoured over other variables because of the availability of detailed sources. Adequate sources for the employment and the number of the establishments in Libya are available in the 1971 Industrial Census. In 1971 the ten administrative regions (muhafadat) contained 2388 industrial establishments with more than two workers, providing jobs for 18,699 workers with an average of about 7.8 persons for each establishment. The study of the industrial distribution offers a typical case of limited industrial concentration among industrial groups and industrial regions. Table 17.3 illustrates that the industries established were mainly those which used local raw materials to meet basic and immediate consumption needs and which do not require a high standard of skill or large amount of capital. Food, building materials and woodwork, the most important industries, actually employ nearly 80 per cent of the total industrial workforce. Food processing easily dominates the industrial structure, employing about half the workforce. It includes olive-oil processing, milk products, fish canning, tomato canning, pasta, fruit juice, icecream, sweets, flour and animal fodder. These industries are mainly dependent on local agricultural products or, in some cases, on raw materials which are imported though they are obviously more economically produced locally. But in general these industries which depend on local agricultural goods are not well

Table 17.3 Distribution of workers and establishments among industrial groups in Libya, 1971

Industrial groups	*Number of establishments*	*Number of workers*	%
Food, soft drinks and tobacco	1,343	9,270	49.6
Textiles and clothes	42	799	4.3
Woodwork and furniture	508	2,774	14.8
Paper and printing	40	893	4.8
Chemical products	35	817	4.4
Building materials	226	2,824	15.1
Metal and engineering	176	1,273	6.7
Others	18	49	0.3
Total	2,388	18,699	100.0

Source: Industrial Research Centre (1973), *Manufacturing Industry in Libyan Arab Republic*, Tripoli, 88.

developed, partly because these commodities are mostly consumed fresh. The building-materials industry includes pipes, tiles, cement, bricks, marble and limestone, while woodwork products were mainly articles such as doors, windows, tables, wardrobes, chairs, desks and some decorative items.

Imported raw materials which are essential for use in the industries are mostly semi-finished commodities, including wheat flour, fat, sugar, wood, timber, cork, cotton, artificial and natural spun silk, wool yarn, plastic, rubber, metal and aluminium sheets, paint, dyes, colouring materials and yeast. The Common Market countries, Lebanon, India, Ceylon, Greece, Sweden, Poland, Romania, Yugoslavia and Czechoslovakia are the main suppliers.

Industrial location

Table 17.4, which indicates the distribution of the industrial establishments among the muhafadat in 1971, shows a marked concentration in certain localities mainly in the north-west and the north-east. Looking at individual muhafadat, Tripoli leads with about 37 per cent of the establishments, followed by Benghazi with 15 per cent, while the least important muhafadat are El-Kaleg (1.2 per cent), Sebha (1.8 per cent) and Gabal Akhdar (2.5 per cent). The muhafadat with the most industrial activities obviously are those with the major urban areas. It is clear from the figure that the roughly triangular area in the north-west of the country, with corners at Misurata, Yefren and Zwara, is the major industrial area in the country, and contains about three-quarters of the total establishments.

Table 17.4 The regional distribution of industrial establishments in Libya, 1971

Muhafadat	*Food & soft Drinks*	*Textiles & clothes*	*Wood & furniture*	*Paper & printing*	*Chemicals*	*Building materials*	*Metalwork*	*Others*	*Total*
Tripoli	364	30	257	23	32	68	95	11	880
Benghazi	92	5	118	16	3	79	55	6	374
El-Zawia	195	5	25	—	—	9	—	—	234
Misurata	98	—	29	—	—	28	13	—	168
Homes	153	2	4	—	—	5	2	—	166
Derna	39	—	37	—	—	22	4	—	102
Gabal Akhdar	36	—	17	1	—	2	3	—	59
Gharian	324	—	4	—	—	5	4	—	337
Sebha	20	—	12	—	—	7	—	1	40
El-Kaleg	22	—	5	—	—	1	—	—	28
Total	1,343	42	508	40	35	226	176	18	2,388

Source: See Table 17.3.

In terms of industrial groups, food processing, building materials and woodwork are seen to be widely scattered and exist in all of the muhafadat, while textiles, paper and printing and chemical industries are concentrated in two or three muhafadat. Tripoli, however, is notable for its high numbers in all industrial groups; it has more than 25 per cent of the total number of the establishments in every industrial group. Benghazi comes second in most of the industrial groups. In El-Zawia, Misurata, Homes, Gharian, Gabal Akhdar, El-Kaleg and Sebha, industry is dominated by the food group.

The data from the 1971 Industrial Census may also be used to give some insight into the size and structure of the establishments. From Table 17.5, which illustrates the distribution of industries among three groups, it is clear that industry in Libya is dominated by small-sized establishments (with five to nineteen workers), which account for about 72 per cent of the total establishments. The food, woodwork and metalwork industrial groups are chiefly characterized by small establishments which account for more than 90 per cent of all the small establishments. Large establishments, on the other hand, account for less than 7 per cent of the total number of establishments.

Table 17.5 Numbers of industrial establishments by size in Libya, 1971

Industry	*Small establishments*	*Medium establishments*	*Large establishments*
Food, soft drinks and tobacco	1,065	221	57
Textiles and clothes	17	12	13
Woodwork and furniture	379	116	13
Paper and printing	9	25	6
Chemical products	9	13	13
Building materials	115	68	43
Metal and engineering	121	38	17
Others	15	3	—
Total	1,730	496	162

Source: Industrial Research Centre (1973) *Manufacturing Industry in Libyan Arab Republic*, Tripoli, 93.

Turning to the spatial distribution of the industrial establishments by size (Fig. 17.1), small-sized establishments are in general more dispersed throughout the muhafadat than the medium and large establishments, which are most prominent in Tripoli and Benghazi and least prominent in Gabal Akhdar, Gharian, Sebha and El-Kaleg. Furthermore, Tripoli is the dominant muhafadat for all categories, for example 103 out of 162 large establishments, about 64 per cent of the total, are located there. It seems that food, soft drinks,

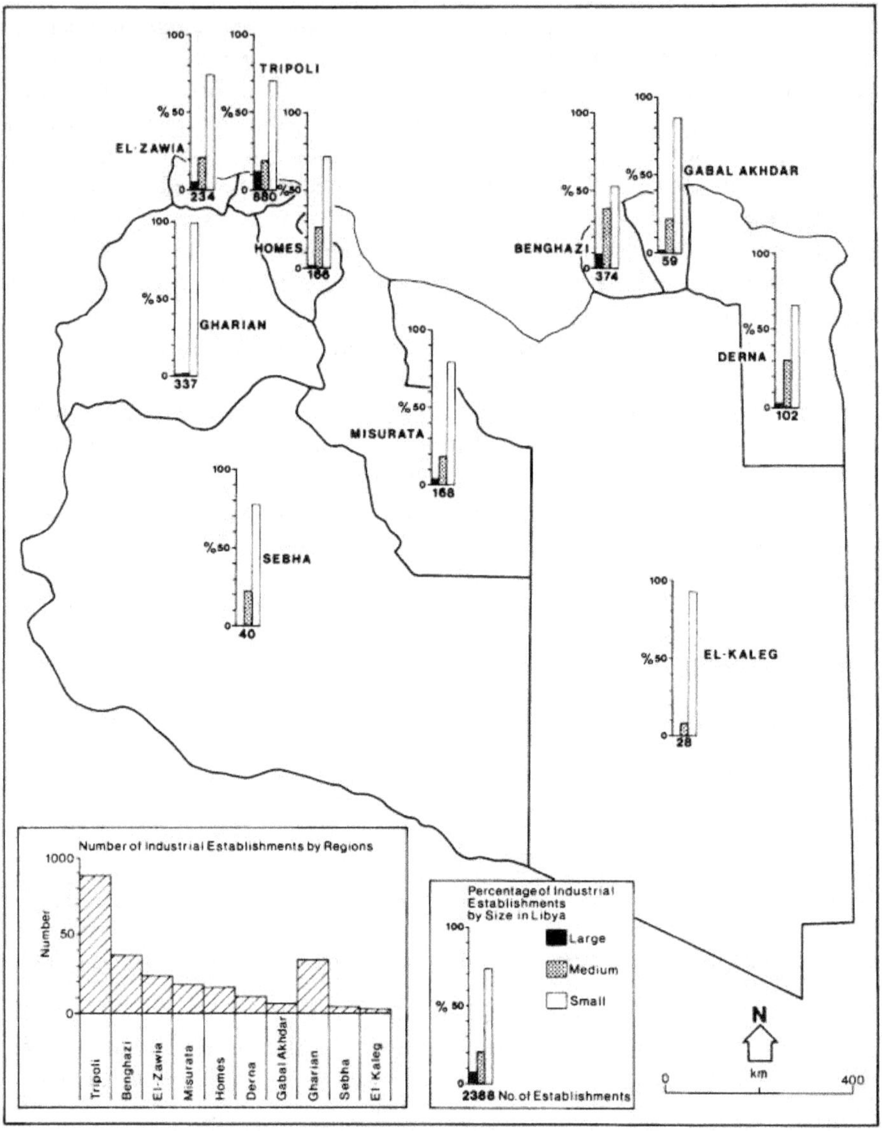

Fig. 17.1 Libyan industrial establishments by muhafadat and size, 1971.

woodwork and building materials are more widely distributed, because these industries producing perishable items, or goods which gain weight, have relatively high transport costs.

Conclusions

From the above characteristics of industrial distribution one can draw a number of conclusions. First, there is a strong correlation between population distribution and industrial location (Fig. 17.2), in particular an obvious contrast between north and south in terms of population and its influence on the market. The north-west and north-east enjoy favourable resources, ports and adequate transportation, and most of the early development-plan projects were concentrated in the north.

Another significant feature of the distribution of industry is that large establishments tend to concentrate in muhafadat which have larger cities. Moreover, industries utilizing local materials tend to be distributed widely among muhafadat, while those using imported materials tend to be concentrated at the ports.

Tripoli muhafada, and in particular Tripoli city, is the most important industrial area in Libya. The main advantages of Tripoli lie in the functions of

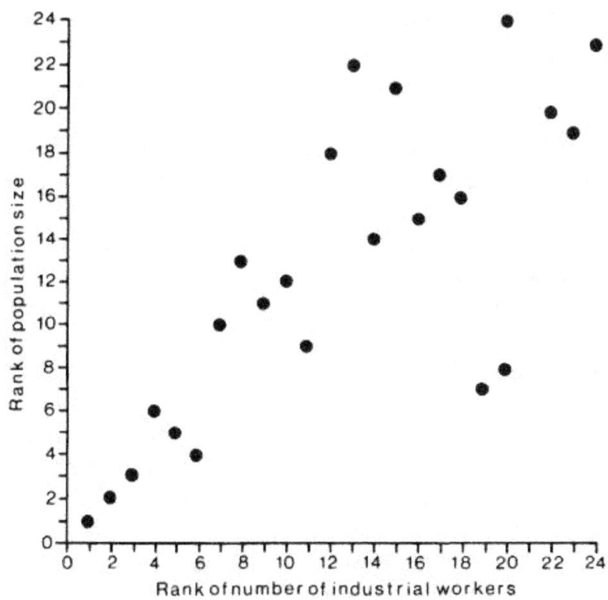

Fig. 17.2 Population size and number of industrial workers of Libyan localities, 1971.

the city as the capital of the country and centre of most government administration, banks, insurance companies, oil companies and of industrial research. The concentration of economic activities carried out by national or international agencies which have moved from Benghazi and Beida also acts as a strong pulling factor in Tripoli. These facts cause it to receive more government investment than any other urban area. It also attracts as the main market, its 1973 population of about 551,000 making it by far the most significant centre of production and consumption in Libya, as indicated by the concentration of skills and of higher incomes. Moreover, as the country's main port and centre of the major transport network, Tripoli has developed as transport node for the whole of western Libya. A large number of buses daily serve areas containing at least three-quarters of the population of the western municipalities.

Despite the industrial development progress and successes achieved in recent years, industry is faced with several problems. The most important lie in the shortage of skilled workers, the limitations of the market and the lack of raw materials apart from agricultural products, building materials and petroleum. The government is trying to solve these problems; if they were solved, Libya's rate of industrial action would be greatly increased. The shortage of skilled workers is temporarily solved by a large influx of foreign workers numbering several hundred thousand in the late 1970s, but this is not without difficulties, particularly their concentration in the major cities and social segregation.

It is impossible to predict with confidence the specific direction of the country's industrial future. It does seem likely, however, that some large and medium-sized establishments for some industries might be profitable in the near future. The oil industry, especially the petrochemical industry which depends on natural gas and crude oil, has a good future with an enormous potential in the country. It can provide other industries such as chemical industries with raw materials. Furthermore, certain industries will achieve a rapid expansion, apart from the food industry, while the increased demand for building materials and the availability of local raw materials such as calcium carbonate, limestone and clay strongly indicate that the building-materials industry, especially cement production, may have a successful future. Manufacture of iron and steel in the large establishment at Misurata will be the basis of the metal industries and engineering equipment.

All these form part of the 1976–80 development plan, which gave high priority to the industrial sector, devoting LD 1090 million or 19 per cent of planned investment. Eighty per cent of this sum is to be allocated to extensions and new projects, 3 per cent to training, industrial research, industrial estates and traditional industries and 17 per cent to industrial and housing credit (Ministry of Planning and Development 1976). The plan is heavily weighted with large investments in industries requiring large capital sums, especially those which will take maximum advantage of the country's resources. Furthermore, an important feature of this plan is that it emphasizes decentralization of industry. More than half of the new projects are located outside Tripoli and Benghazi in nineteen other towns spread out all over the country. Some of these towns, like Misurata, Homes, El-Zawia, Gharian, Jansour, Derna and El-Marj, are destined to become industrial centres. So some of the industrial concentration upon Tripoli and Benghazi will undoubtedly be diminished.

Bibliography and references

ATTIGA, A. (1973) 'The economic impact of oil on Libyan agriculture', in Penrose, E. (ed.) *Agriculture and Economic Development*, London, Frank Cass.

BANK OF LIBYA (1970) 'The Economic Research Division', *Economic Bulletin*, Tripoli, vol. 10.

—— (1974) *Economic Bulletin*, Tripoli, vol. 11.

EL MEHDAWI, M. (1974) 'Real Estate Industrial Bank. A geographical study', *Economics and Business Review*, 10, 45–59.

FAO (1952) *Report to Government of Libya on Agriculture*, Rome, UN.

HIGGINS, B. (1968) *Economic Development, Problems, Principles and Politics*, New York, W. W. Norton & Co.

INDUSTRIAL RESEARCH CENTRE (1973) *Manufacturing Industry in LAR*, Tripoli.

INTERNATIONAL BANK FOR RECONSTRUCTION AND DEVELOPMENT (IBRD) (1960) *The Economic Development of Libya*, Baltimore, Johns Hopkins Press.

LAWLESS, R. (1974) 'Making the desert places bloom', in *Libyan Special Report, Commerce International*, 105, 37–43.

LINDBERG, J. (1973) *A General Economic Appraisal of Libya*, UN Council for Libya.

MINISTRY OF PLANNING AND DEVELOPMENT (1974) *Statistical Abstracts 1974*, Tripoli.

—— (1976) *Social and Economic Development Plan*, Tripoli.

MURABET, M. (1964) *Facts About Libya*, Valetta, Malta Progress Press.

REIB (1977) *Statistical Information of the Bank's Activities 1965–1976*, Tripoli, Share Press.

SECRETARY OF INDUSTRY (1977) *The Achievements of the Revolution in the Industrial Sector*, Tripoli.

CHAPTER EIGHTEEN

Social and economic change in North African medinas: the case of Tunis

RICHARD I. LAWLESS

The impact of European colonialism on pre-colonial urban structures

The individuality of the Arab West, or more properly *Djezira el Maghreb*, within the Middle East region owes much to the marked impress of European colonialism and more particularly to the intense settler colonization experienced by Morocco, Algeria and Tunisia during the nineteenth and early twentieth centuries. French colonization of the Maghreb produced dramatic changes in economic structures, spatial patterns and social organization. In the countryside agricultural colonization transformed many rural landscapes, but the European settlers were always a strong urban minority and their impact on cities and city life was profound.

Like the *Machrek* or Arab East, the Maghreb has a long and venerable urban tradition dating from Phoenician and Roman times. By the seventeenth century, however, the region had entered a period of political and economic decline. The number of towns decreased, and the size of existing cities shrank along with overall population and wealth. Thus, at the advent of colonial rule only a small proportion of the population lived in towns and many of these urban dwellers were Muslims of foreign origins (Andalusians, Turks and Kologlis) or Jews. Of the three political entities in the Maghreb, Algeria was the most weakly urbanized. Continuity of urban occupation was rare; indeed one writer has described its urban history as 'une succession de villes déchues' (Sanson 1974). At the time of the French conquest in 1830 only 5 per cent of the population were urban dwellers. Algiers, the capital of the Ottoman Regency, contained no more than 30,000 inhabitants and Constantine, the second city, 25,000. Tunisia in contrast had known remarkable urban continuity. Its urban system, characterized by a high degree of primacy, was already well established long before the French protectorate. On the eve of colonization in 1881 perhaps 12–20 per cent of the total population were urban; Tunis, the capital, maintained an estimated population of some

120,000 with only 14,000 inhabitants in the second largest city, Kairouan. Morocco, the richest prize, was the last to pass under European control and was not officially partitioned between France and Spain until 1912. At the beginning of the twentieth century between 8 and 9 per cent of the Moroccan population were urban. Fez and Marrakesh, with 100,000 and 60,000 inhabitants respectively, were the major cities and together with Meknès and Rabat-Salé constituted simultaneously the capital cities of the Moroccan Empire under a unique system of court rotation (Escallier 1974).

Colonialism introduced the modern era of urbanization in the Maghreb. The urban system was restructured and extended and its dynamics altered to fulfil new political, economic and social functions. The regional pattern of urbanization was dictated by the colonial extractive economy. Urban growth was concentrated in the major port cities, either old revived ones or new foundations, which rapidly began to dominate the hierarchy. In the interior pre-colonial centres were increasingly bypassed by the imposed economic system, and some experienced marked decline. Thus Fez and Marrakesh were eclipsed by Casablanca, Tlemcen by Oran, Kairouan by Sousse and Sfax; Kalâa, a small pre-colonial town in western Algeria, suffered total de-urbanization (Sari 1970).

Beginning in the nineteenth century, the emergence of a distinct colonial urban pattern in the Maghreb was accompanied by dramatic changes in the internal spatial organization of the region's urban centres. The policies which guided the introduction of modern urban forms varied significantly from country to country and the particular policies adopted by the early colonial planners are of critical importance for any analysis of the impact of modernization on traditional urban structures.

In Algeria, where pre-modern urban structures were weakest and the impact of colonization most brutal, little survived. The Muslim urban population actually declined during the first decades of colonial rule. Massacres, expulsions and the disruption of the pre-colonial economy depopulated the cities, and Algiers and Constantine did not regain their Muslim populations of 1830 until the beginning of the twentieth century. Algiers, the capital, suffered a particularly harsh fate. The existing urban structure of much of the lower Casbah or pre-colonial city was virtually obliterated to make way for new roads and new military and port installations. By the end of the nineteenth century what remained of the Casbah had been almost submerged by new European constructions (COMEDOR 1970), resulting in the dominance both spatially and visually of *L'Alger français*, the colonial city. Only the five small towns of the Mzab in the central Sahara, protected by distance and by the resilience and internal strength of their distinctive Ibadi communities, escaped the ravages of imposed French modernization.

The rapacious *laissez-faire* attitudes which conditioned the French occupation of Algeria had become more restrained by the time French rule was established in Tunisia. At least outward respect for indigenous traditions was observed and for the most part traditional institutions were preserved under the protectorate endowing the colonial regime with a measure of legitimacy. The pre-colonial urban centres or medinas survived more or less intact with limited modern inroads and the French and Italian settlers constructed European-style quarters alongside, creating characteristic dual cities. In Tunis, the historic nucleus remained ecologically distinct, but the presence of an Italian, Corsican and Sicilian proletariat along the eastern edge of the medina prevented the emergence of a sharp boundary between old and new cities.

The approach to urban modernization in Morocco was radically different. Unlike Tunisia, European influences had hardly touched the major cities when direct foreign rule was imposed in 1912. Lyautey, the first French Resident General, was determined to prevent their penetration by modern European constructions and to this end created a system of cultural and religious apartheid which resulted in an extreme form of dual-city development unprecedented anywhere in the Middle East. Existing Moroccan cities were preserved and protected from both foreign and indigenous attempts at modernizing them. Thus the walls surrounding Rabat-Salé were not only left intact but were actually repaired and reinforced by the French. Where possible, new cities for the European settlers were built some distance from the pre-colonial centres, as at Fez and Marrakesh, and separated from them by a *cordon sanitaire* or green belt of open land. In this way Lyautey hoped to separate as fully as possible the European inhabitants from the Moroccans and, while this segregation was never as complete as he had intended, the basic lines of apartheid were remarkably well maintained. The historic Moroccan cities therefore escaped European-inspired modernization, but survived as museum pieces and at a terrible cost to their inhabitants (Abu-Lughod 1975).

Whether they survived complete or in part, untouched by modernism or penetrated by it, as the major element within the framework of the dual city or a small part of the total urban area, the historic centres of the Maghreb experienced profound changes in their social and economic organization. Colonization triggered off a new wave of rural to urban migration – a phenomenon which the authorities had not foreseen and for which no provision had been made. Growing numbers of Muslim migrants therefore simply crowded into the medinas before spilling over into the emerging spontaneous settlements or bidonvilles which made their first appearance during the colonial period. As densities increased and the medinas became increasingly overcrowded, the Muslim bourgeoisie began to desert their traditional homes for new houses in the westernized quarters, and their former palaces and mansions were quickly subdivided to accommodate newcomers from the countryside. This process intensified and accelerated with independence. In this way an almost total proletarianization of the medinas has occurred and overcrowding and extensive subdivision of properties to accommodate several low-income families has contributed to serious physical deterioration. At the same time the old centres have lost most of their former administrative functions and have become increasingly peripheral or marginal to modern economic activity. Because densities have increased many more residents are now striving to gain a meagre livelihood from a much weakened and impoverished economic base.

Overcrowding, proletarianization – even ruralization – and degradation are problems which characterize historic centres not only in the Maghreb but throughout the Middle East (Abu-Lughod 1978). What is remarkable is that there have been few in-depth studies either of the processes which have transformed Islamic medinas from citadels of urbanity to inner-city ghettos or of the present functions which they perform. Paradoxically, more attention has been devoted to rediscovering their ancient properties than to recognizing socio-economic realities. The Tunis medina in an exception. Since 1967 the *Atelier d'Urbanisme* of the *Association Sauvegarde de la Médina* (ASM) has compiled a wealth of data on the causes and consequences of the transformation of Tunis' historic centre (Project Tunis/Carthage 1974). It provides a unique and invaluable case example of the processes of change, present vocation and current problems affecting North African medinas.

The Tunis medina: in search of the problem

For some five centuries before the French occupation Tunis was an intellectual and commercial centre in the great Islamic tradition; its only rival in the Maghreb was Fez. On the eve of colonization the medina and its north and south faubourgs (or *rbats*) (Fig. 18.1) covered an area of 270 ha between the swampy edges of the Lac de Tunis and the vast salt marshes of the Sebkhet es Sedjoumi (Fig. 18.2). It occupied possibly the best microclimate on the Bay of Tunis. The outer walls, belatedly built to circumscribe the two faubourgs, were not strong enough to withstand a military assault but they did provide a tangible symbol of the separation between the city and the surrounding countryside. Within the walls there were few open spaces and most streets offered access to pedestrians, horsemen and pack animals only; wheeled vehicles were unknown. But the complex network of narrow, winding streets and alleyways disguised a definite and logical organization of space (Fig. 18.1). Government was quartered in its own special section of the city, the Kasba, occupying the highest point. Below it was the venerable mosque/university of the Zitouna situated roughly at the city's centre, surrounded by the major suqs or bazaars housing the most prestigious trades and crafts. Each

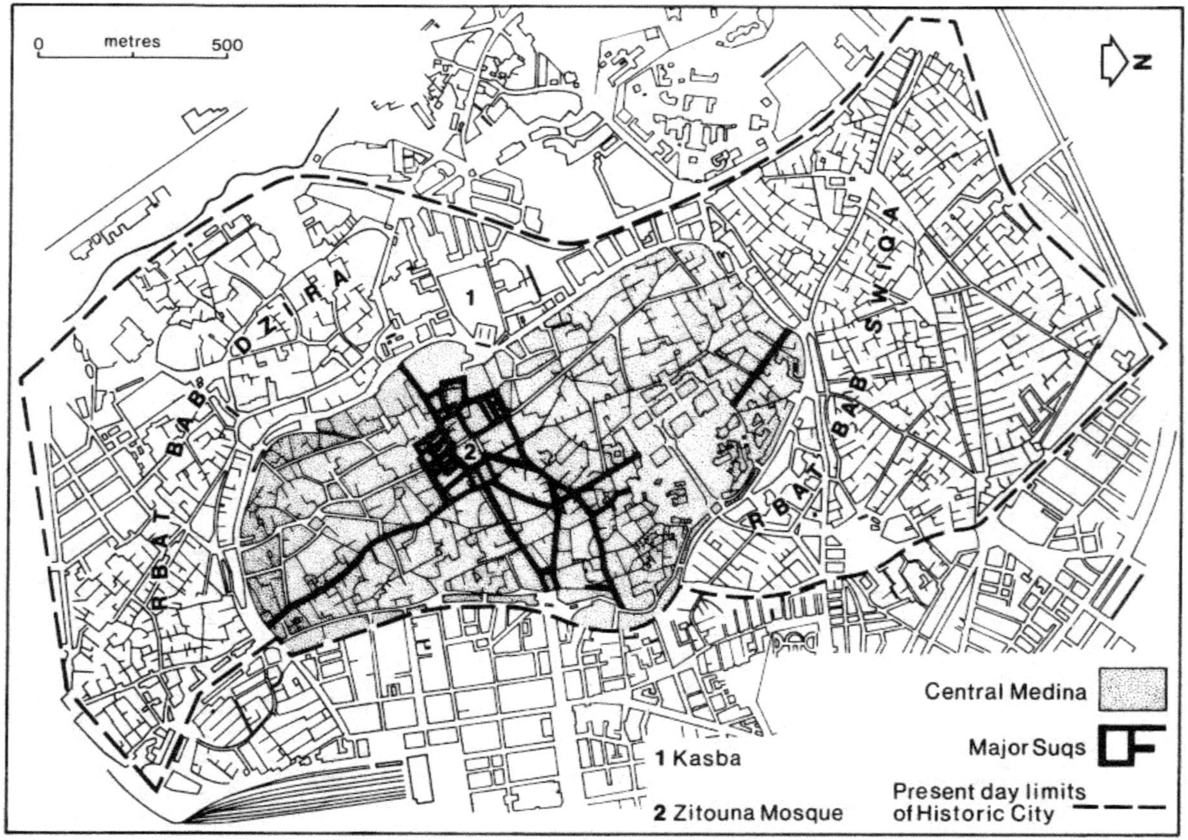

Fig. 18.1 The Tunis medina – traditional morphology.

trade or craft was grouped together and occupied a well-defined section of the bazaar system, its activities regulated by a hierarchy of guilds. Those activities involving unavoidable noise and smells were consigned to locations where they caused least disturbance, usually close to the city walls. Public and private space were clearly differentiated. Most dwellings were built on one or several storeys around an interior courtyard and the rich palaces and mansions of the aristocracy and bourgeois families represented merely a development of this basic structure. For both rich and poor the emphasis was on privacy and security and consequently the houses presented few openings to the outside with solid bare walls devoid of ornament, heavy doors and window shutters. Within the residential quarters the population was organized according to kinship-ethnicity, place of origin and clientage-affinity, and spatial segregation by economic class alone was rare. Each sub-quarter of the medina was relatively self-contained, its residents linked together by bonds of occupational and religious association, ethnic origin, family ties and clientage. The important families lived surrounded by their clients-retainers, or families tied to them through marriages or alliances. Such a system of social units therefore contained citizens of different levels of status and wealth who were held together by a network of mutual interdependence. Predictably the old Tunisois families pre-empted the higher and healthier quarters in the western part of the medina close to the Zitouna mosque and the most prestigious suqs, whereas the newly arrived or the marginal found shelter on the fringes of the city – migrants from the countryside in the north and south faubourgs, minority religious groups such as the Jews in the *hara* and the sizeable European population in the *quartier franc* – both located in the low-lying and malodorous eastern part of the medina which received much of the drainage and effluent from above.

After the advent of French rule and the founding of the orthogonally aligned European city on mud conquered from the lake (Fig. 18.2), the medina remained the place of residence and business for most of the Muslim population. With the exception of the destruction of the inner walls around the central medina in the 1900s, a rehabilitation scheme to rehouse some 400 families in the Jewish *hara* in the 1930s (Danon 1955), and a certain amount of destruction from Allied bombing during the Second World War, its physical structure survived the colonial experience virtually intact. As the colonial period progressed, however, changes began to occur in the social and economic organization of the medina which were to gain momentum and intensify after Tunisia achieved independence in 1956. The aristocratic Livournais Jews and the wealthier Europeans were the first to desert their traditional homes for the new European town which rapidly became the scene of intensive land and property speculation. They were followed by a few upwardly mobile Tunisian Jews, Italians and Maltese who succeeded in

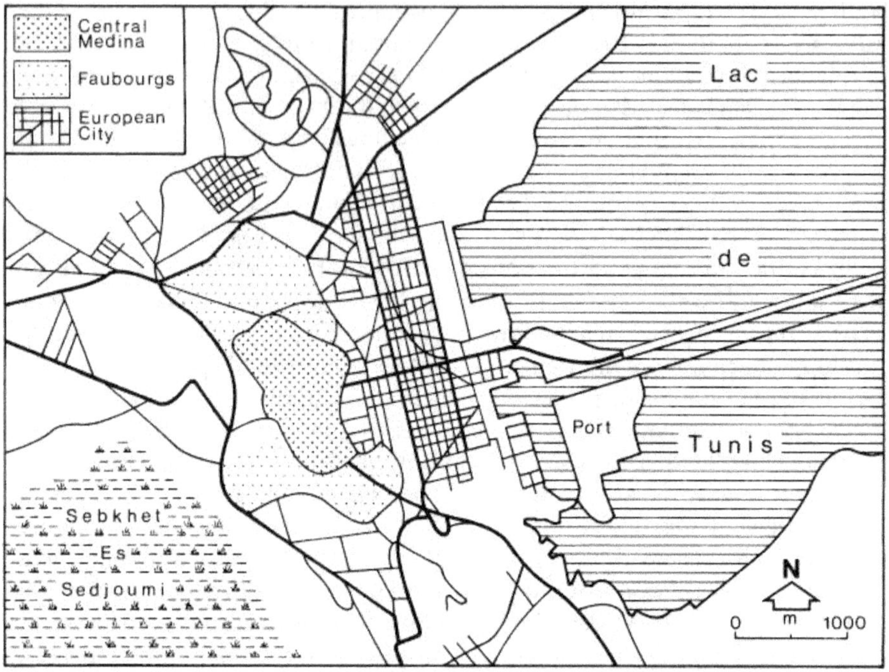

Fig. 18.2 Tunis – dual city

finding employment in the modern quarters. But the majority of the European population, mainly Sicilians and Maltese, and the indigenous Jews formed a miserable proletariat, too poor to escape from their overcrowded, decaying and unhealthy ghettos. From the 1930s many less favoured European immigrants also found accommodation in the medina, aggravating demographic pressure in the lower quarters. Some Spaniards and Italians were eventually forced to seek shelter in the numerous *fonduks* and *oukalas* of the old city, transforming these traditional institutions – caravanserais for itinerant merchants and hostels for temporary migrant workers – into overcrowded rooming houses (Callens 1955). On the eve of independence in 1956 there were 10,000 Tunisian Jews and 19,000 Europeans in the central medina (excluding the two faubourgs), almost half the total population (Sebag 1974).

Throughout the colonial period few Muslims, even among the wealthiest families, deserted the medina for the modern quarters which remained for them an alien environment, *'la ville française'*. Menaced by the assimilationist policies of the French authorities, many Tunisians sought refuge in their traditional institutions, symbols and customs – the Arabic language, Islam, the patriarchal family, and the women's veil – which acquired new meaning as methods of withdrawal from and resistance to the modern styles and values introduced by the colonizer.

In the early 1940s, however, the migration process quickened and acquired new characteristics as the disruptive effect of the French colonial economy on traditional rural systems reached its climax. Because of the powerful polarizing effect of the capital on all other centres in Tunisia's urban system, much of the accelerating migration stream was directed towards Tunis. At first the influx of Muslim migrants was absorbed into the medina and the two faubourgs but not without difficulty as pressure on existing accommodation intensified. Many *fonduks* and *oukalas* were converted into permanent dwellings for these new urbanites, with several single men and sometimes entire families living in one room. Even the stables and often the courtyards of these ancient structures were occupied by migrants in severely overcrowded, insanitary and unhealthy conditions. Other migrants found accommodation in private dwelling houses. The ground-floor rooms, formerly stables and storehouses, of numerous mansions were sublet; elsewhere the whole house was subdivided among several families. As the housing situation reached saturation point, the *Djemaïa des Habous* (Council for Religious Endowments) even allocated a number of secularized *zaouias* (holy tombs) to needy families without shelter (Callens 1955). Eventually, when the medina could no longer absorb the migrant flow new arrivals began to settle in the so-called *gourbivilles*, or spontaneous settlements which sprang up around the edges of the city (Fig. 18.3) at Djebel Ahmar, Taoufik, Mellassine, En-Nejah, Saïda-Manoubia, Zitoun-el-Djerbi and Bel Hassan. They were later joined by those seeking to escape the overcrowded and progressively deteriorating medina (Dardel and Klibi 1955). These spontaneous settlements, chiefly of self-built housing, expanded rapidly in the decade preceding independence – the population of Saïda Manoubia for example quadrupled between 1946 and 1956 (Groupe Huit 1978).

Independence inaugurated far more profound changes in the social organization of the medina than the advent of colonialism. In the space of a few years the vast majority of the Europeans, French, Italian and Maltese, departed for France or Italy and most of the Tunisian Jews emigrated either to France or Israel. Even more dramatic were the internal movements of the Muslim bourgeoisie of Tunis, who quickly abandoned the medina for a more comfortable lifestyle in apartments in the former colonial town or in suburban villas left vacant by the departing Europeans. So long as the French dominated the Tunisians politically, the latter stayed in the medina and adhered to its old value system; once independence was achieved, it was the new government's overwhelming desire to sweep away everything which recalled the colonial period and which seemed to have justified it, the backwardness of the country and above all the cities. Because of this national

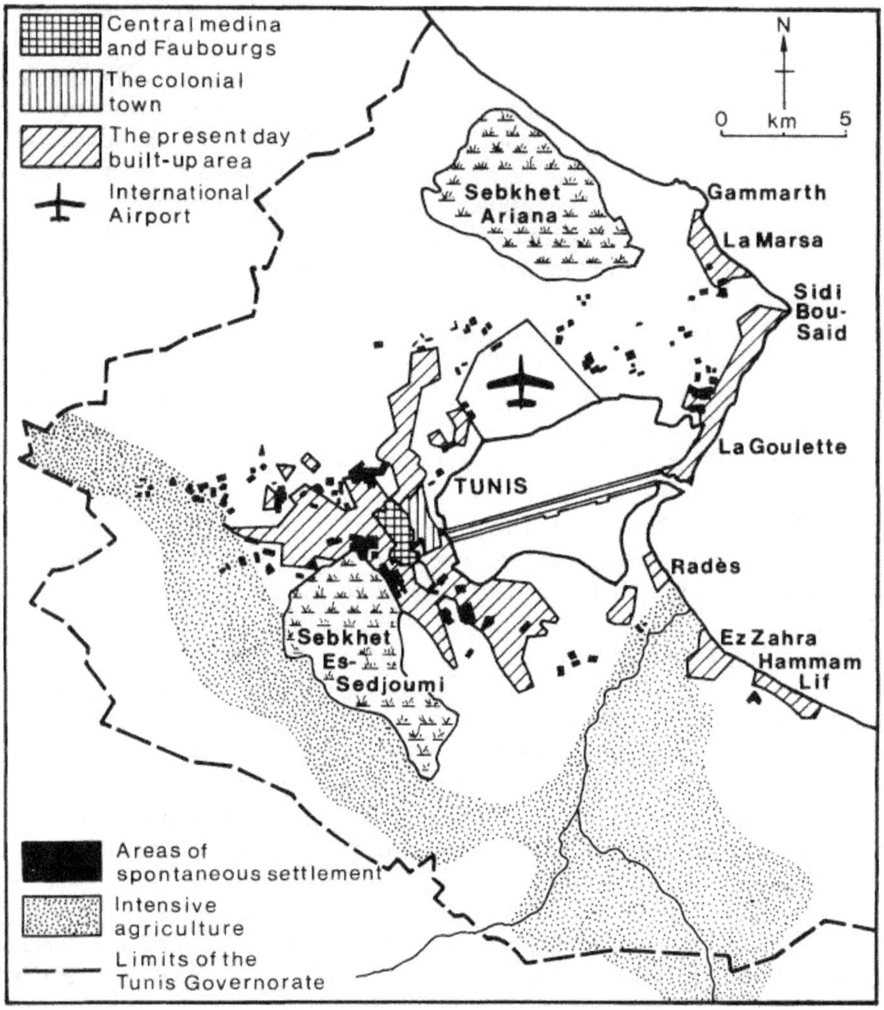

Fig. 18.3 The Tunis agglomeration.

commitment to a modernizing state and a secular society, the medina became a 'haven of archaisms' to a class that rejected it as a place to live, as a place to conduct business and most importantly, perhaps, as a place to learn and worship (Poncet 1974). A move to the modern city and an acceptance of the value systems and the conduct which it imposed was easily accomplished. Paradoxically, the Tunisification of the European city was accompanied by the Europeanization of the Tunisians who occupied it. As the Tunis bourgeoisie abandoned their traditional homes, the medina was invaded by refugees from certain *gourbivilles* destroyed when the new government launched a *dégourbification* programme and by a stream of new migrants from the rest of the country, the vast majority from rural areas. The impact of these population movements on the social structure of the medina has been admirably summarized by Sebag (1974). He writes:

> Not only has the population of the medina become *arabized* to the extent that Tunisian Muslims have replaced Europeans and Jews, but it has also become *provincialized*, because people from the provinces have taken the places left vacant by those born in the capital; *ruralized*, because the newcomers originate from the countryside; *proletarianized*, because a proletariat and sub-proletariat have taken the place of an urban bourgeoisie; and *pauperized*, because the wealthier citizens have been replaced by poor migrants.

Exhaustive studies carried out by the ASM's *Atelier d'Urbanisme* between 1968 and 1974 enable us to examine these processes in more detail (Eckert and El-Kefi 1974). They reveal that the most dramatic changes have occurred in the central medina whereas the two faubourgs, which had traditionally absorbed migrants from the southern oases, now appear to have acquired a relatively stable population (Table 18.1). In 1968 two-thirds of households living in the central medina were migrants compared with less than one-third at the time of independence (Table 18.2). In contrast, between 60 and 70 per cent of the inhabitants of the northern faubourg of Bab Swiqa (excluding the eastern edges of the faubourg, formerly inhabited by Jews and now containing a high proportion of migrants) were born in Tunis. A quarter of all residential

Table 18.1 The distribution of population and the proportion of migrants in the Tunis medina

	Area (ha)	*Inhabitants (1966 census)*	*Density (per ha)*	*Migrants as % of population (1968 ASM Census)*
Medina	93	48,760	520	64.3
Rbat Bab Swiqa	103	59,050	540	45.7
Rbat Bab Dzira	73	35,910	480	51.2
Medina and faubourgs	270	143,720*	513	53.7

Source: Eckert and El-Kefi 1974.

* In 1956 the population of the medina, including the two faubourgs, was 168,000. The decline during the intercensal period 1956–66 reflects the demolition of some 23 ha of residential properties and the extension of commercial activities into areas formerly reserved for residential use. The results of the 1975 Census of Population reveal that the medina's population has remained stable at around 140,000.

Table 18.2 The geographical origin of migrants (heads of household only) enumerated in central Tunis medina (1968 ASM Census)

Region of Tunisia	%
North-east	15.1
North-west	28.0
Sahel	16.3
Centre	7.8
South	32.8

Source: Eckert and El-Kefi 1974.
Note: Since independence migrants from the south have continued to form the highest proportion of new arrivals in the central medina, but between 1956 and 1968 the number of arrivals from the Sahel and the north-east accelerated and achieved the highest *percentage* increase.

properties in the central medina had been subdivided and were rented room by room. Between 1956 and 1968 the number of rented rooms in traditional *oukalas* and in private dwellings converted into rooming houses (*dars oukalisés*) quadrupled.

This process, known as *oukalisation*, was found to be particularly strong in the eastern section of the medina which became the main reception area for migrants following the exodus of the Sicilian, Maltese and Jewish populations. Covering about one-third of the total area of the central medina, these low-lying quarters contained 45 per cent of migrant households in 1968 and 46 per cent of subdivided properties (*structures oukalisés*) whereas in the faubourg of Bab Swiqa the proportion was nearer 70 per cent and the phenomenon of *oukalisation* linked to immigration was less pronounced.

The ASM's enquiries also revealed that almost half the active migrants living in the central medina were unskilled and constituted a poor sub-proletariat of day labourers, manual workers, itinerant traders, porters, odd-job men, prostitutes, domestic servants, caretakers and shoeblacks; many were unemployed; 21 per cent of the active migrants were office workers, mainly clerks, and 17 per cent unskilled workers in the tertiary sector, drivers, waiters and barbers; only 7 per cent were factory workers and less than 5 per cent shopkeepers, civil servants or members of the liberal professions. Data from the northern faubourg, Bab Swiqa, reveal a similar socio-economic structure. A close correlation existed between occupation and the geographical origin of migrants in the central medina (Table 18.2). Some three-quarters of migrants from the south of Tunisia were employed in the more 'traditional' lower-class occupations such as petty traders, bakers, dockers and porters;

three-quarters of migrants from the Tell entered more 'modern' lower-class occupations as manual workers, day labourers, factory workers, second-hand dealers, especially of clothing, and in repair services; while over a quarter of the Sahelians secured employment as clerks. In an urban economy incapable of creating enough jobs to satisfy the increasing demand for work, access to employment for new arrivals is mainly through 'networks of clan solidarity'. In this way it is much easier for a Sahelian to find work in a small office, a Tellian to become a second-hand dealer and a migrant from the Matmata to find work in the port. In 1968 almost three-quarters of all active migrants in the central medina earned less than 26 Tunisian dinars per month, while the average monthly income of nearly half the inhabitants of the entire historic centre was only 20 Tunisian dinars.

Throughout the historic centre, population densities now average 513 per ha compared with 144 per ha for the Tunis agglomeration, rising to 1313 per ha in parts of the central medina. There is severe overcrowding (46 per cent of all households live in one room) and widespread subdivision of properties among poor migrant families possessing few resources. Within some buildings pressure for space or privacy has led to the haphazard construction of more rooms in the patios, which cuts off light and air from the only available source. Newcomers unable to afford the monthly rent for a single room (about 7 Tunisian dinars) squat in the entrance halls. Some housing cells are even rented to one group for the day and another for the night (Micaud 1977).

The transformation of the historic centre into a low-income popular zone has contributed to the rapid physical deterioration in the medina's traditional fabric. Few properties are owned by their occupants, multiple ownership of single dwellings by absentee landlords is common and many houses abandoned by the Europeans and Jews are either occupied illegally or administered by private agents for a fee; factors which have resulted in a lack of investment in the repair and proper maintenance of residential properties. The fine old houses of the medina are not particularly well built, although they are very carefully finished. Therefore, once the façade deteriorates water seeps into the earthen or poorly laid stone walls and decay quickly follows. Palaces and mansions formerly occupied by the high and middle bourgeoisie and more modest dwellings have equally succumbed to neglect and decay. Housing conditions are therefore extremely bad – 50 per cent of dwellings are in a poor state of repair and some lie in ruins. Infrastructures are inadequate and outdated – 60 per cent of houses do not possess piped drinking water and 53 per cent are compelled to use communal lavatories. A sewerage system designed for 40,000 is now expected to serve almost four times as many people.

A continued degradation of historically significant and architecturally irreplaceable structures has also occurred. Not only have the major monuments lost their wealthy patrons with the departure of the upper and middle bourgeoisie, but the nationalization of *habous* (religious endowments) in 1957 was followed by the secularization and sale of many religious buildings (370 between 1957 and 1968), the majority of which have been converted to new uses as dwellings, warehouses or shops. Whereas there were 219 *zaouias* in 1956, only thirty-eight remain today; 181 *mesjids* (mosques) have been secularized and only five of the medina's thirty-one *medersas* (religious boarding schools) retain their original functions.

The historic centre has lost many of its religious functions since independence and the whole of its elaborate educational system has been dismantled. The prestigious university attached to the Zitouna Mosque together with its secondary-level annexes, which enrolled over 15,000 students in the mid-1950s, has been abolished and replaced by new secular institutions located in the modern city. Yet, surprisingly, the medina has retained important administrative functions. The old Turkish administrative area around the Kasba, which was co-opted by the French as their seat of government, now contains a complex of the most important ministries of the Tunisian Republic together with the new headquarters of Tunisia's sole political party – *Parti Socialiste Destourien* – located on the exact site of the former Kasba. Numerous craft industries survive in the old city and still retain, at least in part, their traditional spatial organization and corporate structure, though their role in industrial activity is today dwarfed by large-scale factory production, chiefly located on the outskirts or in the newer parts of the city. The level of activity, however, has declined significantly since the end of the colonial period. Whereas in 1953 there were 5374 *entreprises artisanales* employing over 16,000 workers (Golvin 1957), by 1972–3 the ASM's *Atelier d'Urbanisme* enumerated only 2800 workshops which nevertheless employed 8600 workers and supported at least 34,000 people – a quarter of the medina's inhabitants. Many handicraft activities were long ago undermined by the forced incorporation of Tunisia into the world system of trade as a source of raw materials and as a market for manufactured goods. More recently competition from Tunisia's own expanding 'modern' manufacturing sector and the adoption of western styles and values by a growing number of Tunisians pose a new threat to traditional industrial activities. Certain crafts have experienced a pronounced and seemingly irreversible decline. Such is the fate of the few craftsmen who still manufacture *burnous*, *djellabas* and *kachabias* for a dwindling clientèle faithful to traditional forms of dress. They are all old men and as no young apprentices are being trained their skills will die with them. Some craftsmen have abandoned their trades to become exclusively retailers, while others thrive because they have succeeded in adapting their production to the fast-expanding, if highly vulnerable, tourist market. The coppersmiths and engravers, who now sell three-quarters of their output to tourists, are the most successful example of the latter (Lowy 1976).

Although the major wholesale merchants have disappeared following the reform of Tunisia's traditional marketing systems which are now controlled by national agencies, the medina continues to perform a wide range of retail functions. Between 1967 and 1973 employment in retailing in the medina, excluding the numerous street traders, increased by 5 per cent and the annual turnover by 40 per cent. In 1973 the 3045 retail establishments there (employing 5375 people) realized a turnover of 21 million Tunisian dinars which represented a quarter of the value of all retail transactions carried out in the capital. The appearance of a lucrative tourist market accounts in part for the recent expansion in commercial activities, but even more important has been the increase in the number of low-income consumers (i.e. those earning less than 25 Tunisian dinars a month). Since independence the medina has acquired increasing centrality, emerging as the dominant low-income commercial centre of the entire Tunis agglomeration (Fig. 18.3). Eighty-three per cent of the sales of all textiles, 76 per cent of clothing, 66 per cent of furniture and 64 per cent of kitchen utensils purchased by popular consumers living in Tunis (some 41 per cent of the city's total population) are made in the retail outlets of the medina, compared with only 3 per cent, 6 per cent, 26 per cent and 10 per cent respectively in the modern city's shopping centre. The medina's centrality remains strong, though more selective, for middle-income consumers, but it exerts only a limited attraction for higher-income groups who prefer commercial establishments located in the modern city. The remarkable expansion of those commercial activities serving popular consumers has weakened the traditional links between producer and retailer, so that the medina is increasingly retailing goods which it does not manufacture. The spatial pattern of retailing has also altered in response to these new demands. Traditional retail sellers, partly adapted to tourism, remain around the Great Mosque and new tourist shops have proliferated along the Rue du Zitouna. In contrast, new popular commercial activities have polarized along a north–south axis provided by the Rue des Teinturiers, Souk El Grana and Rue Sidi Mahrez. Pressure on space within the traditional bazaar system has resulted in the expansion of retailing into former residential areas and the conversion of private dwellings and even public buildings to commercial functions. Maintenance of both retail premises and workshops, the vast majority of which are rented not owned by their occupants, is minimal, aggravating the general deterioration and decay of the medina's traditional fabric.

Conclusion

In recent decades the Tunis medina has evoked different and often contradictory perceptions and proposals for intervention among urban planners.

For some it has been regarded as a casualty in the thoughtless drive for modernization; for others an obstacle to such progress. Demolition has been proposed as one solution, preservation in the interests of economic development and international tourism another. Few have recognized that the historic core performs unique and valuable functions as an organic part of the metropolitan structure and many have completely misunderstood its present role. Though its rich and varied architectural fabric is rapidly deteriorating, the medina houses more poor people at less cost than could be achieved elsewhere in the city, and in some branches of retailing the central medina surpasses the adjacent modern city in the importance of its activities. All too often planners have sought to remodel the historic centre in the image of one or another of the current planning ideologies. Planning the past should begin not with revitalization schemes but with the recognition of the medina's actual vitality (Micaud 1978).

References

ABU-LUGHOD, J. L. (1975) 'Moroccan cities: apartheid and the serendipity of conservation', in Abu-Lughod, I. (ed.) *African Themes*, Evanston, North Western University, 77–111.

—— (1978) 'Preserving the living heritage of Islamic cities', in Holod, R. (ed.) *Toward an Architecture in the Spirit of Islam*, Proceedings Seminar One, The Aga Khan Award for Architecture, Philadelphia, 27–35.

CALLENS, M. (1955) 'L'hébergement traditionnel', *Les Cahiers de Tunisie*, 3, 165–79.

COMEDOR (Comité Permanent d'Etudes, de Développement, d'Organisation et d'Aménagement de l'agglomeration d'Alger) (1970) *Etude pour la rénovation et la restructuration de la Casbah d'Alger; Les transformations du tissu de la Casbah pendant la période coloniale*, Algiers.

DANON, V. (1955) 'Le niveau de vie dans la Hara de Tunis', *Les Cahiers de Tunisie*, 3, 180–210.

DARDEL, J. B. and KLIBI, S. C. (1955) 'Un faubourg clandestin de Tunis: le Djebel Lahmar', *Les Cahiers de Tunisie*, 3, 211–24.

ECKERT, E. H. and EL-KEFI, J. (1974) 'L'espace traditionnel de la ville de Tunis: la médina et les deux rbat: faubourg ou gourbiville?' in *Les Influences occidentales dans les villes Maghrébines à l'époque contemporaine*, Etudes méditerranéennes 2, Editions de l'Université de Provence, 211–35.

ESCALLIER, R. (1974) 'La croissance urbaine au Maroc', in *Villes et sociétés au Maghreb: études sur l'urbanisation*, Editions du Centre National de la Recherche, 145–73.

GOLVIN, L. (1957) *Aspects de l'artisanat en Afrique du Nord*, Presses Universitaires de France.

GROUPE HUIT (1978) *Réhabilitation du quartier de Saïda Manoubia*, Ministère de l'Intérieur – Municipalité de Tunis.

LOWY, P. (1976) 'L'artisanat dans les médinas de Tunis et de Sfax', *Annales de Géographie*, 470, 473–93.

MICAUD, E. C. (1977) 'Urban planning in Tunis', in Stone, R. A. and Simmons, J. (eds) *Change in Tunisia*, State University of New York Press, 137–58.

—— (1978) 'Urbanization, urbanism, and the medina of Tunis', *International Journal of Middle East Studies*, 9, 431–47.

PONCET, J. (1974) 'Syncrétisme urbain en Tunisie: la modernisation des villes Tunisiennes: message et problème', in *Les Influences occidentales dans les villes Maghrébines à l'époque contemporaine*, Etudes méditerranéennes 2, Editions de l'Université de Provence, 237–42.

PROJET TUNIS/CARTHAGE (1974) *Sauvegarde et mise en valeur de la Médina de Tunis, Rapport de synthèse: Dossier No. 1 Protection du patrimoine monumentale; Dossier No. 2 Mise en valeur des monuments historiques; Dossier No. 3 Patrimoine immobilier; Dossier No. 4 Descriptif et estimation sommaire des coûts – Hypothèse de financement; Dossier No. 5 Activités opérationelles de l'ASM; Dossier No. 6 Opérations de réhabilitation de l'îlot 111 E-50; Dossier No. 7 Artisanat – propositions de développement; Dossier No. 8 Commerce: principes d'une politique commerciale*, UNESCO, Institut National d'Archéologie et d'Arts, Association sauvegarde de la Medina.

SANSON, R. P. H. (1974) 'La symbolique rurale et la symbolique urbaine du néocitadin algérien', in *Les Influences occidentales dans les villes Maghrébines à l'époque contemporaine*, Etudes méditerranéennes 2, Editions de l'Université de Provence, 136–41.

SARI, D. (1970) *Les Villes précoloniales de l'Algérie occidentale: Nédroma, Mazouna, Kalâa*, Société Nationale d'Edition et de Diffusion.

SEBAG, P. (1974) 'La décolonisation et la transformation des quartiers traditionnels de Tunis', in *Les Influences occidentales dans les villes Maghrébines à l'époque contemporaine*, Etudes méditerranéennes 2, Editions de l'Université de Provence, 247–55.

CHAPTER NINETEEN

Urban planning in Iran

BRIAN D. CLARK

An old saying which states that 'anyone who knows anything at all about Iran must be grossly misinformed' appears particularly appropriate in the post-revolutionary period. For in the field of planning in general, and urban planning in particular, there are few indications of how the new regime will formulate its spatial planning policies. At one level functional changes are occurring such as the conversion of hotels to hospitals and schools, whilst at another level there appears to be greater participation in urban affairs through revolutionary councils. However, no overall strategy for urban development has yet been formulated. For this reason, the chapter must critically review the nature of urban planning during the period before the Shah was deposed.

To appreciate the nature of urban planning and the problems that have been encountered, it is necessary briefly to highlight certain key factors which have been instrumental in shaping the form that it has taken. They include historical, economic, social and political processes which have operated in a land that is large and diverse and contains a range of markedly contrasting physical environments (Fisher 1971). The following factors may be highlighted.

With a total area of 1.65 million sq km, great contrasts exist in the resource base, population distribution and in economic opportunities (Government of Iran 1974). This has given rise to marked differences both between and within regions and cities, leading to much economic and spatial inequality. In theory, if not practice, it has been the aim of national economic planning and regional and urban planning to try to reduce this inequality (Naraghi 1969). The creation of industrial growth poles, establishment of agro-industry, investment in heavy industry and improvement of the national transport infrastructure are all measures which have been attempted in five National Economic Development Plans (NEDPs) and all have had profound effects on urban planning policies.

The Iranian state, particularly in the 1960s and 1970s, was the product of the centralization of political and economic power in the state and in the person of the Shah. Centralization of control has been possible owing to sectoralism, lack of a democratic tradition and state ownership of an increasing oil revenue. Centralization can be seen not only in the way that political power has rested firmly in Tehran, but also in the way that most planning policies in cities and towns throughout the country have been imposed without any real local involvement. This has led to inefficiency in that duplication of effort exists in urban areas and a fragmented range of urban policies have been implemented by competing ministries which are rarely co-ordinated, even when city plans have been prepared (Ghaffarzadeh 1979).

Great opportunities have existed for a range of dynamic urban planning policies as a result of the increases in oil revenue from $1.2 billion in 1970 to $20.9 billion in 1977. In practice, it has only been certain favoured cities which seem to have benefited, and nowhere more than Tehran, whilst small towns and cities have stagnated or declined (Clark 1971).

In sum, whilst opportunities for rational planning, backed by oil revenue, have existed, the approach adopted has been fragmentary and disjointed and this has led to many problems. Any success in urban planning in one particular city is often a function of the influence of those in power who had access to finance and favours in Tehran. The concept of need, whether in physical, economic or social terms, has rarely been the criterion used to formulate urban planning policies.

The growth of urban planning

Origins

The roots of urban planning in Iran are many and various. They include water requirements, Islamic ideology, needs of defence and commerce and grand gestures by the ruling monarch. Scarcity of water, for example, saw the building of the water systems in Isfahan in the fourteenth century and the construction of the Karaj – Tehran canal in the nineteenth. That they are examples of a planned solution to an urban problem is shown by the fact that they required large sums of capital and labour and the consent of the population and decision-makers. Defence, whether in the form of town walls, or internal separation of ethnic quarters, has also been a planned feature of many Iranian towns.

Another example of small-scale urban planning has been the building, extension or embellishment of bazaars and caravanserais. Usually it was the city merchant or landowner who was responsible for their construction. Not only did it increase their rent base, it also gained them prestige, and the building of mosques was considered a pious, religious act. Whilst this type of incremental planning has been the norm there are examples of a more grandiose scale of planning. Certain ruling dynasts, for example, have stamped their imprint on the traditional Iranian city. In the case of Isfahan, the creative genius of Shah Abbas I gave the city its distinctive character with the building of the Maidan-i-Shah, much of the bazaar system, the Chahar Bagh and the Allahverdi bridge across the Zayandeh River (Beaudouin and Pope 1938–9).

Administrative growth of urban planning 1874–1940

The creation in 1874 of city assemblies composed of government officials and distinguished citizens was linked to schemes to improve the health of urban populations in a number of cities. It can be considered as one of the first attempts to create co-ordinated city management. The assemblies had power to raise taxes for city improvement schemes including the building of streets and alleys and for street cleaning. The wide-ranging power of the assemblies was illusory for they were disbanded one year later in 1875. That they failed was largely due to the hostility of the government-appointed city governors, illustrating that even then, as in the 1970s, lack of co-operation between different sections of government has been a major weakness of urban planning.

The Constitutionalist Revolution (1906–9) was essentially an urban-based revolution which challenged the power of the tribal khans. It saw city management in the hands of the elected representatives who formed city assemblies. The assemblies were the first, and perhaps only, example of real public involvement in decision-making in urban Iran in the twentieth century. The Municipal Law of 1913 gave legal force to the assemblies, with elected councils having power to appoint officials to administer the city. Their duties included responsibility for street cleaning, distribution and pricing of food and the production of city maps to guide future growth. The law was not fully understood or implemented and in 1917 the assemblies were dissolved when the Ministry of Interior became responsible for urban affairs. This situation existed until 1949 when the Law of Independence of Municipalities was enacted and which provided the present-day legislative basis for Iranian municipalities (Mozayeni 1974).

Urban planning in the reign of Reza Shah (1925–41)

The most important single physical planning action up to the post-Second World War period was probably the decision of Reza Shah to attempt to modernize the major cities of the country. His solution was simple and brutal but extremely effective, for he decided to drive a major network of straight roads (*khiabans*) through the hearts of the traditional cities (Clarke and Clark 1969). Whilst many cities were inefficient in terms of their ability to cope with the growing number of lorries and motor cars, the new roads created a monotonous townscape, truncated existing residential and commercial areas and bear testimony to the fact that this was essentially single-element planning, predominantly an engineering solution, with little regard being paid to the social and economic disruption which ensued. This road-building programme, however, did encourage many commercial activities to move out of the old core of the city and ultimately led to the development of new commercial centres in many cities of Iran (Clarke 1963). At the same time, new housing quarters were established on the periphery of cities and it was in these areas, together with some of the older wealthier quarters, that street lighting was introduced. Many city walls were demolished and in Tehran most of the fine Qajar architectural heritage was destroyed as part of what many would argue to be ill-conceived redevelopment planning (Bahrambeygui 1977).

These physical changes, which were a deliberate move towards creating a western-looking society accompanied by a policy of secularization, were supported by legislation which in theory should have encouraged more rational city administration and planning. Legislation included laws to allow the compulsory acquisition of land, the introduction of building permits to control the nature and form of growth, and powers to produce city plans to formulate land-use patterns. In practice, little was achieved apart from the establishment of many government departments in towns and cities, each responsible for some aspects of urban development. The reality of urban planning in this period was one of weak mechanisms to control and direct growth, lack of policies as to how the rapidly growing population, encouraged by rural-urban migration and tribal settlement, could be accommodated in cities and limited power over land speculation which was becoming a feature in some urban areas.

By the end of Reza Shah's reign many of the elements which later characterized contemporary planning were present. These included a large, complex and expanding administrative structure, a battery of legislation on ownership and government's right to acquire land, a new landowning group composed of government officials, rural landlords and city merchants who had the power and influence to abuse the legislation, and a public increasingly alienated from the process of decision-making. Urban planning was restricted to some major projects such as road building, the successful planned development of Abadan by the oil company and some rebuilding of small towns damaged by earthquakes. More generally, planning took the form of small-scale developments, such as the provision of electricity, piped water, street surfacing and government housing projects.

Urban planning as part of National Economic Development Planning (1947–78)

Urban planning in the framework of National Economic Development Plans (NEDPs)

Before analysing the development of urban planning from 1945 to the present day, a few brief comments on the economic and administrative status and structure of urban planning will indicate why problems have arisen. One of the key factors is that urban planning has been given a low priority, both in terms of finance for projects and also in manpower training, throughout the whole of the post-war period. The great industrial drive which has dominated all the NEDPs has failed to take into account linked regional and urban needs (Richardson 1975). It has been essentially economic sectoral planning with housing needs, urban infrastructure and other urban requirements not being linked to the major economic objectives. As a result, the vast migration to the urban areas generated by national economic policies has led to overcrowding, shortage of accommodation, the growth of slums and shanty towns and an imbalance in the urban system in the country at large (Clark and Costello 1973).

Another key factor is the administrative structure in which urban planning operates. At the national level the Plan and Budget Organization (PBO) and the Ministry of Housing and Development are the two most important initiators of urban development. The former's role relates to the provision of finance for urban projects, whilst the Ministry has been concerned with implementing housing projects and initiating the production of comprehensive town plans. In the post-war period many changes in the administrative structures and the creation of new agencies concerned with some aspect of urban planning have made it inevitable that uncertainty of objectives, duplication of effort and corruption have led to inefficiency and conflict.

The position has not been helped at the urban level where municipalities, who in theory have considerable powers covering urban social and economic activities, are generally acknowledged to be the poorest and least efficient of all government agencies. Whilst the municipalities face many problems, perhaps the greatest was the very tight control which was kept on their activities by the regional governor (*ostandar*), central-government ministries in Tehran and the PBO. A further weakness of the municipalities has been that they had limited means of raising income and their role as an interface between central government and the population has been very limited. As a result, most urban policies, and this includes projects and the form and content of city plans, have been imposed from Tehran by both government departments and consultants, without any real understanding of the needs and aspirations of the local population.

Urban planning 1947–61

In the first two NEDPs urban development was given a low priority in the overall budget. Emphasis was placed on the provision of potable water, electricity production and street improvements. There was no overall planning strategy and the nature of development was a reflection of the capabilities of individual municipalities. In the first plan it was clear that the poorer municipalities were paralysed because they had to pay 50 per cent of the project cost. In the second plan this was rectified so that the poorer municipalities had to pay a smaller percentage of the total costs (Plan Organization of Iran 1960).

Urban planning 1962–72

In the third NEDP, existing policies were continued. Conflict became apparent, however, when the municipalities attempted to acquire greater funding for project implementation. The PBO was reluctant to do this, believing the municipalities to be inefficient in what they had done and incapable of taking on greater responsibility.

Towards the end of the second plan the PBO appointed consultants to draw up a comprehensive city plan for Tehran. Essentially this was a physical land-use plan and little attention was paid to economic and social criteria. Indicative of the problems faced was a request that the consultants should include their own legislative proposals as to how the plan should be implemented. One of the major recommendations was that a five-year 'service' line should be imposed beyond which no development would be allowed without special permission. This line would move outward every five years until it reached the outer boundary of the plan in twenty-five years' time. Needless to say this policy could not be fully enforced, acted as a stimulus to land speculation, and uncontrolled, sprawling, peripheral development continued.

It was during the third NEDP (1962–6) that steps were taken to allow the preparation of city plans for seventeen cities. Later this was extended to a further nine cities in the fourth plan. There can be no doubt that these city plans have had far-reaching consequences for the development of Iranian cities. The plans have been produced by consultant firms who after studying 'the geography, hydrology, climate, history, demography, sociology, economy, space and traffic' and 'taking into consideration government and private sector plans and intentions' must provide the High Council for City Planning with three alternative plans. The High Council was established in 1965 with powers to guide and advise on the preparation of city plans, set planning standards and approve plans once they are submitted. The High Council normally select the alternative recommended by the consultant. A general plan is then drawn up and sent to the city for comment and then passed to another committee comprising local government officials before being sent to the High Council. From here it is sent to the Ministry of Housing and Development, who then forward it to the municipality for implementation.

A detailed plan is then drawn up which follows the same administrative procedure. Once approved, the plan provides the municipality with a detailed land-use map at the present time and proposed uses ten to twenty-five years ahead, the way that traffic will be incorporated into the land-use system and implementation costs. The consultants also have to provide training for municipal employees who have to implement the plan.

Although it is easy to criticize much that is contained in city plans, including their non-dynamic format, lack of participation by the public and limited social inputs, they have proved to be an important step in controlling the worst excesses of unplanned growth. The implementation of city plans was linked to a further municipality Act which saw the introduction of a new system of housing-construction permits to stop illegal housing. There have, however, been a number of major problems. During the fourth economic plan the first city plans reached their implementation stage. The first impact was a reduction in private investment in housing as a result of delays between commissioning and implementing a plan. Another result was an increase in the price of land within the plan's boundaries and speculation in land outside the boundary but designed to be included later. Some gains, however, did accrue to government as a result of a capital tax on land which was included in the Urban Renovation and Development Act of 1968. The Act also gave greater powers to municipalities compulsorily to acquire land required for renewal and development projects.

It was during the latter part of the fourth plan that an independent renewal organization was created in Tehran to build a new centre for the city at Abas Abad. Conceived as one of the most grandiose planning schemes of all time in Iran, the object was to creat a 5 million square metre administrative, cultural and commercial centre in the north of the city which would also house a population of over 100,000 (Moeinzadeh 1975).

The fifth National Economic Development Plan 1973–8

During this period a new trend can be seen. City plans were being prepared for the majority of the largest urban areas, but it was realized that many of the earliest ones had been too concerned with physical planning.

> Within the framework of the City Plan for each urban centre, development programmes will be based on more comprehensive objectives stressing the

quality of urban life, and as far as possible establishing diverse objectives for social and economic development programmes will be avoided. (Plan and Budget Organization 1974)

As well as intimating that broader social and economic factors would be taken into account in formulating urban plans, another discernible trend has been the attempt to integrate urban planning into broader regional and national planning strategies. To this end a Centre for National Spatial Planning was established in 1975 with wide powers of research, personal training and formulation of planning policy. A National Spatial Strategy Plan has been produced and in comparison with many earlier reports on planning it is a critical and realistic document about alternative options for urban development. It notes that inequalities are increasing and that a high target rate of economic growth will lead to a deterioration in the quality of life (Centre for Research and Training in Regional Planning 1976).

Conclusions

At the present time it is only possible to draw conclusions about a system of urban planning that probably no longer exists! As has been demonstrated, urban planning has evolved in a complex and at times conflicting manner. First, it has seen the allocation of funds by central government to provide a range of urban utilities including streets, piped water, electricity, sewerage systems and urban parks. Secondly, it has attempted to co-ordinate land-use activities, including housing provision, by the introduction of comprehensive city plans. Both approaches have been to provide cities with services for which demand has been created or expanded as a result of population growth, physical expansion or technological change. The results have not always been entirely satisfactory because of the administrative structures, conflicting objectives as to ends and means and the low priority given to urban problem-solving. Urban planning policies, including housing strategies, have mainly been conceived to solve problems created as a result of national economic development strategies. As one author has put it, 'urban policies have been basically corrective and ad hoc rather than an integral component of a nationally coordinated spatial economic plan' (Ghaffarzadeh 1979). This has led to urban policies being introduced during a crisis; policies such as the housing programme of 1975, controls on land speculation and attempts at co-ordination of different levels of planning.

Currently the country is facing a crisis of a far greater magnitude than it has faced for many years. It can be argued that recent events in the country bear devastating testimony to the fact that one of the causes of the revolution was that there had been little attempt to improve facilities in the poorer quarters of the cities. As Iran enters a new political era it will be interesting to note the form that urban planning policies take. That they will be required is not in question; what is at stake is whether the new regime, with its professed objective of enhancing the quality of urban life for all the population, has learned anything from the successes and failures of previous urban strategies.

References

BAHRAMBEYGUI, H. (1977)*Tehran, an Urban Analysis*, Tehran, Sahab.

BEAUDOUIN, E. E. and POPE, A. U. (1938–9) 'City plans', in Pope, A. U.(ed.) *A Survey of Persian Art from Prehistoric Times to the Present*, vol. 2, *City Plans*, London, Oxford University Press.

CENTRE FOR RESEARCH AND TRAINING IN REGIONAL PLANNING (1976) *National Spatial Strategy Plan*, Tehran, Government of Iran.

CLARK, B. D. (1971) 'Iran: changing population patterns', in Clarke, J. I. and Fisher, W. B. (eds) *Populations of the Middle East and North Africa*, London, University of London Press.

CLARK, B. D. and COSTELLO, V. F. (1973) 'The urban system and social patterns in Iranian cities', *Transactions of the Institute of British Geographers*, 59, 99–128.

CLARKE, J. I. (1963) *The Iranian City of Shiraz*, Dept of Geography Research Paper Series No. 7, Durham, University of Durham.

CLARKE, J. I. and CLARK, B. D. (1969) *Kermanshah. An Iranian Provincial City*, Centre for Middle Eastern and Islamic Studies, Pub. No. 1, Durham, University of Durham.

FISHER, W. B. (1971) *The Middle East: A Physical, Social and Regional Geography*, 6th edn, London, Methuen.

GHAFFARZADEH, H. (1979) 'Hamadan. An evaluation of the urban planning process', unpublished PhD thesis, Dept of Geography, University of Aberdeen.

GOVERNMENT OF IRAN (1974) *Iran National Report to the United Nations Conference on the Human Environment*, Tehran, Govt of Iran.

MOEINZADEH, M. (1975) 'Urbanization in Tehran: problems and opportunities', *International Union of Local Authorities Newsletter*, 9 (9/10), 12.

MOZAYENI, M. (1974) 'City planning in Iran: evolution and problems', *Ekistics*, 38, 264–7.

NARAGHI, E. (1969) 'Regional studies in Iran', in *Multidisciplinary Aspects of Regional Development*, Paris, OECD.

PLAN ORGANIZATION OF IRAN (1960) *Second National Development Plan*, Tehran, Govt of Iran.

PLAN AND BUDGET ORGANIZATION (1974) *Iran's 5th Development Plan 1973–1978 Revised. A Summary*, Tehran, Govt of Tehran.

RICHARDSON, M. W. (1975) 'Regional planning in Iran', *Growth and Change*, 3, 16–19.

CHAPTER TWENTY

The political geography of Cyprus

MICHAEL P. DRURY

It would seem that not only is no man an island; neither, within the spectrum of political geography, can any island be so regarded. This is peculiarly true of Cyprus which, despite the rich copper deposits which gave it its name, the well-watered Troodos Massif which helps to make it the most cultivated of all Middle-Eastern states and the scenic attractions which lure tourists who provided more than £30 million in 1978 or some 20 per cent of its annual budget, has proved irresistible to a succession of foreign powers, not because of what it possesses but because of where and what it is.

As the only island of significance to be located in both the Middle East and the 'Middle Sea', Cyprus possesses unique locational characteristics which have combined to give it a place in the annals of conflict and international manipulation out of all proportion to its size. Its 9251 sq km of territory supports a population which grew rapidly during the first three-quarters of this century from 235,000 to some 650,000, maintaining throughout that period a ratio between Greek Cypriot and Turkish Cypriot of around 5:1. This bipartisan population *is* the 'Cyprus Problem'. Whereas other Mediterranean islands evoke images ranging between the extremes of hedonistic affluence and deep-rooted poverty, Cyprus's problems are not noticeably economic, but are connected with the search, so far in vain, for solutions to the issue of conflicting ethnic interests which has raised its head frequently if irregularly ever since Turk began to join Greek on the island in 1571. This population, though regarded, particularly since independence in 1960, by the non-Graeco-Turkish world as 'Cypriot', has seldom felt able to regard itself in such a homogeneous light.

Two amplifications are necessary. First, 'ethnic' is an ambiguous word. It is here used as the Cypriots use it, having inherited from their two motherlands and such twentieth-century treaties as that of Lausanne (1923) the following dangerously simple rule: adherents to the Orthodox faith and to Islam are regarded as Greeks and Turks respectively. There is usually, but not necessarily, a linguistic distinction; converts change their ethnicity and therefore there is no automatic hereditary link. All citizens of the new republic were, for legal reasons, required to declare themselves in one category or the other following independence. Secondly, Cyprus was without Turkish inhabitants until the Ottoman conquest. No previous invaders had settled, other than as an administrative ruling class, since the arrival of the Greeks in pre-classical times. Immediately after 1571, however, some 30,000 settlers, either the soldiers involved in the invasion and their families, or agricultural colonizers, particularly from the Konya region, were introduced. Thereafter the Greek population declined from an estimated 150,000 in 1570 to as little as 25,000, but by the eighteenth century Greek numerical ascendancy had been re-asserted.

The people of Cyprus, if they have had the dubious privilege of owning a home in such parts of the island as Nicosia's large northern suburb of Omorphita, have been in sight of or overtly disrupted by active conflict for twenty-one out of the twenty-six years 1953–79. During this period the population distribution pattern has been totally transformed. In the early 1950s the pattern was not significantly different from that of the preceding hundred years: clearly the total population had grown greatly and, as elsewhere, this growth was chiefly concentrated in the urban centres, which accounted for 37 per cent of the population in 1960. Despite these trends, the vast majority of the population was to be found in ethnically homogeneous nucleated villages, but without, on an island-wide scale, any clearly defined large areas exclusively inhabited by one ethnic group. In none of the island's twelve administrative divisions – the six towns and their accompanying districts – did the ratio of Greek to Turk exceed 7.4:1 or fall below 2.2:1. This complex settlement pattern was complemented by the pattern of land ownership in which, in a Mediterranean agricultural context yet to experience holding consolidation (which did not begin until the early 1970s), the problems of small fragmented plots were exacerbated by an ethnic admixture in which the Turkish Cypriots possessed a widely scattered 18 per cent of the cultivated lands. Yet, as a consequence of respective accretion and erosion of the pre-1974 Greek and Turkish Cypriot population distribution following intercommunal tension peaks in 1964, 1967 and ultimately in 1974 itself, these two elements are now firmly separated by a swathe of territory isolating from the remainder the 39 per cent of northern Cyprus which forms the Turkish Federated State of Cyprus (TFSC). Proclaimed in February 1975, the TFSC was regarded four years later as illegal by all except Turkey. It 'exercises all powers except those expressly given (by it) to the Federal Republic of Cyprus' (part of the constitution of the TFSC, approved by referendum within northern Cyprus in June 1975). All Greek Cypriots except some 3000 and all

Turkish Cypriots except some 300 now live south and north respectively of that divide, which itself consumes 3 per cent of the island's territory; so that an almost total polarization has been achieved, not only in the physical sense but also in the economic, social and political spheres. These changing trends are summarized in Figs 20.1 and 20.2.

Whilst the Cypriot population of whatever ethnic or political colour can in no sense of the word be described as passive – the EOKA (National Organization of Cypriot Fighters, founded in 1953) campaign against the British colonial presence gave ample evidence of that – it is equally clear that many if not most of the events which come under the heading of Cyprus's political geography are a consequence of factors outside the control of the indigenous population. One can go further: although not devoid of physical attractions, it is evident that the conventional physical resources of Cyprus have been as innocent as have been its inhabitants in influencing its fate; what has been crucial has been its location.

Though islands peripheral to the major continents may, with justification, have gained comfort from the strategic advantage which their very insularity imparted, a similar status within the basin of the Mediterranean has produced no such accompanying guarantee of a peaceful or unmolested existence. Through a succession of Eurocentric eyes, the Mediterranean was seen not as the edge of three continents but as a convenient and central corridor for all forms of advance, economic, military or cultural. For at least 3000 years successive peoples, whether native to the basin or not, have come to regard control of bases within the Mediterranean as crucial to the advancement of their own interests, and the string of islands which serve as moated forts along the main west – east axis of this sea have seldom avoided for long the attentions of rival powers all too well aware of the advantages to be gained from possession of one or more of these core sites. In the Mediterranean's three European backwaters, the Tyrrhenian, Adriatic and Aegean seas, the incidence of such potential fortresses is far greater than in the main body of the Middle Sea, but this very frequency – there are some 120 inhabited islands in these three seas, half being in the Aegean – coupled with their smaller average size and relatively poorer location makes them, individually, less desirable than those few which act as stepping stones along the main basin.

The Balearics, either wholly or in part, changed international hands seven times in the eighteenth century; Corfu, commanding the Strait of Otranto and thus the Adriatic, experienced five such changes in the nineteenth. Such pointers only begin to indicate the intricacies of the changing political status and affiliation of these islands: often occupation has been partial and ownership divided, or the position has been left deliberately ambiguous with a limbo being created to appease rival foreign factions. This was the case with Crete during much of the nineteenth century and until its incorporation within

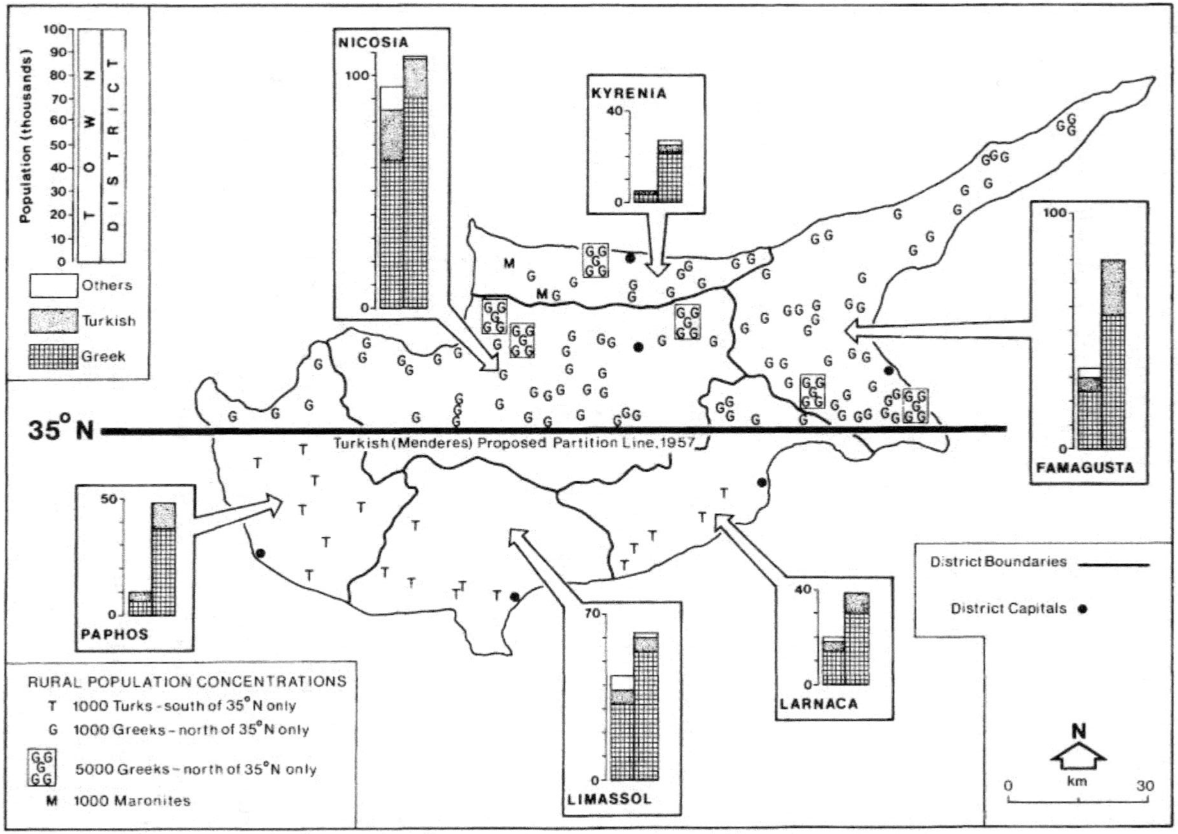

Fig. 20.1 Cyprus: the pattern of communal distribution in the year of independence, 1960.

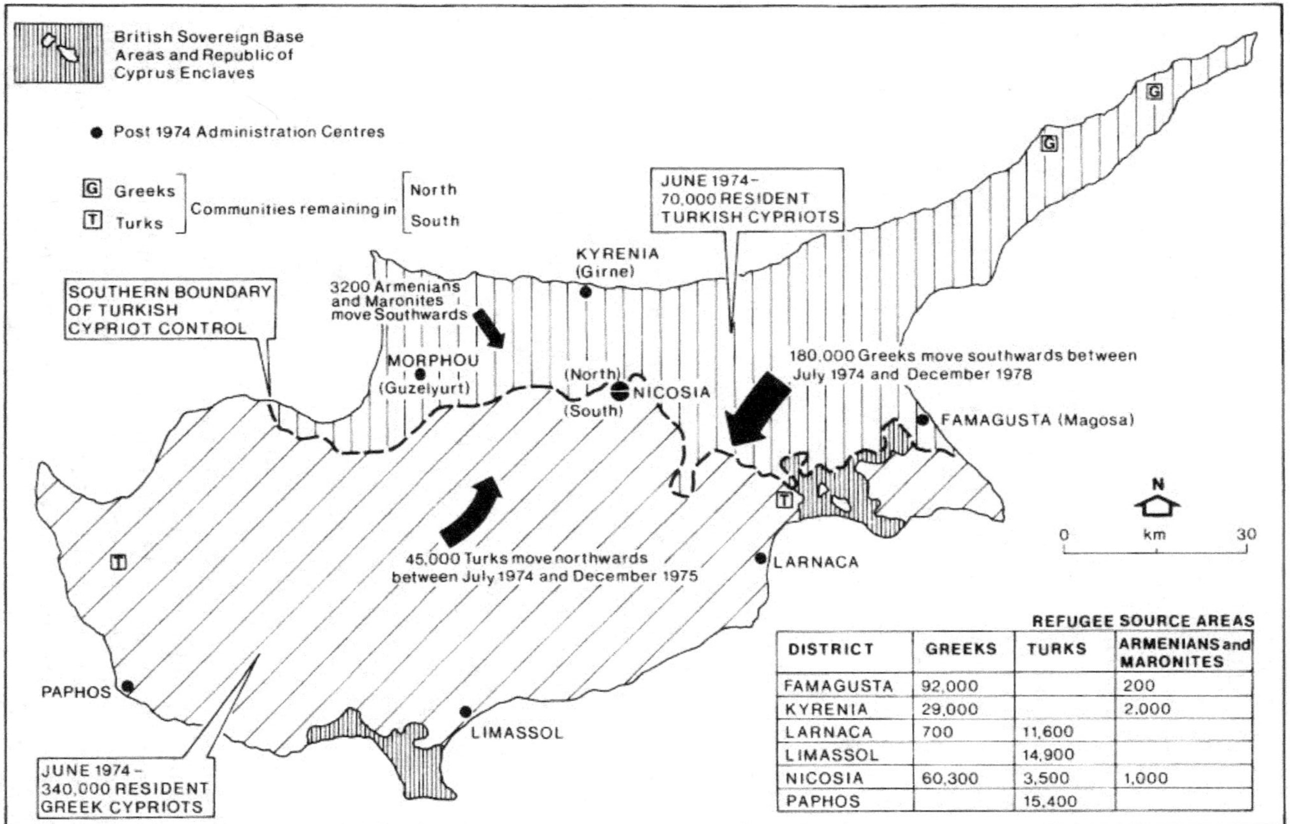

DISTRICT	GREEKS	TURKS	ARMENIANS and MARONITES
FAMAGUSTA	92,000		200
KYRENIA	29,000		2,000
LARNACA	700	11,600	
LIMASSOL		14,900	
NICOSIA	60,300	3,500	1,000
PAPHOS		15,400	

Fig. 20.2 Cyprus: the 1974 *de facto* partition and subsequent population movements.

Greece in 1913, when a fluctuating degree of autonomy was maintained under Egyptian–Albanian rule whilst still legally a part of the Ottoman Empire; with Malta, owned from 1530 to 1798 by the peripatetic Knights of the Order of St John who still maintain an embassy in Valletta; and with Rhodes which, owing to the Italian occupation between 1912 and 1947, has continued to enjoy tax concessions as compensation for its earlier sufferings, despite having emerged as one of Greece's few prospering peripheral areas.

Into this pattern Cyprus fits all too well. As the easternmost link in the island chain it has had additional rarity value, particularly during times when the eastern Mediterranean seaboard has had attention focused on it: such times have been frequent and prolonged and range from Phoenician expansion of trade links through Crusader activity and the Suez Canal to the deployment of the NATO defensive system. The history of Cyprus has thus been a tempestuous one, not least since the changeover from a Turkish to a British administration in 1878. In confirmation of these tendencies, events leading up to and since independence have indicated the continuing relevance of location and have left the government of Cyprus with a degree of control over the island which has been debatable, fluctuating, but never total. 'Never total' because, epitomizing this vital locational role, when the British government granted independence to Cyprus it retained full sovereignty over two areas, the Dhekelia and Akrotiri Sovereign Base Areas which together comprise 3 per cent of the island. Having 'lost' Suez in 1956, official British thinking revolved around Eden's view that 'no Cyprus – no certain facilities to protect our supply of oil. No oil – unemployment and hunger in Britain. It is as simple as that' (as quoted in *Keesing's Contemporary Archive*, June 1956, item 924). The only concession which the Cypriot government could secure was that, if Britain 'in view of changes in [her] military requirements, should at any time decide to divest [herself] of the aforesaid sovereignty . . . it is understood that such sovereignty . . . shall be transferred to the Republic of Cyprus' (HMSO, Cmnd. 1093 1960).

Being a pawn in such processes is not, however, entirely without its compensations. A pivotal position in the eastern Mediterranean (Fig. 20.3) has drawn Cyprus into many of the upheavals on the encircling mainland. Whilst, inevitably, this involvement centres around Graeco-Turkish rivalries, it has also meant that Cyprus has provided throughout its history a refuge, meeting ground and base for the changing rulers of its neighbouring coasts and the populations which they have displaced or disturbed. Thus, on more than one occasion Cyprus has welcomed such minorities as the Maronites and the Armenians, both of which continue to maintain distinctive communities within the island, even though in Turkish Cypriot eyes they are, as Christians, identified with the Greek Cypriots and have shared the latter's recent fate. The Armenians number some 3700, 99 per cent of whom lived in the four largest

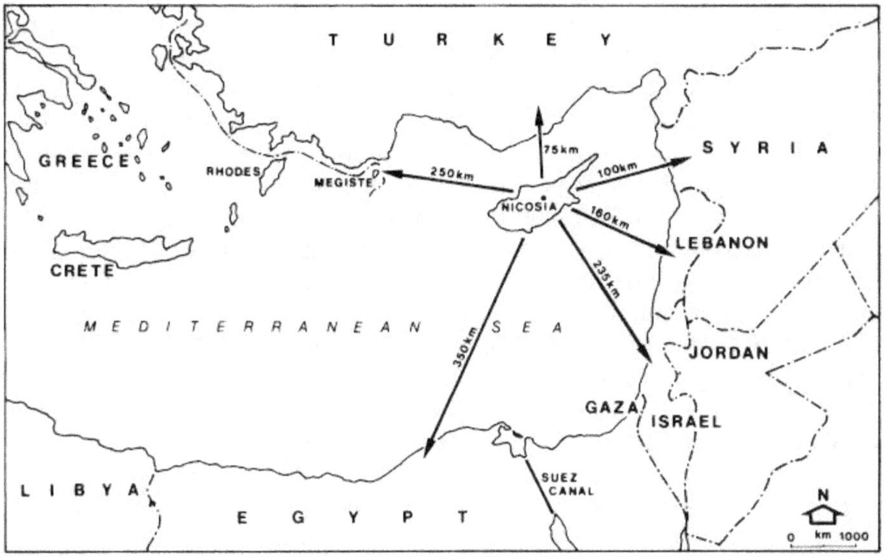

Fig. 20.3 The location of Cyprus within the eastern Mediterranean.

towns, but those living in Famagusta and northern Nicosia quit their homes in 1974. The Maronites number some 2700, 80 per cent of whom lived in a cluster of villages north of Morphou (Figs 20.1 and 20.2); all of this element again became refugees in 1974, their villages falling entirely within the TFSC. Similarly, in the past it meant that during the centuries of the Crusades the Lusignan dynasty established themselves as the island's feudal overlords in compensation for failures on the mainland further east. It also meant that Venice, representing the south-eastern bastion of European Christendom during its rule in Cyprus (1489–1571), was regarded as an intolerably close irritation below the underbelly of a dynamic Muslim state emerging out of Anatolia, thus generating the island's conquest by the Turks. In the nineteenth century it meant that when the opening of the Suez Canal realigned the world's shipping routes after 1869, thus re-establishing the eastern Mediterranean's nodality, Britain was only too eager to take over the island's administration from an ailing Ottoman Empire which had twice previously had its offer of Cyprus in return for other favours turned down by the British. More recently, it has encouraged the emergence of Cyprus in the role of safe neutral territory to which the conflicting interests in the Lebanese and Israeli – Arab struggles might escape either separately or, occasionally, to confer. Thus, until 1978 Cyprus was Israel's only neighbour prepared to recognize its existence and welcome visiting Israeli citizens; undeterred, Arab interests have continued to visit and invest in the island whilst, particularly since 1974, Palestinian elements have met with some success in equating their struggle with that of the Greek Cypriots. Finally, just when the Government of Cyprus was faced with its greatest crisis, that of absorbing almost 200,000 Greek Cypriots into the 60 per cent of the island which remained under its control, events in the Lebanon began to drive Christian Lebanese to find temporary refuge on Cyprus. In the peak year of 1976, 78,000 such Lebanese escaped to Cyprus and stayed until granted a visa to go elsewhere. Their arrival, usually with visible assets immediately to hand, gave the flagging Cypriot economy a crucial fillip and in particular rescued the tourist industry from the doldrums into which it had sunk when cautious western tourists, who had contributed £24 million to the Cypriot economy in 1973, thereafter stayed away rather than have their holidays spoilt by thoughts of invasion and refugees.

Clearly, then, there are some advantages to be gleaned for an island state within the Middle East which has no direct connection with issues of political sensitivity peculiar to that region. Equally crucial to the economy, if damaging to the government's pride, have been those two British bases which by their presence prevented the Republic of Cyprus from being synonymous with the island of Cyprus. The bases have consistently employed some 2 per cent of the Cypriot workforce (4800 Greek Cypriots and 1900 Turkish Cypriots in the last 'normal' year of 1973) and have regularly spent within the island, through official channels, a sum equivalent to 10 per cent of the island's GNP. In addition to this are the annual rents but these have not been paid since 1964 because of legal complications over the fact that payment was to be made to the Cypriots as a whole; the Turkish Cypriots having not participated in the government since that date, the British have withheld payment. In 1979 the Cyprus government claimed £400 million in rent arrears from the British. The casual expenditures of the British (some 24,000 in 1973) were also significant, as were private arrangements to rent properties outside the bases which brought considerable wealth and expansion to the adjacent cities of Limassol and Famagusta until 1974.

Since 1974 the existence of the bases has been crucial in other respects relevant to Cyprus's political geography. As British territory, the bases were presumed to be safe from attack and became sanctuaries for large numbers of both ethnic elements. Dhekelia was within easy reach of much of the Greek Cypriot population of Famagusta and its district as they fled in anticipation of the arrival of the Turkish army; Akrotiri was equally accessible to the Turkish Cypriots of parts of Paphos and Limassol districts who, anticipating reprisals from their Greek Cypriot neighbours, fled there. It took fifteen months for the British government to negotiate the transfer of the latter to new homes in the north of the island, whilst Dhekelia (or, more accurately, its enclaved Greek villages of Xylotymbou, Ormidhia and, to a lesser extent,

Xylophagou) in 1979 remained packed with displaced Greek Cypriots from further north. Presumptions about safety were indeed correct. The Turkish army did not advance into British territory, with the result that Dhekelia Base has a northern boundary adjoining the TFSC for 48 km (Fig. 20.4), allowing many Greek Cypriots the opportunity to see their former homes from the safety of a divide which, at least on the British side, is not mined. Further, through the keeping open of one frontier crossing on the road to Famagusta, those Turkish Cypriots in possession of a work permit are able to move south to work on the base alongside their Greek Cypriot counterparts and thus Dhekelia has since 1974 formed the only notable example of an inter-ethnic employer within the island.

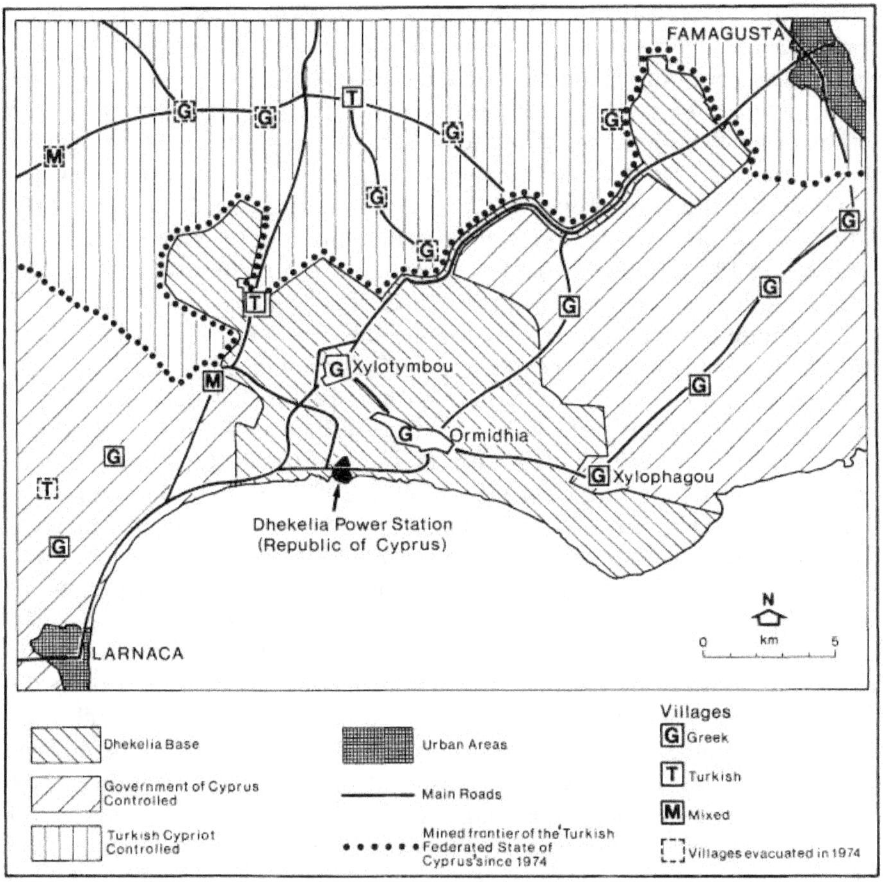

Fig. 20.4 The Dhekelia Sovereign Base Area.

These bases were one of several unusual consequences of the island's Independence Constitution. Independence was not the culminating dream of the majority of the population, most Greek Cypriots having supported the idea of union (Enosis) with Greece. The British government had offered Cyprus to the Greek government in 1915 in return for the latter's entry into the First World War on the Allied side. This Greece refused to do, but that offer, along with the evidence of the Greek state's steady advance to embrace, with the exception of the two at the entrance to the Dardanelles, all of the Mediterranean islands with a significant Greek population, convinced Greek Cypriots that Enosis was not only proper but feasible and ultimately acceptable to the British. Independence was conceived and came into fruition as a compromise between the three interested parties – Greece, Turkey and Britain – all of which were allowed to station troops on the island. It purported to give some satisfaction to the anti-colonial urges of the Greeks, some guarantee of internal security to the Turks – understandably fearful of the consequences of a reduction of their minority status from 18 per cent within Cyprus to under 1 per cent within Greece – and some saving of international face and security for the British. Above all, the constitution went to elaborate lengths to guarantee maintenance of the status quo which it established, notably by providing the Turkish element with 30 per cent of the fifty parliamentary seats and similar representation throughout the civil service and the right of veto over all aspects of governmental policy not deemed to be exclusively relevant to the Greek Cypriots.

The failure of the constitution to achieve this object was hardly surprising, since such a legalistic tightrope required profound balancing skills. Not for the first time, the hopes of the party relinquishing power that those receiving it should live in harmony were dashed because of the overwhelming centrifugal forces (in this case the pull of the two 'motherlands') which re-emerged once the euphoria surrounding the achievement of self-determination had waned. Tensions escalated into violence in December 1963, as a result of which the Turkish element 'withdrew'. This withdrawal had a spatial dimension as well as a socio-economic and political one. Not only did the allotted 30 per cent of seats remain empty from then on, but there was a withdrawal by the Turks from all aspects of corporate Cypriot life. Those whose home areas were regarded as both vulnerable and undefendable abandoned them, seeking the security of ethnic enclaves, either in the five major towns or in rural areas where large single villages or clusters of smaller settlements made defence seem possible. In all, besides the five towns – Kyrenia was the exception – fifty discrete rural areas emerged, amounting to some 10 per cent of the island's territory. The area administered by the central government thus became akin to a landscape riddled with holes (Fig. 20.5). That these holes survived for a decade was a consequence of three circumstances.

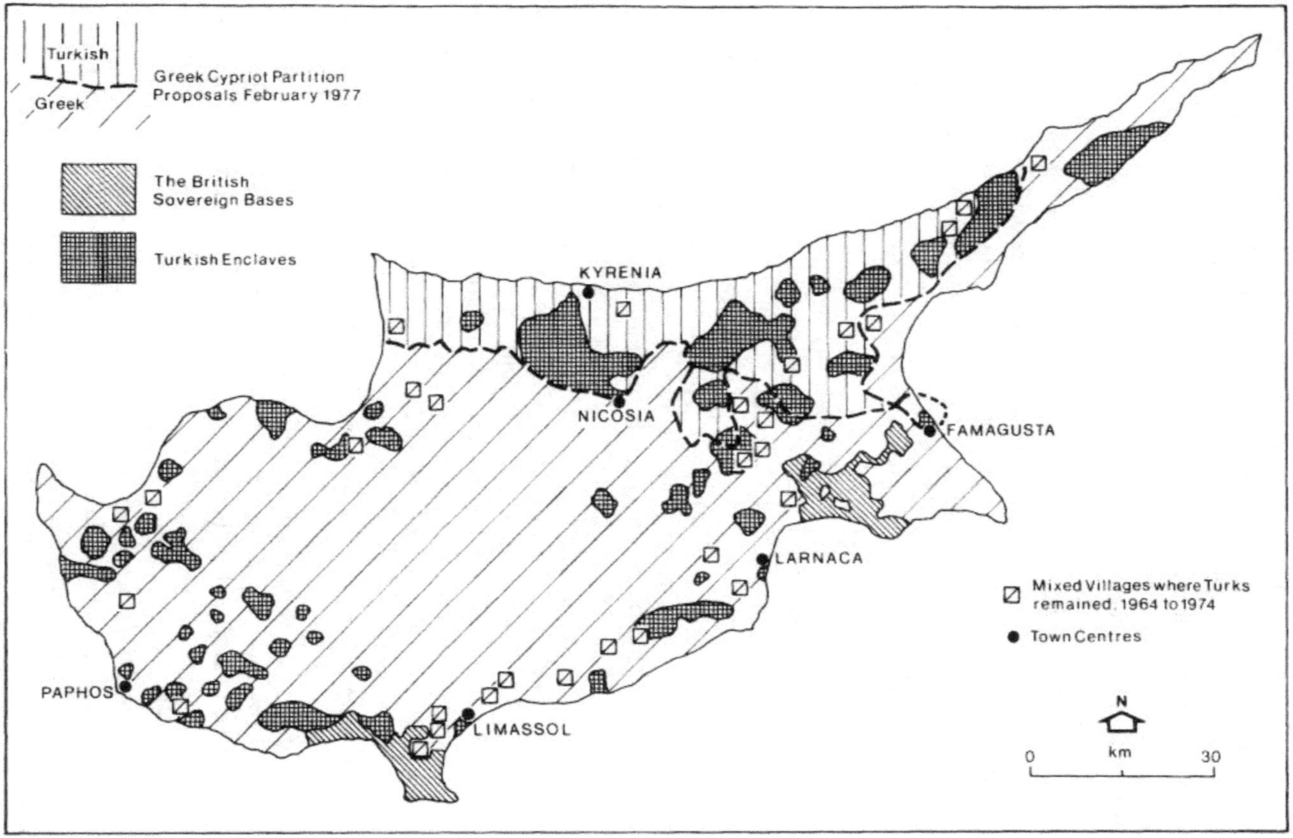

Fig. 20.5 Cyprus: the 'Turkish Enclaves', 1964–July 74, with the Greek Cypriot partition proposal of 1977.

First, the Turks were determined to defend them and helped by the threat of intervention from mainland Turkey were able to do so, despite two major callings of the Turkish bluff in 1964 and 1967, until 1974 when, armed with both superior military strength and moral indignation, Turkey finally intervened. The identification of minority consciousness with specific and clearly demarcated territory was of crucial psychological importance. Just as the Vatican City had been created in 1929 for the express purpose of allowing the Holy See to possess territory and thus encounter other territorial states as a lay equal, so the Turkish Cypriots, who had been granted by the 1960 Constitution separate municipal status but without territory in the five main towns, felt that the legal recognition of two communities was inadequate, at least for the less numerous one, unless a spatial dimension was added. This, as well as genuine fear of the Greek majority's intentions, drove the Turks to establish and maintain these enclaves, despite the obvious economic and strategic difficulties involved.

Secondly, the Greeks, though clearly disliking the new situation, could do little about it, except by covert means. What they did was to ignore them. Thus, for ten years during which the Cypriot economy boomed and standards of living rose visibly on all Greek sides, the Turkish element was left to stew in its own juice. New roads, electricity supply lines and water distribution systems were constructed around the enclaves so that, as the 1964–74 decade advanced, not only did the material gap between the two communities widen, but also the fact that many of the enclaves consistently refused to admit Greeks caused decreasing inconvenience to the Greeks. Conversely, particularly after 1967, the Turks were not prevented from moving freely throughout the island – though relatively few did so – with the result that their awareness of the improving Greek lot was as great as was the latter's ignorance of, and indifference towards, the Turk's mounting frustrations and sense of divorce from any centralized authority. The overwhelming majority of the Cypriot population thus came to live, be educated, work and relax in an ethnically homogeneous environment; *de facto* partition had already taken place.

To see that the spatial distribution of these two environments did not alter a third element was introduced: the United Nations. Introduced in March 1964 as a neutral third party wedged between the sights of the two combatants, the United Nations presence in Cyprus cost the world some £150 million during its first fifteen years and has, besides being led and administered by a variety of nationals, involved the troops and police of eight nations: Australia, Austria, Canada, Denmark, Finland, Ireland, Sweden and the United Kingdom. Prior to 1974 the contingent averaged some 5000 men, since when it has dwindled to below 2500, and during that first decade was dispersed to over fifty points close to the ethnic interfaces. This presence was sufficient to achieve its prime objective – that of the spatial status quo – so long as both sides were relatively quiescent, but it was neither sufficiently strong nor empowered to intervene should a total breakdown in intercommunal relationships occur. The United Nations thus provided the all-important face-saving excuse for peaceful coexistence rather than the reason for it. In addition, it provided a further source of income for the island's economy.

By 1974, however, events were moving to a head, not so much in Cyprus but in Greece where a tired and unpopular regime needed a triumph to bolster its image and, in selecting Cyprus as the stage for that triumph, did no more than its many predecessors in regarding the island as an appropriate pawn. The temporary overthrow of President and Archbishop Makarios in favour of Nicos Samson provided Turkey with an excuse for physical intervention. The combining of lay and spiritual leadership in one man was perhaps an inevitable development on Cyprus where the church had been autocephalous for many centuries; the Turks had utilized this fact and had dealt with the Greek population through their archbishop for most of the Ottoman period. The outstanding personality of Makarios meant that church and state continued to be regarded as synonymous, a situation which obviously proved unacceptable to the Turkish Cypriots. Nevertheless, by the early 1970s, the Turks had come to regard Makarios as a 'known devil', considerably safer than some untried but blatantly militant and pro-Enosis former terrorist. According to the Treaty of Guarantee which accompanied the Independence Constitution, Greece, Turkey and Britain were obliged to intervene should the constitutional status quo be threatened. Britain being unwilling and Greece being temporarily disqualified by her own actions, only Turkey was left to carry out this obligation.

The background to and details of these events have been widely discussed and speculated upon: it is sufficient simply to state that, as a consequence, when a ceasefire was agreed between Cypriot government and Turkish forces, some 180,000 Greek Cypriots and 45,000 Turkish Cypriots found their homes on the 'wrong' side of the new divide (Fig. 20.2). A chaotic spatial pattern which nevertheless meant that 99 per cent of the island's population was housed in communities controlled by its own ethnic group was thus transformed into a simple bifurcation which, by the end of 1975, had obliged 38 per cent of the island's population to abandon their homes. This partition, although not as simple as that of the 35 °N parallel, suggested by the Turkish premier Menderes, in 1957 (Fig. 20.1), bore almost as little relation to communal distributions. Typically, Kyrenia, least Turkish of the six towns, was the first to fall to the Turkish advance, whilst Paphos, the most Turkish, remained the most divorced from these operations. Clearly, strategic and economic motives rather than communal integration had been the dominant influences. The 39 per cent of the island that was occupied had previously contributed an estimated 58 per cent of the GNP; 75 per cent of the tourists had stayed in hotels now in the north; 80 per cent of the citrus orchards, which produced the island's leading agricultural export, were taken, as were 56 per cent of the mining and quarrying outputs and 50 per cent of the manufacturing capacity. Turkey had clearly exceeded its treaty obligations.

Except for one checkpoint in Nicosia and the crossing from the Dhekelia Base the new frontier had become, by the end of the 1975, a completely closed and heavily defended strip of territory, not only rendering sterile the 3 per cent of the island upon which it lay, but also causing severe infrastructural indigestion to both sides. With the line running through the heart of the island's capital, its only international airport, the sole stretch of (mineral) railway, and immediately to the south of the main port (Famagusta) and through the most prosperous area of irrigated agriculture (Morphou citrus region), economic upset certainly matched the social chaos that it created. Nevertheless, the powers of economic recuperation have been equally great. New air and sea ports have been constructed on both sides, roads have been re-routed and the population re-housed, often in satellite towns in close proximity to the industrial estates developing around Limassol, Larnaca and Nicosia. Within the Middle East the climatic and nodal attractions of Cyprus stand out, as does the reputation for efficiency and speed in the conducting of its mercantile and administrative affairs. These aspects have helped to rescue the island from the economic brink at a time when not only had such disastrous internal events taken place, but political decisions had to be taken concerning the redirection of Cypriot production and trade. Cyprus, for instance, was obliged to relinquish its preferential trading links with its main trading partner, Britain, following the latter's accession to the EEC, and had to seek new outlets for goods, many of which, such as sherry, fresh fruit and vegetables, had been specifically geared to British requirements. Such readjustments have been accomplished, with trade links being widened and strengthened throughout the Middle-Eastern and eastern European spheres in particular. Commerce itself has been specifically emphasized; the Cypriot merchant fleet is now three times the tonnage of that of any other Middle-Eastern state, amounting to some 10 tons deadweight for every inhabitant of government-controlled Cyprus, whilst the island's International Trade Fair, first held in 1976, has established itself rapidly as a symbol of the new Cyprus. The success of the government's economic measures and the relative failure of the TFSC which, despite an absolute decline in its population following 1974 and the acquisition of the greater share of the island's resources, has not recaptured the north's former economic well-being, repeats the economic dichotomy which prevailed from 1964 to 1974. Such a dichotomy is as dangerous now as it was then. In addition, there is the added problem of complacency born out of economic success. The ability to restore material standards has been identified, both within and without the island, with a solution to the basic problems.

Yet the political problem remains. The post-1974 years of sporadic intercommunal talks have produced nothing beyond a preparedness to continue with such talks, whilst discrete economies develop corresponding to the political realities of a divided island.

The political geography of Cyprus thus emerges as a multifaceted issue. Clearly, the intransigence of the two communities has been great. Proposals from one side have only become acceptable to the other long after, in the light of a changed situation, they have been withdrawn. Thus, the Greek willingness to accept a federal or cantonal solution, as shown in their proposals of 1977 (Fig. 20.5), came two decades after their rejection of similar proposals which led the Turkish side to make demands for 'Taksim' (partition) in retaliation against 'Enosis'. Yet such intransigence is perfectly understandable: on only the most infrequent occasions have the two communities been encouraged to think as one; usually their distinctive traits and aspirations have provided fertile pasture for external manipulation. Perhaps, having been used so often, this realization is gaining credence, but such bitter animosities, nurtured for so long, are not swept away with ease. Should the Cypriots be tempted to forget that it is their island rather than themselves which has made them such a popular target for international attention, they should be reminded of this by the American decision to supply arms again to Turkey because of the latter's strategic role within the western alliance and by the Graeco-Turkish wranglings over ownership of the Aegean seabed and its possible oil deposits, in both of which incidents Cyprus has emerged as a key bargaining point. Unless alone and unaided they put their own house in order, they are as unlikely to be consulted by the international community about their fate now as they were when Britain, Greece and Turkey arranged for them to emerge into independence on such disastrously weak foundations in 1960.

Bibliography and references

CYPRUS GEOGRAPHICAL ASSOCIATION (1976) *International Symposium on Political Geography*, Nicosia, Cyprus Geographical Association.

DRURY, M. P. (1972) 'Cyprus: ethnic dualism', in Clarke, J. I. and Fisher, W. B. (eds) *Populations of the Middle East and North Africa*, London, Univ. of London Press.

HARBOTTLE, M. (1970) *The Impartial Soldier*, London, OUP.

HMSO (1960) *Cyprus*, London, HMSO, Cmnd. 1093.

IERODIAKONOU, L. (1971) *The Cyprus Question*, Stockholm, Almquist & Wiksell.

KAROUZIS, G. (1976) *Proposals for a Solution to the Cyprus Problem*, Nicosia, Cosmos.

—— (1977) *Land Ownership in Cyprus*, Nicosia, Strabo.

KOLODNY, E. Y. (1971) 'Les Turcs de Chypre', *Revue de Géographie de Lyon*, 46, 5–56.
MINORITY RIGHTS GROUP (1977) *Cyprus*, Report No. 30, London.
POLYVIOU, P. G. (1976) *Cyprus in Search of a Constitution*, Nicosia, Nicolaou & Sons.
VASSILIOU, G. V. (1976) 'Trade agreements between the EEC and . . . Cyprus', Shlaim, A. and Yannopoulos, G. N. (eds) *The EEC and the Mediterranean Countries*, Cambridge, CUP.

CHAPTER TWENTY-ONE

Provincial and district delimitation in the Kingdom of Saudi Arabia

NASSIR A. SALEH

Although the delimitation of areas for administrative purposes has been known in Arabia for at least the past fourteen centuries, the present pattern of provincial and district delimitation of the state of Saudi Arabia has distinct spatial characteristics which stem from the special structure of its central and local government, the geographical personality of its land and people, the foreign influences upon its different parts and the organization of its society.

The essence and authority of the Saudi provinces

The establishment of the Saudi provinces goes back to the first years of the state's history when up to 1932, the year the country was named the Kingdom of Saudi Arabia, it was called the Kingdom of al-Hijaz, Najd and its Dependencies. The kingdom was then divided into two major administrative units, al-Hijaz and Najd, but with the rapid introduction of government services to the different parts of the kingdom and the increasing responsibilities of the central government, these units gradually increased in number and were further divided into smaller units. The need to set an elaborate law to define the functions and responsibilities of the provincial governors (*amirs*), the subordinate governors (also called *amirs*) and the local councils was felt as early as 1939, when the Regulations of the Provincial Governors and Administrative Councils were sanctioned (Document No. 1). This law was clearly a step towards the maintenance of full supervision, control and co-ordination of government services, as well as organizing the functions and responsibilities of the highest administrative authority in the provinces and their subdivisions. It did not, however, state the number of provinces actually in operation and neither did it establish lines or limits of authority for the different levels of administrative units. For example, no mention was made of the precise authority of the provincial governors over their subordinate governors, that is the governors of the subdivisions which fall within their jurisdiction.

According to the law of 1939, the provincial governors are the highest administrative as well as executive power in the provinces, and are responsible to the highest authority in the state, that is the king. Their main functions are to deal with the affairs of their provinces in such a way as to refer them to the appropriate government agencies and departments, to supervise the implementation of the laws and instructions in the government departments in their provinces, to report any corruption in the government departments, to be alert to and directly responsible for any movement against public morality and security, to seek justice for all members of the community, regardless of any consideration except that it is within the authority invested in them, and to execute all judgments passed by the Sharia courts and all orders by the state high authority. The law states persistently that the authority of the provincial governor is confined to his province, whose limits should be known, and he is not to interfere with the affairs of other provinces.

The law also stresses the fact that all heads of government agencies and the departments must abide by the orders and instructions of the governor in the administrative affairs of the units concerned, whether provinces, districts, towns or villages, according to the authority invested in him by law and instructions and to the extent that his instructions are in harmony with the laws and instructions organizing and administering the affairs of those agencies and departments. The governors are also responsible for the welfare and economic development of the communities and they must advise or draw the attention of the other government departments to take the appropriate action to achieve this goal.

Each provincial administrative council consists of four to eight elected members according to the status and size of the province. The bases upon which this evaluation is judged are not defined. The governor of the province is always the chairman of the council, but meetings can be held under the chairmanship of his deputy in the council. The members are to be elected for two years, but they may be re-elected for more than one term.

The responsibilities of the provincial governors and the administrative councils as prescribed in the law of 1939 are overlapping, unspecified in certain aspects, obscure and vague in others. The general features of size, shape, population of and relation between the different levels of the administrative units are ignored. The law itself was never applied at the national level, and hence administrative councils are not to be found outside the western part of Saudi Arabia. Provincial, district and village governors held the responsibilities and authority assigned to them, but they became varied in this respect according to their personalities. As years passed, their authority became traditional but vague and obscure, except when specific responsibilities and authority are invested in them individually.

In 1963 a step was taken to reorganize the provincial system in the form of the Provincial Regulations, sanctioned by Royal Decree No. 12 on 21. 5. 1383 AH, 9 October 1963 AD (Document No. 2). Although the Royal Decree states in its second paragraph that the Regulations should be implemented within nine months of their proclamation, none of the items has been applied or adopted except those concerning the responsibilities of governors at all levels, which are identical to those stated in the law of 1939.

Present territorial administrative divisions

The state of Saudi Arabia is estimated to be 2.2 million sq km in area, with most of its southern boundaries being undefined. The foundation of the state goes back to the middle of the eighteenth century when the religious reformation movement of Sheikh Mohammad Abd Al-Wahhab was adopted by the family of al-Saud, the governors of Diriyah village in the heart of Najd. Until the beginning of the twentieth century the state first expanded and later collapsed. The real political architect of the state, however, was King Abd-al-Aziz Al-Saud (*c.* 1881–1953), who restored the state capital al-Riyadh in 1902 and brought the state into its present territorial shape and unity.

By 1928 two territorial administration divisions were recognized, namely al-Hijaz and Najd with a variety of smaller divisions attached to each. The state is now composed of fourteen first-order administrative divisions called *amarat* or *muqataat* (provinces) depending on the status of each province. Their emergence is largely the outcome of a blend of historical, geographical and social factors (Saleh 1975:14–199).

The provinces of Saudi Arabia vary greatly in the size of their areas and the number of their inhabitants (Table 21.1). This is a direct consequence of the areal variations between the different parts of the country in terms of their physiographic conditions. In countries like Saudi Arabia with predominantly arid land, the most useful variable that can explain the differences in the areal size of administrative divisions is the size of the cultivated or uncultivated land in each unit as related to the total area of the cultivated or uncultivated land of the country. Generally speaking, the areal size of the Saudi provinces is dependent on the size of their cultivated land. In other words, the size of the desert proper usually determines the areal extent of the administrative divisions, whether at the provincial or district level. For example, the size of first-order administrative divisions (provinces) in the south-western part of the country is relatively small because of the higher proportion of cultivated areas and higher population densities (Fig. 21.1). As seen in Table 21.1, the provinces of Jazan, Asir, and al-Bahah account for 38.8 per cent, 6 per cent and 2 per cent of the total cultivated area of the country respectively, while their areas are only 0.63 per cent, 4.5 per cent and 0.68 per cent of the total area respectively. This is in contrast to other parts of the country, where

Table 21.1 Population, size and cultivated area of Saudi Arabian provinces

Provinces	*Total area* *Sq km* *	%	*Cultivated area* *Donum*[a]	%	*Population 1974* *Thousands*[b]	%	*No. of settlement centres*[b]
Al-Riyadh	418,835	19.09	579,671	14.615	1,272	18.90	1,992
Makkah	119,573	5.45	947,528	23.90	1,754	26.10	4,088
Asir	98,734	4.50	237,642	6.00	681	10.12	4,597
Al-Bahah	14,919	0.68	77,762	2.00	186	2.80	1,296
Eastern	708,000	32.27	101,593	2.56	770	11.40	667
Hail	169,596	7.73	54,722	1.38	257	3.81	540
Al-Jauf	74,596	3.40	10,979	0.28	65	0.97	85
Jazan	13,822	0.63	1,537,136	38.80	403	6.00	4,537
Al-Madinah	143,049	6.52	29,123	0.73	519	7.70	1,742
Najran	87,760	4.00	29,911	0.75	148	2.20	242
Northern Boundaries	115,624	5.27	200	0.005	129	1.90	130
Al-Qaseem	76,790	3.50	319,635	8.00	317	4.70	509
Al-Qurayyat	53,972	2.46	5,228	0.13	31	0.50	98
Tabouk	98,730	4.50	33,753	0.85	194	2.90	472
Total	2,194,000	100.00	3,964,883	100.00	6,726**	100.00	20,995

Source: [a]Miscellaneous agricultural data from the Statistical Division, Ministry of Agriculture and Water; [b]Ministry of Finance and National Economy, *The General Population Enumeration, 1934*, Riyadh.

* Area is estimated using the grid method and calculated using the planimeter.

** The official census adds the following figures to the total population of the Kingdom of Saudi Arabia: bedouins along the boundaries 210,000; Saudis living outside the country 73,000.

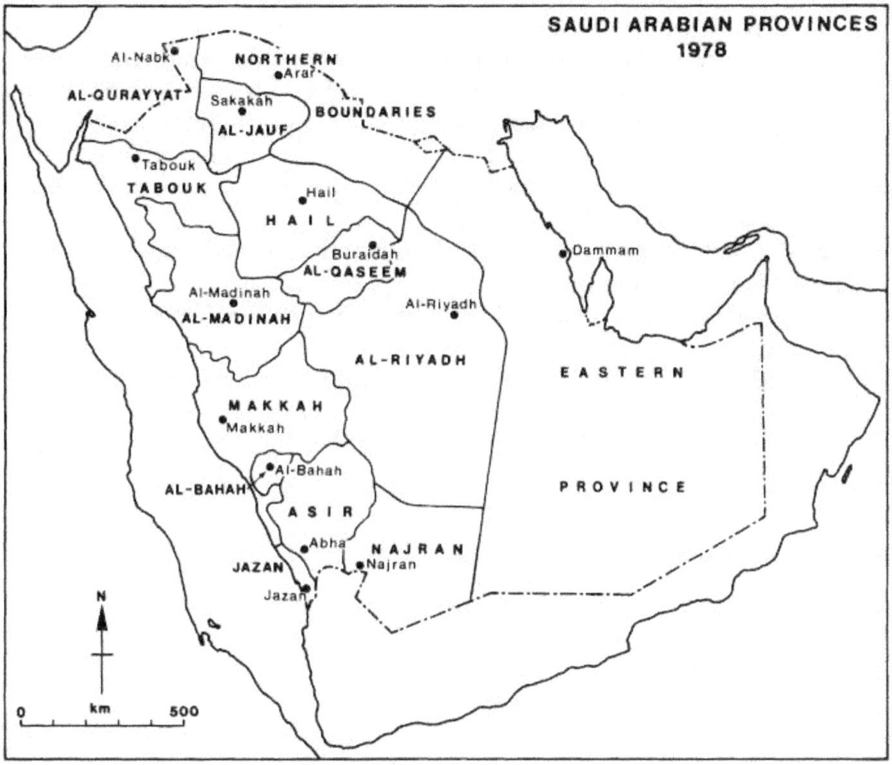

Fig. 21.1 Saudi Arabian provinces, 1978.

large provinces are associated with the dominance of the desert proper. An example of this association is the Eastern province, which accounts for about 32 per cent of the total area of Saudi Arabia, but only 2.5 per cent of the total cultivated land of the country, with a low average population density. The provinces of Hail, the Northern Boundaries, al-Jauf, al-Qurayyat, Tabouk and al-Madinah reveal the same pattern.

A similar relationship exists in other countries of the Middle East and North Africa where arid land predominates. For example, in Iraq, Egypt, Libya and Algeria the small-size first-order administrative divisions are located in the densely cultivated and inhabited areas of these countries; that is, in the eastern part of Iraq, along the Nile valley in Egypt, in Jabal al-Akhdar and the north-west of Libya, and in the coastal plain and Tell mountains in Algeria (UNESOB 1970; Office of the Geographer 1969). However, there are exceptions to this general pattern in Saudi Arabia, where there exist provinces with large areas as well as large cultivated areas, such as the provinces of al-Riyadh and Makkah. This is largely the outcome of the presence of densely cultivated parts within these two provinces, such as the al-Sarah area in Makkah. In fact, within these provinces we find that the pattern mentioned earlier exists at the second or third level of administrative divisions. For example, the second- and third-order divisions in the widely cultivated areas of al-Sarah are small in area compared to other divisions of the same level in the arid parts of Makkah province, such aş the subdivisions of Tarabah, Ushairah and al-Mahani (Saleh 1975: Fig. 6.1).

The same pattern exists elsewhere in the different provinces of the state. For example, the most favourable conditions for sedentary life in the Kingdom of Saudi Arabia are to be found in the upland plateau of al-Sarah where part of al-Bahah province is located. Because of its high population density and the existence of a mass of tiny settlements, this region contains 80 per cent of the total number of second-order administrative divisions of the province. The other 20 per cent is divided between the semi-arid Tihama proper in the west and the inhospitable interior plateau in the east (Saleh 1975:297–301).

Patterns of provincial and district delimitation

The diverse effects of geographical, historical and social factors on the areal extent and delimitation of the Saudi provinces and their subdivisions have produced diverse and clearly identifiable patterns as far as their spatial distribution and characteristics are concerned.

One of the major aspects of provincial administration is that the number of first-order administrative divisions has decreased steadily, giving some indication of a growing decentralization tendency. At the beginning of the reign of King Abd-al-Aziz every major town and village in Najd considered itself as a separate administrative unit with the king as the only authority to be consulted and obeyed. The same was true with the different towns and villages in al-Hijaz when the late King Faisal was its governor in the late 1920s and early 1930s. This situation was, of course, impossible to continue as the affairs of the different parts of the country have become complex and the services improved. The first-order administrative divisions were thirty-six in 1958, nineteen in 1973 and fourteen in 1977 (Table 21.2). Many unofficial sources suggest that the number of provinces will decrease even more to become finally seven or eight, which would be a suitable number if the 1962 provincial law was to be implemented. The same sources suggest the idea of attaching al-Jauf, Northern Boundaries and al-Qurayyat provinces to Hail to form one province. It is also suggested that al-Bahah, Jazan and Najran provinces be part of Asir province.

The second major aspect is that the limits of provincial divisions and subdivisions cannot be called boundaries in the strict sense. This is because there exist no boundary lines that separate one provincial division or

Table 21.2 First- and second-rank provinces in Saudi Arabia, 1958, 1973 and 1977

1958	*1973*	*1977*
First-rank provinces		
Al-Riyadh	Al-Riyadh	Al-Riyadh
Hail	Makkah	Makkah
Al-Qaseem	Al-Madinah	Al-Madinah
Eastern (Al-Hasa)	Eastern	Eastern
Northern Boundaries	Asir	Asir
	Hail	Hail
	Al-Qaseem	Al-Qaseem
	Northern Boundaries	Northern Boundaries
	Northern (Tabouk)	Northern (Tabouk)
Second-rank provinces		
Makkah	Al-Jauf	Al-Jauf
Al-Zaima	Al-Qurayyat	Al-Qurayyat
Wadi al-Jumum	Najran	Najran
Al-Kamil	Jazan	Jazan
Al-Khurmah	Al-Bahah	Al-Bahah
Afif	Bishah	
Raniah	Al-Khasirah	
Al-Khasirah	Afif	
Al-Taif	Raniah	
Tarabah	Yonbu	
Al-Lith		
Al-Birk		
Al-Rith		
Sharurah		
Northern (Tabouk)		
Al-Jauf		
Al-Muwaih		
Dhulum		
Midrikah		
Jeddah		
Rabigh		
Yonbu		
Al-Madinah		
Taima		
Bishah		
Baljourashy (now Al-Bahah)		
Al-Qunfudhah		
Jazan		
Najran		
Abha (Asir)		
Al-Qurayyat		

Note: All provinces are first-order administrative divisions but may be classified as being of first or second rank.

subdivision from another. Most of the areal limits are known only to a few officials working in the areas concerned. These limits mostly coincide with natural features which do not extend over the whole boundary zone. They are also associated with tribal limits which are well defined and precisely known only in a few localities such as al-Sarah area. The only criterion used by officials and researchers to decide the limits of provincial divisions and subdivisions is the attachment of the various towns and villages to the centres of the different provinces and districts. This is the procedure which was followed in the field studies made by the author in different administrative areas of the kingdom. It enabled him to draw the limits between the second-order administrative divisions in al-Bahah, one of the smallest provinces (first-order administrative divisions) in the state.

Another major aspect of Saudi provincial administration is the absence of a uniform spatial hierarchical order of administrative subdivisions within the provinces and even within individual provinces. As can be seen from Table 21.3, six patterns can be distinguished bearing in mind that the term 'province' refers to the first-order administrative divisions, although the official terms for them might vary according to their status and rank. To examine these variations in the hierarchical order of administrative subdivisions and the problems they indicate, a closer look at the relationship between tribal life and settlement patterns in Saudi Arabia is necessary.

Apart from the south-western part of the country, settlement centres, mainly agricultural oases or trade centres, are located in predominantly arid lands where pastoralism is practised. Although most of the inhabitants of these oases and trade centres are of tribal origin, once they are settled they no longer identify themselves with tribal institutions. In other words, tribalism as a political and social concept ends at the village boundaries. Thus a distinction between tribal nomadic organization and settled life has developed to the extent that two territorial political units can be recognized. The first is the village territorial political sphere of influence, which is limited to the village boundaries, and the second is the tribal territory, whose size depends upon the power and wealth of the tribe.

The animosities that took place between tribes and villages in eastern, central, northern and north-western Saudi Arabia constituted a major part of the history of these areas before the consolidation of the Saudi state during the first quarter of this century. Within these areas, however, certain villages were the residence places of the tribal chiefs and they therefore became tribal centres whose territorial political sphere of influence extended to the boundaries of the tribal territories. Hail, al-Wajh, al-Jauf, Yonbu, Tabouk, Khaibar, al-Ula and Diba are only a few examples of these tribal centres. During the formation period of Saudi provinces and governorates, these tribal centres became centres for either provinces or governorates, with the result

Table 21.3 Six patterns of hierarchical order of the administrative subdivisions of the Saudi provinces

(1)
Muqatat al-Riyadh

- Amarat Mantiqah (Governorate)
  - Amarah (Town)
  - Markaz (Village)
- Amarah (District)
  - Markaz (Village)

(2)
Al-Mantiqah al-Sharqiyyah
(Eastern Province)

- Amarah (District)
  - Markaz (Subdistrict)
- Amarah (District)
  - Qura (Village)
- Amarah (Town)

(3)
Mantiqat al-Madinah

- Amarah (Governorate)
  - Markaz (Subdistrict)
    - Qura (Village)
- Amarah (District)
  - Qura (Village)

Other Province: Tabouk

(4)
Mantiqat Asir (Province)

- Amarah (District)
  - Qura (Village)

Other Provinces: Al-Bahah, Jazan, Bishah and Yonbu

(5)
Mantiqat Makkah

- Amarat Mantiqah (Governorate)
  - Amarah (District)
    - Qura
  - Markaz (District)
    - Qura
- Amarah (District)
  - Markaz (Village)

(6)
Mantiqat al-Jauf

- Amarah (Village)

Other Provinces: Al-Qaseem, Najran, Northern Boundaries Province and al-Qurayyat that their limits coincide with the limits of the tribal territories which they controlled. Examples of this trend are Hail province (the Shammar), al-Jauf province (the Ruwala), Tabouk (Bani Atiya), al-Qaseem province (the Harb of Najd), al-Ula governorate (the Awlad Ali), Diba governorate (the Huwaitat), al-Wajh governorate (the Beli), Yonbu province (the southern sections of the Juhainah) and Rabigh governorate (the Masrouh section of the Harb). With the absence of a strong bond at the subtribe level, and the confinement of the village's territorial political sphere of influence to its boundaries, delimitation of second-order administrative units within those provinces and governorates developed a unique pattern in which administrative subdivisions emerged around villages with their territorial extent being confined to the village boundaries. In other words, most of the territorial limits of these subdivisions are confined to the boundaries of single villages, with the result that, although villages become centres of second-order administrative units, their territories could hardly be termed 'districts'. For example, the provinces of al-Qaseem, Hail, al-Jauf, al-Qurayyat, Tabouk, the Eastern province and most of the governorates of al-Riyadh province are divided into administrative subdivisions with most of their territories totally confined to single villages.

With the absence of an intermediate territorial administrative unit between the provincial and village level, certain disadvantages and problems crop up. One of these problems is that villages constituting administrative subdivisions try to extend their administrative catchment areas beyond the village boundaries to include villages which have not yet acquired administrative status. This might produce conflict between these villages and confusion among the inhabitants of the latter type of villages as to what authority they should recognize for their administrative affairs. Another disadvantage of this system is that the emergence of an administrative subdivision in a village involves the establishment of a government office which is normally supplied with employees, furniture, transportation facilities and official entertainment expenses. All this proves to be very costly if one realizes that in each of the aforementioned provinces and governorates the number of administrative subdivisions evolved around single villages is unjustifiably great and that tens of villages apply every year to be the centres of administrative subdivisions. For example, al-Qaseem province had eighty-three administrative subdivisions of this type in 1393 AH (1973 AD) and applied for twenty more to be considered in the fiscal year of 1393–4 AH.

As one moves south-westward to al-Sarah and Tihama proper, the whole pattern of settlements changes radically. Man-to-land attachment and tribal bonds are stronger. Villages feel and acknowledge ethnic association with each other. Ethnic groups, whether at the tribe, subtribe or clan level, have developed distinctive territorial feelings to the extent that each village is tied to a hierarchy of territorial units from the village to the tribal level, with mutual recognition of their respective territorial limits. An example taken from al-Sarah should serve to illustrate this hierarchy of territorial units. The village of al-Qarn is inhabited by members of the Bani Salim clan, which is a subdivision of the Bani Dhabian, a subtribe of the Ghamid. The emergence of provinces and governorates in this part of Saudi Arabia has been completely based on and in line with the distribution of tribal territories where a group of related tribes has been incorporated into one province or governorate, as in the case of the territorial limits of the provinces of Asir, Jazan, al-Bahah, Najran and Bishah, and the governorates of al-Taif, al-Lith and al-Qunfundhah (Makkah province). When delimiting second- and third-order administrative units within these provinces or governorates, the same principle was adopted, whereby related subtribes were grouped to form administrative subdivisions which can be termed 'districts' in the true meaning of the word. One advantage of this system is that district limits are fairly well known and acknowledged, a situation which does not leave any ground for conflicts or disputes between districts over their territorial limits, since most district limits are drawn so as to coincide with tribal or subtribal territorial limits. Another advantage is economic, since the number of administrative subdivisions in this part is very small compared to those in the provinces of eastern, central, northern and north-western Saudi Arabia; hence the amount of money spent to maintain their administrative functions is much less. If we combine the provinces of al-Bahah and Asir, whose area is roughly equal to that of al-Qaseem, we find that the number of subdivisions in these two provinces is only fifty, compared to 103 in al-Qaseem, bearing in mind that the population of the two provinces of al-Bahah and Asir is almost three times that of al-Qaseem.

Another significant difference between south-western Saudi Arabia and the rest of the country in the distribution pattern of provincial subdivisions is that a large proportion of the centres of administrative subdivisions in the central, northern and north-western parts of the country emerged in the peripheral areas of the provinces or governorates for the sole function of observing the international boundaries or inter-provincial limits in order to prevent any clashes, illegal passing or trading along the international boundaries, or inter-tribal conflicts along the provincial limits. Most of these centres emerged in tiny villages or hamlets and it is not unusual to find the government offices of some of these centres stationed in tents in the middle of completely arid land, isolated from the rest of the province, not to mention the provincial centre. Many examples could be cited to illustrate this trend in al-Qaseem, Eastern province, Northern Boundaries, al-Qurayyat, al-Jauf, Tabouk, al-Madinah, Yonbu and Makkah province. But the point to be stressed here is that unnecessary duality and duplication in the functions of these centres is encountered, since they emerge along both sides of the inter-provincial limits.

For example, the centres of the administrative subdivisions of Hadhah, al-Asaihir, al-Harrah (al-Madinah province) and al-Furai (Makkah province) are located along the territorial limit between two provinces with the identical function of preventing inter-tribal clashes between the three tribes of Utaibah, Mutair and Bani Sulaim, while they are located within a 10 km radius from each other in a very rugged and isolated part of Harrat Rahat. The problems and implications of such territorial administrative arrangements have been dealt with previously, but it remains to be said here that elimination of such duality of function, and therefore minimization of government expenditure, could only be done by amalgamating and incorporating such centres into one district, to be attached to one of the provinces to which those centres were originally attached. By doing this, districts of this type will assume more efficient and firm control over inter-tribal conflicts with the aim of integrating tribal elements by fostering the feeling of territorial belonging among them, a situation which will undoubtedly enhance the improvement of their social welfare and standard of living. Parallel to this, we find that in the south-western part of Saudi Arabia the basic aim of district delimitation is to foster a common feeling of territorial belonging among the inhabitants at the district level. District offices function as venues through which the demands of the inhabitants for more effective public services are carried to the provincial office and from there to the different central government departments. Districts engage in healthy competition to apply for such services as schools, roads, health centres, social security offices, agricultural units, communication offices and the like.

Therefore, the author emphasizes the need for such a feeling of territorial belonging at the district level, which cannot be fostered by the present pattern of district delimitation in the greater part of Saudi Arabia. This implies the need to reconsider the distribution and territorial extent of the existing administrative subdivisions within most of the Saudi provinces, with the aim of achieving a uniform pattern of territorial hierarchy as far as the provincial administration is concerned. This will undoubtedly provide a solid base for further unification of the different patterns of territorial hierarchy of the territorial administrative units of central government departments.

Another major point which is largely underestimated is the variations among the Saudi provinces as far as the rank, status and authority delegated to their governors are concerned. Provincial governors with higher ranks have more power and connections to persuade the central government departments, to meet the demands of the inhabitants of their provinces for public services and to implement the development programmes within their provinces – a situation which results in unbalanced welfare. Furthermore, the authority delegated to these provincial governors of higher rank provides them with adequate power to decide upon certain cases and to execute certain decisions without referring them to their higher authority (the Ministry of Interior); while such cases and decisions are subject to a strict procedure in other provinces. In other words, the inhabitants of the different provinces are subjected to different patterns of decision-making process. This could leave regrettable consequences on the perception and mental attitude of the inhabitants towards the viability and justification of the authority of the local administrative units.

In conclusion, a wide variety exists among the Saudi provinces, whether in their rank and administrative status, in the authority delegated to their governors, or in the hierarchical order of their administrative subdivisions. These variations indicate clearly a continued lack of co-ordination and planning at the national level. They might also provide obstacles against future adjustment. For example, they could well be the main factor behind the continuous delay in implementing the Provincial Regulations of October 1963, which were drawn up to standardize existing procedures of provincial government with the aim of creating uniform territorial administrative units for provincial administration as well as the administration of the central government departments and ministries.

Bibliography and references

AL-HAMDAN, Y. A. (1971) 'Development of local government in Saudi Arabia', unpublished MPA thesis, University of Pittsburgh.

DOCUMENT NO. 1 *Regulations of the Provincial Governors and Administrative Councils*, issued on 13.1.1359 AH.

—— 2 *The Provincial Regulations*, issued by the Royal Decree No. 12 on 21.5.1383 AH.

—— 3 *Detailed National Budget Plans of the Fiscal Years of 1375, 1377, 1383/84, 1385/86 and 1393/94 AH*, Budget Department, Ministry of Finance and National Economy.

MINISTRY OF FINANCE AND NATIONAL ECONOMY, GENERAL DEPARTMENT OF STATISTICS, CENSUS UNIT (1971) *Guide to the Towns, Villages and Hamlets of the Saudi Provinces*, al-Riyadh.

OFFICE OF THE GEOGRAPHER (1969) *Geographic Report; Africa; Civil Divisions*, No. 15, US State Department.

SALEH, N. A. (1975) *The Emergence of Saudi Arabian Administrative Areas: A Study in Political Geography*, PhD thesis, University of Durham.

UNITED NATIONS ECONOMIC AND SOCIAL OFFICE IN BEIRUT (UNESOB) (1970) *First and Second Order Administrative Divisions in Selected Countries in the Middle East*, Beirut.

Index

www.ingramcontent.com/pod-product-compliance
Lightning Source LLC
LaVergne TN
LVHW011112300926
849160LV00008B/354
9780415825481